Biology Under the Influence

Biology Under the Influence

Dialectical Essays on Ecology, Agriculture, and Health

Richard Lewontin and Richard Levins

MONTHLY REVIEW PRESS
NEW YORK, NY

Monthly Review Press
New York Copyright © 2007
All rights reserved

Library of Congress Cataloging-in-Publication Data

Lewontin, Richard C., 1929– Biology under the influence : dialectical essays on ecology, agriculture, and health / Richard Lewontin and Richard Levins. p. cm. ISBN 978-1-58367-157-3 (pbk.) — ISBN 978-1-58367-158-0 1. Biology—Philosophy. 2. Biology—Social aspects. I. Levins, Richard. II. Title. QH331.L5355 2007 570.1—dc22 2007033070

Design Terry J. Allen

Monthly Review Press
146 West 29th Street – Suite 6W
New York, NY 10001
http://www.monthlyreview.org

10 9 8 7 6 5 4 3 2 1

Table of Contents

Part Two:

Part Three:

Dedication

Five Cubans are now serving long sentences in U.S. prisons because they were monitoring the activities of Cuban émigré terrorist groups in Miami. From their cells they have been active both in helping to make prison life more bearable for the other inmates in their immediate community and continuing to be full participants in the life of the Cuban revolution. We admire their steadfastness and creativity in resistance, and we dedicate this book to Antonio Guerrero, Fernando González, René González, Gerardo Hernández, and Ramón Labañino, and to people all over the world struggling for their release.

Introduction

Biology Under the Influence is a collection of our essays built around the general theme of the dual nature of science. On the one hand, science is the generic development of human knowledge over the millennia, but on the other it is the increasingly commodified specific product of a capitalist knowledge industry. The result is a peculiarly uneven development, with increasing sophistication at the level of the laboratory and research project, along with a growing irrationality of the scientific enterprise as a whole. This gives us a pattern of insight and blindness, of knowledge and ignorance, that is not dictated by nature, leaving us helpless in the big problems facing our species. This dual nature gives us a science impelled both by its internal development and the very mixed outcomes of its applications to understand complexity as the central intellectual problem of our time. But it is held back by the philosophical traditions of reductionism, the institutional fragmentation of research, and the political economy of knowledge as commodity.

This means we have to be engaged on two fronts: 1) we stand against the obscurantist anti-science, which ranges from direct manipulation of the EPA and FDA by the government and the hype of the drug companies, to creationism and the mystification of mathematical chaos; 2) we also reject scientism, the claim that other people's ideas are superstition while ours are uniquely objective knowledge verified by numbers. We reject the postmodern view that, still reeling from having discovered the fallibility of science, comes to deny the validity of knowledge or, overwhelmed by the uniqueness of the particular, refuses to see patterns even of uniqueness. Scientism focuses mostly on the last stages of research, hypothesis testing, thus ignoring the questions of the origins of the hypotheses to be tested and of the source of the rules of validation. We challenge both the mystical holism that sees everything as so much "all one" that it becomes a shifting blur without parts, and the reductionism which claims that the most fundamental truths are found in the smallest parts of things. We trace how this works out in agriculture, health, ecology, and evolution. Then we step back and look at the processes of abstraction and model building, and return to examining the present-day obstacles to an integral, complex, dynamic view of the world.

We come to this project as participant observers. Both of us have been active in overlapping though somewhat different areas of population genetics, ecology, evolution, biogeography, and mathematical modeling. As participants we have been engaged in the nuts and bolts of our sciences in lab and field and before the computer. In our scientific work we have attempted to apply the insights of dialectical materialism that emphasizes wholeness, connectedness, historical contingency, the integration of levels of analysis, and the dynamic nature of "things" as snapshots of processes. Although we have variously worked with enzymes, fruit flies, corn, ants, gene frequencies, and orange trees, our point of view was always influenced by how we see the world as a whole.

We also step outside of the specific scientific problems to become observers and examine the nature of science and the uses of mathematics and modeling. In this, we step into what usually fits within philosophy of science. Occasionally we have worked jointly. At other times our separate work was strongly or loosely influenced by our ongoing dialog of almost forty-eight years.

We have also been political activists and comrades in Science for the People; Science for Vietnam; the New University Conference; and struggles against biological determinism and "scientific" racism, against creationism, and in support for the student movement and antiwar movement. On the day that Chicago police murdered Black Panther leader Fred Hampton, we went together to his still bloody bedroom and saw the books on his night table: he was killed because of his thoughtful, inquiring militancy. Our activism is a constant reminder of the need to relate theory to real-world problems as well as the importance of theoretical critique. In political movements we often have to defend the importance of theory as a protection against being overwhelmed by the urgency of need in the momentary and the local, while in academia we still have to argue that for the hungry the right to food is not a philosophical problem.]

The essays in this book were written over a 20-year period and were directed at different audiences, some academic colleagues and some activists with little technical knowledge. Not all chapters will be equally relevant to everyone. Redundancy is usually undesirable in books, but here it is justified by two considerations: the removal of repetition would destroy the coherence of some chapters, and since the approach is largely unfamiliar, its repetition in different contexts may not be amiss.

Some of the entries are short essays from our column "Eppur´ Si Muove" that ran in the journal *Capitalism, Nature, Socialism*. These include "Are We Programmed?" about genetic determinism, "The Politics of Averages" about statistics, "Schmalhausen's Law" about vulnerability, "Life on Other Worlds," "Evolutionary Psychology," as well as several others. Longer pieces, some previously published, discuss uncertainty, the political economy of agriculture, Cuba, systems theory, model building, the organism/environment relation, and chaos. And, of course, there is Isador Nabi's contribution, "Greypeace," through which in a jest we spend our rage.

There are also important topics we do not discuss. We do not have any essays specifically on feminist analysis, cultural criticism or the role of subjectivity in social life, design plans for a better world, or questions about how to get there. Here we are consumers of the work of our comrades. This can lead some critics to the mistaken conclusion that we are indifferent to these questions and are mechanistic materialists.

We have followed the same rule as in the previous book, *The Dialectical Biologist*: we do not say anything where we have nothing to add.

PART ONE

1

The End of Natural History?

Biologists in the present century find themselves in a deeply contradictory position on questions of diversity and change. They are the inheritors of a nineteenth-century natural historical and evolutionary tradition in which the immense diversity of organisms and the long-term change that has occurred in the living world were at the very center of interest. There are millions of distinct species now extant representing less than 0.1 percent of all the species that have ever lived, and these too will become extinct. Yet only a minute fraction of all the kinds of organisms that might be imagined have ever or will ever exist. No two individuals within a species are identical, the species composition is always changing, population sizes vary markedly from year to year, and the physical conditions of life are in constant flux.

In the late eighteenth and nineteenth centuries, the ideology of change, central to the bourgeois revolutions and the social upheavals necessary for the growth of capitalism, was transferred easily onto the natural world. Herbert Spencer declared change to be "a beneficent necessity," and although it made Tennyson sad, he heard nature cry, "I care for nothing, all shall go." But the bourgeois revolutions succeeded and the Whig interpretation of history has become Whig biology. We are at the End of Natural History. The world has settled down, after a rocky start, to a steady state. Constancy, harmony, simple laws of life that predict universal features of living organisms, and the self-reproduction and absolute dominance of a single species of molecule, DNA, are the hegemonic themes of modern biology. Biologists suffer from a bad case of physics envy, and no branches of biology have been more ruthless in their search for a Hamiltonian, a single equation whose maximization will character-ize the entire biosphere, than ecology and evolution. Indeed, the price of admission into "real science" for these natural historical fields has been to give up their concentration on change and contingency and to prove their status as science, rather than mere butterfly collecting, by producing some universal predictive laws. If there must be change, at least let it be caused by some sim-ple law-like force.

On the model of Newtonian physics, change and diversity, rather than being the natural state of things, become deviations from the natural state of rest or regular linear motion, deviations that must be explained by externali-ties. But there's the rub. In classical physics, systems are sufficiently isolated from each other so that their ideal motion can be studied in isolation, taking into account the effect of an external impetus. The moon will continue in its utterly predictable course around Earth unless some very large object intrudes from outer space. But every population, species, and community, indeed the whole damned biosphere, is constantly changing in what appear to be unpre-dictable ways. Nor are the boundaries between the system and outer space so clear. How are we to explain system change as a result of unpredicted external-ities if we are not sure what is external?

There have been two responses, one from a prescientific tradition, and one from the bowels of physics itself.

The first kind of response denies the constant turnover and instability of living organisms while it alienates the human species from the rest of nature and reasserts the reality of the distinction between artificial and natural. Human technological society, disturbing the natural world from its normal state of harmony and balance, becomes the externality. In a transformation of quantity into quality, what was in the early stages of its evolution just another part of the harmonious balanced whole, escapes into another sphere of action and becomes an autonomous actor dominating and exploiting the rest of nature from the outside. It does this, of course, at its peril, since, like any exploiter, it may extinguish both itself and the system that supports it by imprudent exploitation. Under this model, the task of science is to uncover the laws of behavior of the unperturbed natural world and to use these laws to hold in check the effects of the external perturbing force.

The other response does not attempt to identify externalities that cause
unpredictable irregularities in an otherwise simple law-like system, but denies
the very existence of the irregularities and asserts the predictability of the biosphere from simple generating principles. There have been three such attempts
in the last twenty-five years, whose names are metaphors for the anxiety of
meaninglessness that has engendered them. They are catastrophe theory,
chaos theory, and complexity theory. All are attempts to show that extremely
simple relationships in dynamical systems will lead to what seem at first sight
to be unpredictable changes and extraordinary diversity of outcome, but which
are, in fact, utterly regular and law-like.

Catastrophe theory—developed in the 1960s by the mathematician Rene
Thom—shows that in some systems, which are changing in time according to
quite simple mathematical laws, the changes observed may be continuous and
gradual deformations of the state at a previous instant, and at a critical point the
entire shape of the system will undergo a "catastrophic" change and then continue its development along a totally new pathway. Many physical deformations
under continuously increasing force will reach a critical point at which they
will break like a bent branch. The classic example, known by sometimes
painful experience to the denizens of the Malibu beach, is the breaking wave.
As a swell develops into a deep convex curve there is a continuous deformation
of shape whose tubularity is suddenly and catastrophically lost at a critical
point in its roll, and the wave comes crashing down. The practitioners of catastrophe theory hoped that it would provide the explanation of changes in shape
during the development of individual organisms, and of the extinction of
species, among other things, but there is currently no trace of this theory in
biological practice. Indeed, the externalities view has more recently triumphed
in the claim that truly "catastrophic" events, meteor impacts, rather than mathematical catastrophes, have been responsible for a major part of species extinc-
tions. The fascination with the possibility of these external catastrophes has
resulted in a complete neglect of the question of why every species goes
extinct, with or without meteors.

In the 1980s, chaos theory was introduced to show that some very simple
dynamic systems may go to equilibrium or undergo regular oscillations in one
range of parameters and in other ranges will pass from one state to another in
what appears to be a totally random fashion, but which in fact can be exactly
predicted, moment by moment, from the equations of motion. So an uncertain

and diverse world is really the solution to a trivially simple equation. In particular, mathematically chaotic regimes were offered as an explanation for the
unpredictably varying population sizes that species typically display from generation to generation. Where chaos theory reigns, historical contingency disappears. The entire demographic history of a population from its initial condition is already immanent in the deterministic equation of its growth and is completely fixed by processes internal to the organisms that make up the population. No reference need be made to historical processes in the outside world or
to random variation that arises from the finiteness of real populations. Thus far,
biologists have been unable to make use of chaos theory outside of the specu-
lative realm, because no one knows how to reconstruct these hypothetical ahistorical equations of motion from data that appear as random.

Most recently the thinkers at Santa Fe Institute have begun to develop a theory of complexity which, they promise us, will generate the dazzling variety of
life histories from the behavior of networks of simple entities with lots of simple connections. Not wanting to break with previous speculations, they also
claim that living systems are "at the edge of Chaos." There will be "laws" of
complexity of which life will be one example, but only one. Complexity theory is yet another attempt to produce a theory of order in the universe, though
one that is vastly more ambitious than astrophysics. Not only was the entire
history of the stars immanent in that millionth of a second when the universe
began but the history of life as well. It is not simply that we have reached the
end of history, there never was any history to begin with.

None of these theories, all meant to tame diversity and change, and most
important, to expunge historical contingency, envisions the alternative, that living beings are at the nexus of a very large number of weakly determining forces
so that change and variation and contingency are the basic properties of biological reality. As Diderot said, "Everything passes, everything changes, only
the totality remains."

2

The Return of Old Diseases
and the Appearance of New Ones

A generation ago, the commonsense position of public health leaders was that infectious disease had been defeated in principle and was on the way out as an important cause of sickness and mortality. Medical students were told to avoid specializing in infectious disease because it was a dying field. Indeed, the Epidemiology Department at the Harvard School of Public Health specialized in cancer and heart disease.

They were wrong. In 1961, the seventh pandemic of cholera hit Indonesia; in 1970, it reached Africa, and South America in the 1990s. After retreating for a few years, malaria came back with a vengeance. Tuberculosis has increased to become the leading cause of death in many parts of the world. In 1976, Legionnaires' disease appeared at a convention of the American Legion in Philadelphia. Lyme disease spread in the Northeast. Cryptosporidiosis affected 400,000 people in Milwaukee. Toxic shock syndrome, chronic fatigue syndrome, Lassa fever, Ebola, Venezuela hemorrhagic fever, Bolivian hemorrhagic fever, Crimean-Congo hemorrhagic fever, Argentine hemorrhagic fever, hanta virus, and, of course, AIDS, have confronted us with new diseases. The doctrine of the epidemiological transition was dreadfully wrong. Infectious disease is a major problem of health everywhere.

Why was public health caught so completely by surprise?

Part of the answer is that science is often wrong because we study the unknown by making believe it is like the known. Often it is, making science possible, but sometimes it is not, making science even more necessary and surprise inevitable. Physicists in the late 1930s were lamenting the end of atomic physics. All the fundamental particles were already known—the electron, the neutron, and the proton had been measured. What more was there? Then came the neutrinos, positrons, mesons, antimatter, quarks, and strings. And each time, the end was declared.

But the explanation demands something more than the obvious fact that science will often be wrong. Before we can answer why public health was caught by surprise, we have to ask: What made the idea of the epidemiological transition seem so plausible to the theorists and practitioners of health?

There were three main arguments:

1. Infectious disease had been declining as a cause of death in Europe and North America for nearly a hundred and fifty years, since the causes of mortality were first systematically recorded. Smallpox was almost gone, tuberculosis was decreasing, malaria had been driven out of Europe and the United States, polio had become a rarity, and the childhood scourges of diphtheria and whooping cough were on their way out. Women were no longer dying of tetanus after giving birth. Just look ahead: the other diseases would go the same way.

2. We had ever better "weapons" in the "war" against disease: better laboratory tests to detect them, drugs, antibiotics, and vaccines. Technology was advancing, while the germs had to rely on their only ways of responding— by mutations. Of course, we were winning.

3. The whole world was developing. Soon all countries would be affluent enough to use the advanced technologies and acquire a modern health portrait.

Each of these arguments was loosely plausible, and each of them wrong. The problem is that although they seem to be historical arguments, they completely lack an understanding of historical contingency or the way in which historical changes alter the conditions of future change.

First, public health professionals had too short a time horizon. If instead of counting only the last century or two they had looked at a longer period of human history they would have seen a different picture. The first confirmed eruption of plague—the Black Death—hit Europe in the time of the Emperor

Justinian when the Roman Empire was in decline. The second plague spread in fourteenth-century Europe during the crisis of feudalism. What the relation of economic and political events was to these outbreaks is unclear, but when the historical record is more complete the causal paths are easier to follow. The great plague of northern Italy at the beginning of the seventeenth century was directly consequent to the famine and widespread movement of armies during the dynastic wars of the period. And the most devastating epidemiological event we know of accompanied the European conquest of the Americas, when a combination of disease, overwork, hunger, and massacre reduced the Native American population by as much as 90 percent. The Industrial Revolution brought the dreadful diseases of the new cities that Engels wrote about in relation to Manchester in his *The Condition of the Working Class in England.*

So instead of the claim that infectious disease is in decline forever, we have to assert that every major change in society, population, use of the land, climate change, nutrition, or migration is also a public health event with its own pattern of diseases.

Waves of European conquest spread plague, smallpox, and tuberculosis. Deforestation exposes us to mosquito-borne, tick-borne, or rodent-carried diseases. Giant hydroelectric projects and their accompanying irrigation canals spread the snails that carry liver flukes and allow mosquitoes to breed. Monocultures of grains are mouse food, and if the owls and jaguars and snakes that eat mice are exterminated, the mouse populations erupt with their own reservoirs of diseases. New environments, such as the warm, chlorinated circulating water in hotels, allow the Legionnaire's bacteria to prosper. It is a widespread germ, usually rare because it is a poor competitor, but it tolerates heat better than most, and it can invade the larger but still microscopic protozoa to avoid chlorine. Finally, modern fine-spray showers provide the bacterium with droplets that can reach the furthest corners of our lungs.

Second, public health was narrow in another way: it looked only at people. But if veterinarians and plant pathologists had been consulted, new diseases would have been frequently seen in other organisms: African swine fever, mad cow disease in England, the distemper-type viruses in North Sea and Baltic mammals, tristeza disease of citrus, bean golden mosaic disease, leaf-yellowing syndrome of sugarcane, tomato Gemini virus, and the variety of diseases killing off urban trees would have made it obvious that something was amiss.

The third way public health was too narrow was in its theory: not paying any real attention to evolution or the ecology of species interactions. Theorists of public health did not realize that parasitism is a universal aspect of evolving life. Parasites usually don't do too well in free soil or water and so they adapt to the special habitats of the inside of another organism. They escape competition (almost) but have to cope with the partly contradictory demands of that new environment: where to get a good meal, how to avoid the body's defenses, and how to find an exit and get to somebody else. The subsequent evolution of parasites responds to the internal environment, external conditions of transmission, and whatever we do to cure or prevent the disease. Large populations of crops, animals, or people are new opportunities for bacteria and viruses and fungi, and they keep trying.

A deep problem is the failure to appreciate the evolutionary change that occurs in disease organisms as a direct consequence of the attempts to deal with them. Public health theorists did not consider how the bugs would react to medical practice, even though drug resistance had been reported since the late 1940s and pest managers already knew of many cases of pesticide resistance. The faith in magic bullet approaches to disease control and the widespread use of military metaphors ("weapons in the war on . . . "; "attack"; "defense"; "come in for the kill") made it harder to acknowledge that nature, too, is active, and that our treatments necessarily evoke some responses.

Finally, the expectation that "development" would lead to worldwide prosperity and major increases in resources applied to health improvement is a myth of classical development theory. During the Cold War, challenges to the World Bank/IMF approach to development were marginalized as communist. In the actual world of dominance of already formed rich economies, the poor nations obviously could not close the gap with the rich, and even when their total economies grew it did not mean that the mass of people prospered or more resources were devoted to social need.

More deeply, social processes of poverty and oppression and the actual conditions of world trade were not the stuff of "real" science that deals with microbes and molecules. So a cholera outbreak is seen only as the coming of cholera bacteria to lots of people. But cholera lives among the plankton along the coasts when it isn't in people. The plankton blooms when the seas get warm and when runoff from sewage and from agricultural fertilizers feed the algae. The products of world trade are carried in freighters that use seawater as ballast that is discharged before coming to port, along with the beasts that live in that ballast water. The small crustaceans eat the algae, the fish eat the crustaceans, and the cholera bacterium meets the eaters of fish. Finally, if the public health system of a nation has already been gutted by structural adjustment of the economy, then the full explanation of the epidemic is, jointly, *Vibrio cholerae* and the World Bank.

So, at one level of explanation, the failure of public health theory identifies mistaken ideas and too narrow a vision. But these in turn require further explanation. The doctors who looked only at the last 150 years were educated people. Many studied the classics. They knew that history did not begin in nineteenth-century Europe. But earlier times somehow did not matter to them here. The rapid development of capitalism led to ideas about the unique novelty of our own time, immortalized by Henry Ford as "History is bunk." They share American (and less extremely, European) pragmatism, an impatience with theory (in this case evolution and ecology). Therefore they did not see the commonality of plants and people as species among species. Ministries of health do not talk to ministries of agriculture. Agriculture schools are rural and state supported, their students often drawn from farm communities. Medical schools are urban and usually private, and their students come from the urban middle class. They do not fraternize or read the same journals. The pragmatism of both groups is reinforced by the sense of urgency to meet an immediate human need.

The development of a coherent epidemiology is thwarted by the false dichotomies that permeate the thinking of both communities: the either/ors of biological/social, physical/psychological, chance/determinism, heredity/environment, infectious/chronic, and others that we will discuss in other chapters.

One more level of explanation helps us understand the intellectual barriers that led to the epidemiological surprise. Narrowness and pragmatism are characteristic of the dominant ways of thought under capitalism, where the individualism of economic man is a model for the autonomy and isolation of all phenomena, and where a knowledge industry turns scientific ideas into marketable commodities—precisely the magic bullets that the pharmaceutical industry sells people. The long-term history of capitalist experience encourages those ideas that are reinforced by the organizational structure and economics of the knowledge industry to create the special patterns of insight and ignorance that characterize each field and make inevitable its own particular surprises.

3

False Dichotomies

Our understanding of nature is deeply constrained by the language we need in talking about it, a language that is itself the result, as well as the replicator, of long-standing ideological practice. All of science, even "radical" science, is plagued by dichotomies that seem unavoidable because of the very words that are available to us: organism/environment, nature/nurture, psychological/ physical, deterministic/random, social/individual, dependent/independent. A remarkable fraction of the radical reanalysis of nature that we ourselves have engaged in has revolved around a struggle to cut through the obfuscations that have arisen from those false oppositions.

One aspect of the dichotomies of general/particular and external/internal is the relation between averages and variations around those averages. A major divergence in explanation, especially in political struggles over the causes of disease and social dislocation, concerns the determinative importance of over-all average conditions as opposed to the role of preexistent individual variation. Where one locates the causes of tuberculosis or domestic violence—whether in social and environmental stresses or in intrinsic physical and psychic variation among individuals—has powerful political consequences.

All environments vary in space and time, from the widespread and long lasting to the extremely local and transitory events that we often call random. All organisms vary, both in response to the intricate patterns of environment and because of their own internal dynamic. For most medical, epidemiological, and social research, that variation is a nuisance, and much ingenuity goes into removing the variation experimentally or statistically in order to detect average or "main" effects. For understanding the processes of evolution, in contrast, variation between organisms within a species is the necessary ingredient for evolution by natural selection and an object of interest in its own right. Ecology, a science that developed in part as an extension of physiology, and in part as an aspect of evolution, has been somewhat confused about the importance of average conditions affecting "typical" individuals, as opposed to variation in those conditions and in the responsive properties of individuals. We need to consider the relation among the population average, its range of variation, and the extreme values that occur within the population, all aspects of the interpenetration and mutual determination of variation in organisms and their environments.

First, different traits of the same organism differ in the consequences of variation. For some traits, such as body temperature, blood sugar, or the oxygen supply to the brain or heart, a constancy of the trait itself is critical. When internal or external fluxes displace them, mechanisms come into play that bring them back within the tolerable range. For these traits, increased variation may mean either that they have been subjected to more environmental buffeting or that the self-regulatory mechanisms have been weakened. Individuals differ in their self-regulating systems, but under the "normal" conditions in which individuals have evolved outcomes are essentially the same—all the temperatures of blood sugars or brain oxygen levels are within the tolerable range. Under more extreme conditions of temperature or nutrition or elevation, the individual differences become more important, as some manage to keep the physiology in the tolerable range but for others a critical threshold is crossed resulting in death. Finally, in even more extreme conditions, none of the individuals have enough regulatory capacity and variation disappears along with the population.

Other traits are part of the regulatory system itself, and therefore are themselves varying. Changing metabolic rates stabilize temperature. Varying food intake and insulin levels buffer blood sugar. Redistribution of blood keeps the brain breathing. Varying activities seem to be important for human well-being. For these traits, variation indicates that things are working well. If malnutrition prevents us from raising metabolic rates, if labor discipline prevents us from varying our activity or eating as part of self-maintenance, then our physiological state can move out of the tolerable range, and we have the heart disease, muscle pains, headaches, and depression of alienated labor. We avoid here the added complexity that the same traits are both regulated and regulators.

Second, although many traits are continuously variable, often critical thresholds distinguish between good and bad outcomes. But the numbers of individuals who are across the threshold changes as a consequence of the average level of conditions and, as a result, the manifest variation in the trait changes. Differences in susceptibility to disease, and especially mortality, are magnified at low nutritional levels. Measles, a disease that consumes protein, did not kill students in New York City elementary schools when we were children, although everyone contracted the disease. During the same era measles was the leading cause of child mortality in already malnourished West Africa, so that differences in individual metabolism and resistance would have been of the greatest importance.

The same phenomenon applies to the incidence of casual violence or the prevalence of rape. Not everyone who watches TV violence commits murder; not all sexist men are rapists. But if the average systemic validation of violence increases, then perhaps 1/1,000 instead of 1/10,000 will so act. A serious error in the analysis of causes arises when we fail to take into account the dialectic of average conditions and variations in response to those conditions, and instead take variability as an independent causal force having an intrinsic magnitude. When urban rebellions broke out in American cities in the 1960s, one response was to say that when people are sufficiently deprived by others of social power and economic security while the consciousness of their deprivation becomes heightened, they will rebel. The reaction by the right to this explanation was to point out that everyone in the inner cities did not burn and loot, but that these activities were the work of a small group. This group, it was claimed, had a biological predisposition to violence. Thus the explanation is relocated from the average level of conditions to an intrinsic preexistent variability among individuals. Putting the issue of biological causes aside, it is certainly true that individuals differ in their willingness to put up with insult and injury, and also in how they choose to express their unwillingness. But whether a significant number will find inaction intolerable surely depends on the level of that insult and injury. So the level of oppression that leads to rebellion depends upon the pattern of variation in response among individuals, but that variation in response depends upon the level of the challenge.

[Third, quite aside from the effect of average level on the proportion of individuals falling over a threshold, differences in average conditions have a magnifying or reducing effect on the quantitative response of organisms to small variations in environment. An old problem in plant breeding is whether the difference between new varieties and old ones is most easily observed under stress conditions or under the optimal conditions of growth. The arguments were partly a reflection of *a priori* ideological views about social relations. Is the true test of individual merit one's behavior "under fire," in the most challenging circumstances that separate the sheep from the goats, or will the conditions allowing the greatest flowering of intrinsic abilities magnify differences that are small in depauperate circumstances? Partly, the argument is about which traits of the organism are at issue. Consider, for instance, infant deaths in poor communities. They are not spread out uniformly in the community but tend to cluster in those households with a low educational level, little social support, poor nurturing skills, etc., whereas in an affluent community these deficits may be merely inconvenient, rather than leading to mortality.

But the analysis of the causes themselves continues in the same way. Illiteracy or poor skills are not givens[Perhaps a slight visual deficiency made the blackboard blurry in a poorly lighted, overcrowded school room. Poor vision leads to a learning deficiency, discouragement, and dropping out. The individual variation was a consequence of the lack of means (attention, lightbulbs, glasses) and the predominance of a deviation-enhancing mechanism, low vision, that would mobilize the restorative (deviation-reducing) self-regulation in more fortunate circumstances. At the next level we come back to the individual variation. After all, not all children arrive in school with poor vision, a personal misfortune. Ah, but poor vision is often associated with vitamin A deficiency in poor communities. True, but not everyone. . . . Thus, we cycle back and forth between a focus on the systemic, average conditions that make people vulnerable and the range of variation that guarantees that some will fall over some critical value. A correct analysis and program for action demands that average and variant, systemic and individual explanations, are not seen as mutually excluded alternatives, but as codeterminants of the same reality.]

4

Chance and Necessity

Since the major breakthroughs of quantum physics in the 1920s and 1930s and the discovery of random mutation as an evolutionary force, people have been asking whether the world is determinate or random. The usual implication of *random*, whether it be a "random" number or a "random" mutation, is that some event has arisen that could not have been predicted no matter how much information was available about the prior state of the world. The spontaneous disintegration of a radioactive nucleus is said to be "random" because there is no difference in state between the nucleus and other nuclei up until the instant that it disintegrates. Randomness has been associated with lack of causality, and with unpredictability and thus of irrationality, a lack of purpose, and the existence of free will. It has been invoked as the negation of lawfulness and therefore of any scientific understanding of society. It then becomes a justification for a reactionary passivity. As the bumper sticker says, "Shit happens." So stop complaining.

For the most part, however, randomness and causation, chance and necessity, are not mutually exclusive opposites but interpenetrate.

First, the fundamentalist approach to randomness that equates it with lack of any causation excludes a large domain of events to which the notion of randomness applies. If, hurrying to a meeting, you rush out into the road and are struck by a "random" car whose driver was on his way to work, it is nevertheless clear that both your path and that of the car were determined and even planned well in advance. What makes the encounter "random" is that the causal pathways of the colliding objects were independent of each other. Opponents of the Darwinian mechanism of evolution have sometimes accused evolutionists of believing that complex organisms have come into existence by purely random processes. After all, don't biologists claim that all mutations are random? But this confuses the two concepts of randomness. It may indeed be true that some mutations are the result of indeterminacy at the quantum mechanical level, but that is beside the point. The essence of Darwinism is that the processes that produce the variation among organisms in the first place, the mutations, are causally independent of the processes that lead to the incorporation of these variations into the species. Mutations are random *with respect to* natural selection. Unless we are dealing with phenomena at the deepest level of quantum mechanics, randomness means causal independence, not the lack of causation.

Randomness by causal independence has powerful implications in biology. Biological objects differ from other physical systems in two important respects. They are intermediate in size and they are internally functionally heterogeneous. As a consequence their behavior cannot be determined from a knowledge of only a small number of properties, as one can specify the orbit of a planet from the planet's distance from the sun, its mass, and its velocity, without being concerned about what it is made of. Biological objects are at the nexus of a very large number of individually weak forces. Although there are indeed interactions among these forces (and the interactions are often of the essence), it is also the case that there are very large numbers of subsystems of causal pathways that are essentially independent of one another, so that their effects on an organism appear as random with respect to one another. Variations in nutrients over a meadow are causally independent of genetic variations among windborne seeds that fall in different parts of the meadow, so the interaction between environment and genotype that determines the growth of the plant is an interaction of factors that are random with respect to one another.

Individual local events that are the intersection of large numbers of specific causal pathways impinge on society as if they were random. The death of Franklin Roosevelt was surely not an accident with respect to the president's own body, circulation, and general state of health. But it was an accident at the level of international politics.

Second, determinacy can arise out of randomness, even the abyssal randomness of quantum physics. The most accurate clocks in the world, measuring time to nanoseconds with no cumulative error, are based on random radioactive decay. Whether individual events are random in the quantum sense or only in the sense of independent causes, the cumulation of large numbers of independent occurrences in averages, sums, and probabilities allows extremely accurate and repeatable prediction. Moreover, the statistical regularities can be altered by determinate processes. Although we cannot predict which mutation will occur in a gene when we change the temperature or expose an organism to a mutagenic chemical, we know the average effect of increasing temperature, of ionizing radiation, of toxic chemicals, and even of the presence of other genes, on both the average mutation rates and on how drastic those mutations may be in their effect.

The Chernobyl meltdown was both an accident and a caused event. Some months before the catastrophe the director of that nuclear power plant gave a reassuring interview in which he said that the safety backup system was so good that we would not expect a serious accident more often than once in 10,000 years. The chilling aspect of this is not that he was wrong, but that even if he overestimated his own plant's safety, he was right. There are more than 1,000 reactors in Europe, so the chance of something happening to one of them is about 1 in 10 years. It happened to happen at Chernobyl. For the director it was an unlikely accident, but for Europe it was not so improbable. A chance event with low probability becomes a determinate certainty when there are a large number of opportunities.

Third, randomness can arise from determinacy. A standard technique in the computer simulation of real world processes is the generation of so-called random numbers. But these numbers are more properly called *pseudo-random* numbers because they are generated by some extremely simple deterministic numerical rule: for example, by using the middle 10 digits of the successive powers of some starting number. If I know the starting number I can exactly reproduce the pseudo-random sequence. Nevertheless the numbers are "random" as far as the process I am simulating is concerned, because the rule of generating them is utterly unrelated to the rest of the process.

Fourth, random processes are causally constrained. "Random" does not mean "anything goes." Random changes in organisms are nevertheless changes in the neighborhood of the preexistent state. A mutation in green peas or in fruit flies results in the alteration of the development of green peas or fruit flies. The flies will not produce vines that climb trellises nor will the peas fly around and lay eggs. The dangerous "mutants" of early science fiction are fictional precisely because they are impossible in the light of the organization of the body in which they occur, not because they are rare. Random changes are then unpredictable only within the domain of the allowable, and one of the major unsolved problems of ecology and evolution is how to delimit the allowable domain for organisms and communities within which random processes can operate. It is precisely the problem of historical materialism: Where can you get from here?

The interpenetration of chance and determination bears on the problem of how there can be a scientific approach to society when individual human behavior and consciousness seem unpredictable. Those who despair point out that people are not machines, that there are subjective processes in the making of decisions, that it is not classes but individuals who make choices. Terms such as "the human factor" or "subjective factors" with their implication of chance and unpredictability are invoked as the negation of regularity and lawfulness. And indeed it is true that individual behavior and consciousness are the consequences of the intersection of a large number of weakly determining forces. But it does not follow that where there is choice, subjectivity, and individuality there cannot also be predictability. The error is to take the individual as causally prior to the whole and not to appreciate that the social has causal properties within which individual consciousness and action are formed. While the consciousness of an individual is not determined by his or her class position but is influenced by idiosyncratic factors that appear as random, those random factors operate within a domain and with probabilities that are constrained and directed by social forces.

5

Organism and Environment

[Nothing is more central to a dialectical understanding of nature than the realization that the conditions necessary for the coming into being of some state of the world may be destroyed by the very state of nature to which they gave rise. As it is in nature, so it is in the study of nature. Darwin's most powerful contribution to the development of modern biology was not his creation of a satisfactory theory of evolutionary mechanism. Rather, within that theory, it was his rigorous separation of internal and external forces that had, in previous theories, been inseparable. For Lamarck, the organism became permanently and heritably transformed by its willful striving to accommodate itself to nature and so incorporated that outer nature into itself. By totally confounding inner and outer forces in an unanalyzable whole, premodern biology was in fetters that made further progress impossible. Darwin's division of forces into those that were completely internal to organisms and determined the variation among individuals and those that were external, the autonomous forces molding the environments in which organisms found themselves accidentally, "burst those fetters asunder." For Darwinian biology the organism is the nexus of the internal and external forces. It is only through natural selection of internally produced variations, which happen to match by chance the externally generated environmental demands, that what is outside and what is inside confront each other. Without such a separation of forces the progress made by modern reductionist biology would have been impossible. Yet for the scientific problems of today, that separation is bad biology and presents a barrier to further progress.]

The development of an organism is not an unfolding of an internal autonomous program but the consequence of an interaction between the organism's internal patterns of response and its external milieu. Many experiments have demonstrated and a great deal has been written about codetermination of the organism by the interplay between gene and environment in development. Even there, however, the environment is treated as external impingement on an autonomous program or as necessary resources for its realization. But aspects of the environment that are regular occurrences become themselves part of the developmental process. When a seed germinates only after a soaking rain, it is not merely responding to a signal that conditions are suitable. The rain becomes a factor of development as much as the proteins of the seed coat. The development of our ability to see presupposes light, the development of our muscles presupposes movement.

What has received far less attention, both in concept and in practice, is the reciprocal codetermination, the role of the organism in the production of the environment. Darwinism represents the environment as a preexistent element of nature formed by autonomous forces, as a kind of theatrical stage on which the organisms play out their lives. But environments are as much the product of organisms as organisms are of environments. The Darwinian alienation of the environment from its producer, though a necessary condition for the formation of modern biology, stands in the way both of the further development of the sciences of evolution and ecology, and of the elaboration of a rational environmental politics.

There is no organism without an environment, but there is no environment without an organism. There is a physical world outside of organisms and that world undergoes certain transformations that are autonomous. Volcanoes erupt, the earth precesses on its axis of rotation. But the physical world is not an environment, only the circumstances from which environments can be made. The reader might try describing the environment of an organism that he or she has never seen. There is a noncountable infinity of ways in which the bits and pieces of the world might conceivably be put together to make environments, but only a small number of those have actually existed, one for each organism. The notion that the environment of an organism preexists the organism is embodied in the concept of the "ecological niche," a kind of hole in ecological space that may be filled by a species, but it may also be empty, waiting for an occupant. Yet if one asks an ornithologist to describe the "niche" of, say, a phoebe, the description will be something like, "The phoebe flies south in

the fall, but returns to the northern mixed forest early in the spring. The male
marks out a territory that it patrols and over which it forages for insects, while
the female, arriving two weeks later, builds a nest of grass and mud on a horizontal ledge into which she deposits four eggs. Usually insects are caught in
flight but nestlings are fed by regurgitation of insects caught near the ground."
[The entire niche is described by the sensuous life activities of the bird, not by
some menu of external circumstances. Organisms do not experience or fit into
an environment, they construct it.]

[First, organisms juxtapose bits and pieces of the world and so determine
what is relevant to them. The grass growing at the base of a tree is part of the
environment of a phoebe that uses it to make a nest, but not of a woodpecker
who makes an unlined nest in a hole in the tree. A stone lying in the grass is part
of the environment of a snail-eating thrush that uses it as an anvil, but is not part
of the world of the flycatcher or woodpecker. Temperature would seem like an
externally given, fixed condition, but every terrestrial organism is surrounded
by a shell of warm moist air produced by its own metabolism, a shell that constitutes its most immediate "environment." When we ask, "What is the temperature tolerance of an ant?" we discover many different meanings. What temperature can an ant tolerate for a few minutes or hours while foraging? What temperature can an ant nest in a tree tolerate for a complete life cycle? What temperatures allow for sufficient vegetation and prey to permit a population of ant
colonies to persist in contact with other ant species?

Even the relevance of fundamental physical phenomena is dictated by the
nature of the organism itself. Size is critical. Although gravitation is an important force in the immediate environment of large objects like trees and human
beings, it is not felt by bacteria in a liquid medium. For them, because of their
size, Brownian motion is a dominant environmental factor, while we are not
buffeted to and fro by bombarding molecules. But that size disparity is a consequence of genetic differences between life-forms, so just as environment is a
factor in the development of an organism, so genes are a factor in the construction of the environment.

[Second, organisms remake the environment at all times and in all places.
Every organism consumes resources necessary for its survival, and produces
waste products that are poisonous to itself and others. At the same time organisms create their own resources. Plant roots produce humic acids that facilitate
symbiotic relations and these change the physical structure of the soil in ways

that promote absorption of nutrients. Ants farm fungi and worms construct their own housing. Many species change the conditions of their surroundings in such a way as to prevent their own offspring from succeeding them. That is what it means to be a weed. Every act of consumption is an act of production and every act of production is an act of consumption. And in the dialectic of production and consumption the conditions of existence of all organisms are changed. At the present no terrestrial species can evolve unless it can survive an atmosphere of 18 percent oxygen. Yet that oxygen was put into the atmosphere by early forms of life that lived in an atmosphere rich in carbon dioxide that they made unavailable to later forms by depositing it in limestone and in fossil hydrocarbons.

Third, organisms by their life activities modulate the statistical variation of external phenomena as they impinge on the organisms. Plants average their productivity over diurnal and seasonal variation in sunlight and temperature by storing the products of photosynthesis. Potato plants store carbohydrate in tubers. We appropriate that storage in our body fat, in warehouses, and in money.

Finally, the organism transduces the physical natures of the signals from the outer world as they are made part of its effective environment. The rarefaction of the air that strikes my eardrums and the photons that strike my retina when I hear and see a rattlesnake are transformed by my physiology into elevated levels of a chemical signal, adrenaline, and that transformation is a consequence of my mammalian biology. Were I a rattlesnake a very different transformation would occur.

A consequence of the codetermination of the organism and its environment is that they coevolve. As the species evolves in response to natural selection in its current environment, the world that it constructs around itself is actively changed. At present, because of the narrow problematic of both evolutionary biology and ecology that envision a changing organism in a static or slowly changing autonomous outer world, we know little beyond the anecdotal about the way in which changing organisms lead to changing environments. We know rather more, but still far too little, about how, through their life activities, organisms are the active makers and remakers of their milieu. But a rational political ecology demands that knowledge. One cannot make a sensible environmental politics with the slogan "Save the Environment" because, first, "the" environment does not exist, and second, because every species, not only the human species, is at every moment constructing and destroying the world it inhabits.

✳ How does this address the material agency question?

✳ If all organisms are destroying the environment why are humans 'bad' (is a Marxist view of the 'social' assumed?)

6

The Biological and the Social

Struggles for legitimacy between political ideologies eventually come down to struggles over what constitutes human nature. At present, in its starkest form, the struggle is between a vulgar biological determinism, typified by sociobiology, and an extreme subjectivity. For determinism, all social phenomena are merely the collective manifestation of individual fixed propensities and limitations coded in human genes as a consequence of adaptive evolution. At the opposite pole, subjectivity claims that all human realities are created by socially determined consciousness, unconstrained by any prior biological and physical nature, all points of view being equally valid. At best, liberal thought attempts to combine the biological and the social in a statistical model that assigns relative weights to the two, allowing for some component of interaction between them. But the division of causality between distinct biological and social causes that then may interact misses the real nature of their codetermination.

Like any other species, human beings clearly have certain biological properties of anatomy and physiology that both constrain and enable them, properties that are partly shared with other organisms as a consequence of being living systems, and that are partly unique as a consequence of the particular genes possessed by our species. We all have to eat, drink, and breathe; we are all susceptible to attack by pathogens; [there are limits to the external temperatures that our naked bodies can survive; and we will all die. No historical contingency or change in consciousness can remove those necessities. But at the same time, the central nervous system of human beings, combined with their organs of speech and manipulative hands, leads to the formation of social structures that produce the historical forms and transformations of those needs. Whereas human sociality is itself a consequence of our received biology, human biology is a socialized biology.]

At the individual level our physiology is a socialized physiology. The time course of blood pressure or serum glucose with age, the integrity of the epithelial interfaces between the insides and outsides of our bodies, the ways in which we perceive distance or pattern, the availability of our immune systems for confronting invasions by other organisms, and the formation and disruption of linkages in our brains—all are variably dependent on class position, the nature of work, the social status of our ethnicity, the commodities that circulate in our society, and the techniques of their production.

At the next level we select our environments actively or they are selected by others for us, sometimes on a moment-to-moment scale as when one is forced to work in the heat of the midday sun, or sometimes through less frequent decisions about where to live, what work to do, with whom to associate, when and how to reproduce. But an environment for settlement or work has many more properties than those that guided the selection. A site on a river may be chosen as a political center for the ease of collecting tribute there, but can also be a breeding place for snails that transmit schistosomiasis.

The socially conditioned construction and transformation of our environments determine the actual realization of biological limits. The boundaries of human habitation do not correspond to the geographical extremes of temperature or oxygen or food availability that could support us in a socially untransformed world, but correspond to those places where economic activity and political power provide the means to regulate our temperature, provide oxygen, and import food. In so doing we also change the determinants of the boundaries of other organisms. The northern boundary of wheat in North America is not the limit of where wheat plants can mature successfully, but where the profitability of wheat in good harvest years makes up for the poor ones so that an average profitable return on wheat is greater than for alternative crops.

As technology provides cultural mediations between ourselves and physical conditions, new environmental impacts are created. A severe winter in an urban environment does not produce frostbite but hunger—when the poor divert resources from food to fuel. Racism becomes an environmental factor affecting adrenals and other organs in ways that tigers or venomous snakes did in earlier historical epochs. The conditions under which labor power is sold in a capitalist labor market act on the individual's glucose cycle as the pattern of exertion and rest depends more on the employer's economic decisions than on the worker's self-perception of metabolic flux. Human ecology is not the relation of our species in general with the rest of nature, but rather the relations of different societies, and the classes, genders, ages, grades, and ethnicities maintained by those social structures. Thus, it is not too far-fetched to speak of the pancreas under capitalism or the proletarian lung.

The socialization of the environment also determines which aspects of individual biology are important for survival and prosperity. Melanin metabolism, no longer of much relevance for heat balance, has become a sign of social location that affects the way in which people have access to resources and are exposed to toxicities and insults. But an organism under stress along one axis of its conditions of existence will be more vulnerable to stresses along other axes as its conditions of homeostasis are taxed. Thus, there will be a clustering of harmful outcomes to health and well-being in households or families under deprivation or stress, even when the conditions that precipitate the cluster seem physiologically trivial. It is the social mediation of individual biological phenomena that turns a single day's incapacity from the flu into the loss of a job for an already marginalized worker, with consequent catastrophic economic failure and a disintegration of health and the general conditions of life.

Beyond the transformation of biological needs into forms that are specific to different times and places, the kind of social interaction that is biologically possible for the human species has an even more powerful property, the property of negating individual biological limitations. No human being can fly by flapping his or her arms, nor could a crowd of people fly by the collective action of all flapping together. Yet we do fly as a consequence of social phenomena. Books, laboratories, schools, factories, communications systems, state organizations, and enterprises are the means of production for airplanes; fuel, airports, pilots, and mechanics make it possible for any of us to do what

Leonardo could not. Nor is it "society" that flies, but individual human beings who go from one place to another through the air. No human being can remember, unaided, more than a few facts and figures, but a social product, the *Statistical Abstract of the United States,* as well as the library that contains it, constitutes a negation of that limitation. But the social process leading to such a negation begins only when a condition of existence is perceived as a *limitation,* that is, when an alternative world is deemed possible. Although it may indeed be a generalized biological property of the human central nervous system to be able to make mental constructs of things that do not exist and to plan actions in advance of their willful realizations, the domain of what we imagine to be changeable is socially constructed. Indeed, the vulgar reductionist claim that human beings are inevitably driven by their biology to behave in certain ways is self-fulfilling for it takes those behaviors out of context and places them in the domain of unquestionable "facts of life," part of the substrate of unexamined conditions of existence. That is why the present ideological struggle over the biological and the social is the elementary political conflict between those who wish to change the nature of human existence and those who prefer to keep it in its present state.]

7

How Different Are Natural and Social Science?

A caricature of the study of "nature" and "society" sees social science as deeply corrupted by the subjective elements introduced by the observer, whereas natural science is carried out by objective means. And it is not only the positivist natural scientist, scornful of social science, who propagates this view.

It is often argued, especially by social scientists, that dialectics is fundamentally different in natural science than it is in social science. The difference is said to come from the active participation of human beings in the dynamics of society and especially from the unique role of subjectivity. It is not fruitful, however, to debate whether nature and society are different despite similarities or similar despite differences. Much of the dispute depends on the level of analysis. Obviously, each domain of study is different. In particle physics, quantum mechanical randomness is a central feature. In most ordinary chemistry vast numbers of relatively few kinds of atoms allow for a statistical averaging that masks micro-scale randomness. But macromolecules such as DNA are represented only once or a few times in each cell and behave mechanically. The physiology of individual organisms can be understood in part as goal directed, while the metaphor of the organism is misleading in the study of ecological communities. Societies also have their unique properties, not the least of which is the emergence of labor, culture, ideology, and subjectivity. But the question remains: Is the uniqueness of the social different in kind from the uniquenesses of other domains?

Those who argue that it is point out that the observation of social processes is itself a social process. They emphasize that social processes involve the subjectivity of the objects of study, and sometimes talk casually about "the human factor," which presumably makes uncertainty inevitable and duplication impossible. (The terms "human factor" or "human condition" are not analytic terms. They do not refer to the role of labor in our formation, to the use of language or symbols, or to sexual reproduction. Rather it is most often a term of exasperation or despair.) They add that in natural science we can design experiments and observe a large number of repetitions that cancel out many sources of error. Therefore, they claim, natural science can be objective in ways that social science cannot. They add that it would be futile to expect to have predictive equations for society, whereas even the complex patterns of the earth's atmosphere can, in principle, be thought of as obeying a very large set of as yet unspecified equations. Who could even conceive of writing equations that would predict the emergence and content of postmodernism?

This argument is fallacious for a number of reasons. First, it accepts too much of the natural scientists' self-description. Writing equations, and even prediction from them, is only one activity of science. Formulating a problem, the definitions of relevant variables, the choices of what to include or leave out, the decision as to what is an acceptable kind of answer, the interpretation of results, the rules of validation, and the linking of the conclusions from different studies into a theoretical framework are all the results of social processes, some very idiosyncratic ones, interacting with the natural phenomena being studied. Science has become very sophisticated in correcting for the idiosyncratic subjectivities of its practitioners but not for the shared biases of communities of scholars. A long tradition of the Marxist study of the scientific process is lost when Marxists take scientists at their word and accept the self-description of scientific objectivity, or indeed when postmodernists imagine that the critique of science began with Thomas Kuhn.[1]

Second, science is not the same as quantification or experiment. There have been situations where numerical results have been vital in making theoretical decisions. In tests of relativity theory, the overthrow of parity, the confirmation of the Mendelian 3:1 ratio in genetics, and the prediction of the existence of the planet Neptune from anomalies in the orbit of Uranus precise measurement has been critical. But even here the important conclusions have not been quantitative but rather qualitative or semi-quantitative: that gravitation can affect light, that genetic traits segregate, that there is something else out there beyond Uranus. Statistical tests are often used to decide that some phenomenon has or has not had a "significant effect" on some process or is more or less important than some other phenomenon. Such tests can be used to demonstrate the relation between health and class, the association between poverty and the suicide rate, and the growing concentration of wealth.

But in other discoveries, numerical results played a much smaller role: the recognition of the *Australopithecine* fossil Lucy as close to human ancestry, the formulation of the structure of DNA, the confirmation that mosquitoes transmit pathogens, the role of plaque formation in coronary heart disease, the patterns of continental drift and the expanding universe. The various roles of precise measurement separate different branches of natural and social science rather than natural and social science from each other.

Large-scale computer programs can simulate important aspects of a process, but in the end what we are left with are more numbers. These are often useful for projections as long as nothing important changes. And they are certainly essential in design, where quantitative precision can be crucial. But there is no substitute for qualitative understanding, the demonstration of a relation between the particular and the general, that requires theoretical practice distinct from the solving of equations or the estimation of their solutions.

Nor is experimentation a necessary ingredient of science. Though processes of the very small can be duplicated in the laboratory, we certainly cannot replicate supernovae or epidemics or species formation or continental drift. Here we need other methods of verification. The study of large-scale social phenomena shares with ecology, evolution, epidemiology, and biogeography the characteristic that the number of examples of each kind available for study is small compared to the number of relevant *kinds* of objects that actually exist or are possible. Therefore, replication is not possible. We cannot compare 150 socialist revolutions stratified by the degree of sexism in their societies to compare outcomes, or fifty isolated continents with and without large mammals to see how they affect the development of agriculture or the evolution of birds.

In contrast, there are relatively few kinds of atoms or fundamental particles or stars, each present in extraordinarily large numbers of essentially identical replications. But there are a reasonably large number of small businesses and local officeholders for comparative study. Here prediction is performed not on controlled experiments but on sets of data not used in making the prediction. These differences certainly affect the methodologies of the sciences and the kinds of questions they deal with, but they do not separate natural from social science.

Thus the lack of equations or of controlled experiments in social science does not make it fundamentally different from natural science as such. Nor does the question of predictability. While the classical examples of physical science showed the glorious confirmatory power of accurate prediction, the modern theory of dynamical systems reveals many situations, even rather simple ones, in which precise prediction is not possible. (Weather prediction is a notorious example. The modern interest in "chaos" was stimulated by Lorenz's attempt to solve a model of the atmosphere with only three variables and his discovery that even arbitrarily small changes in a variable could result in drastically different outcomes, and arbitrarily small errors of estimation could make the predictions extremely uncertain.) Yet even chaotic systems have regular as well as seemingly random aspects. We may not be able to project the trajectory of a self-regulating population and yet we know that it will most likely oscillate between certain bounds, and that on one of the downswings it might become extinct. The structure of capitalism makes class struggle inevitable; the uniqueness of each historical configuration makes the particular forms of class struggle and even the outcome uncertain from the perspective of that structure alone.

Thus there are two kinds of uncertainty in science: all systems, no matter how complex, have an outside from which influences not included in the theory may penetrate and have major effects; and the dynamics of complex systems themselves may result in chaos, a combination of predictable and unpredictable aspects of the process. Neither of these is unique to social science.

Subjectivity is subjective only from the inside; our theories do not describe how it feels. But subjectivity can also be studied objectively. Beliefs and feelings have causes, and are themselves causes. They may become more or less common. We can, for example, include fear or despair as links in the progress of an epidemic, responding to the prevalence of a deadly disease and the availability of effective treatment, and affecting the contagion rate that feeds back into prevalence. Changing subjectivities must be included for any realistic assessment of the AIDS pandemic. The study of many different subjectivities reveals patterns of subjectivity that make psychosocial therapies possible.

Thus there is no basis for arguing that dialectics is all right in natural science, where predictability and lawfulness prevail, but not in social science, where the erratic operation of capricious subjectivities thwarts science. Or alternatively, that dialectics is all right in social science where contradictions play themselves out before our eyes, but not in natural science where nature is deterministic and mechanical or statistical. Both dialectical materialism and the more limited insights of systems theory are relevant to understanding both natural and social processes.

8

Does Anything New Ever Happen?

The tired, discouraged author of the Book of Ecclesiastes, writing in the second or third century BCE, assures us that "there is nothing new under the sun" and that "all is vanity." More recently, the arch-Whig Francis Fukuyama allowed that perhaps things *used* to happen, but now history has ended. In the time between the two, many quaint sayings repeated the same theme, including "You can't change human nature" and "Plus ça change, plus c'est la même chose."

Claims that phenomena are radically new or only the same old story do not arise from some general ideology but are meant in each instance to do specific work. In some instances, those who prefer that there be no change, as well as those who have tried to promote change only to see their efforts frustrated, join in picking out ways in which different times are similar in order to deny difference. For instance, to support an argument that entrepreneurship is a basic and unchangeable aspect of human nature, any kind of exchange of goods is seen as "trade," and all trade is interpreted as a form of capitalist exchange. So, a Stone Age male corpse found in the Alps with more flints than he could use himself, or a couple of Cubans exchanging rationed goods to meet their different needs, are lumped into a universal human propensity for commerce (presumably on the same chromosome as the genes for the propensity to cheat on exams and distrust strangers). From such a perspective the Soviet Union was merely a continuation of the czarist empire with superficially changed rhetoric, and all revolutions are alike in that they merely replace one group of rulers with another. Yet, in a seeming reversal of ideology, bourgeois apologists have asserted that capitalism has undergone a revolutionary change, replacing domination by the owners of capital with that of technocrats as a result of "the managerial revolution," while it falls to Marxist theorists to remind us that "plus ça change, plus c'est la même chose."

It is always possible, of course, to find similarities and differences among phenomena. Darwin's tracing of evolution depended on both: the similarities betrayed common ancestry and accounted for the constraints within which divergences occurred, while the differences indicated historical divergence. If there were only differences, with each kind of organism unique and showing no common features with any others, then special creation would be a better explanation of the observations than evolution.

Depending upon the work to be done, it is appropriate to stress similarity or change. In looking at contemporary capitalism, we see the continuation of exploitation, the extraction of profits, and the changing means of production as the main source of wealth, the commodity relation penetrating everywhere. From the perspective of challenging the whole system these elements of continuity are more important than the new: the rise of information industries, the increasing independence of financial instruments several steps removed from production seen as major investment opportunities that offer the highest rate of turnover of capital, the endemicity of unemployment, and the rise of the transnational corporation. But when we plan strategies, then we have to increase the magnification and examine the novel features that affect organizing, the need for across-the-board solidarity, the increasingly dangerous position of the Untied States as a declining economic power with a first-rate military, now facing the problem of how to use its military in the service of the economy.

In looking at similarities it is important to note that two "similar" objects or events may have quite different significance and may be on quite different trajectories of development because they are in different contexts. For instance, voting is now widespread in many societies. But voting has had quite different roles: the confirmation of an existing power relation (all Germans were allowed to vote in Hitler's 1934 plebiscite, giving him the authority to rule by decree); the choice between political parties, within which the populace at large has little voice; a referendum ratifying the results of extensive prior popular consultation, as in voting on budgets in New England town meetings; a popularity contest driven by advertising technicians—all are "votes."

When conservatives emphasize the absence of change they speak of "ethnic conflict" and "ancient enmities" rather than nationalistic conflict, which represents a political choice. But if conservatives underline similarity when it is spurious, anarchist thought often emphasizes continuity, as in the belief that the catharsis of revolution creates "new people," ready and willing to live collective lives in equality and solidarity, freed from previous consciousness. The real experience of building socialism shows otherwise. Some social relations are extraordinarily tenacious, and as Rosa Luxemburg pointed out, we are attempting to build a future with the materials of the past.

The claim that nothing new is happening is a common device for opposing social and political action, either on the grounds that no action is possible because the present situation is an unchangeable constant of nature or that no new action is required because things are not materially different than they have always been. The most active current manifestations of these conservative moves oppose demands for radical action in two spheres where public consciousness has been raised to a threatening degree—social inequality and environmental deterioration. The problem of inequality has been a dominant social agony of bourgeois life since the revolutions of the eighteenth century, revolutions claiming equality as their legitimizing principle. The response to a demand that is unrealizable within bourgeois society has been to claim that really new social relations are biologically impossible because human nature is continuous with a competitive, aggressive, self-oriented, and self-aggrandizing nature built into our nonhuman ancestors by evolution. Nothing really new arose in the evolution of the human species. We are simply "naked apes" possessed of our own species-specific form of unchanged and deeply entrenched animal natures, so attempts to change social arrangements are delusory.

Our anxiety that the present form and scale of transformation of resources will soon make a materially decent life untenable for human beings has been met with the claim that nothing new is happening. Doesn't Marx remind us in the *Grundrisse* that every act of production is an act of consumption and every act of consumption is an act of production? And not only for human beings. Every species of organism consumes the resources necessary for its life and, if unchecked by predation or competition, would undergo unlimited growth. Every organism produces waste products that are poisonous to itself. And why all the fuss about extinction? After all, 99.999 percent of all species that have ever existed are already extinct and, ultimately, none will escape extinction. Time and chance happeneth to all. Moreover, no species of vertebrate or flowering plant has become extinct in Britain in the last hundred years despite the toxic outpouring from the "dark satanic mills." The Greeks had already completely deforested their land in Classical times and there hasn't been any prairie in North America for more than a century, but that didn't stop either the Greeks or the Americans from becoming dominant in their time.

Both of these arguments emphasize the present operation of the same basic forces that were the motors of past history, and the continuity of the present with the past. But that emphasis misses essential features of dynamical systems that allow the occurrence of novelties despite continuity and a uniformity of underlying processes. First, domains of the world that had not previously been touched by the process may become incorporated. All species use resources, but human beings are unique among species in placing nonrenewable resources like fossil fuels and minerals at the center of their consumption. Second, domains of the world that were not previously in contact may be juxtaposed and interact. Most of the chemical reactions produced by humans have never before taken place because the reactants have never been in contact. Third, dynamical systems change their shape at critical values of the continuous variables, so-called catastrophe points, as when a stick, increasingly bent by continuously increasing forces, suddenly breaks. So, even for renewable resources, low rates of production and consumption of these resources may lie within a range of values that allows for a dynamic stability of the system, though exploitation outside this range may result in a collapse. But a mathematical "catastrophe" may also be a constructive novelty. As the central nervous system of human primate ancestors grew larger, with connections multiplying, parts of the brain began to perform new functions, among them linguistic functions that have no analogue in nonhuman primates. Fourth, nonlinear dynamical systems behave smoothly and predictably for some ranges of their parameters, but outside these ranges oscillate wildly and without any obvious predictability (so-called chaotic regimes). An economy of petty money-lending local producers, supplying a local market, does not have the same dynamic as globalized finance capital.

It is said that when Galileo, confronted by the nasty tools of the Inquisition, retracted his claim that the earth, like other heavenly bodies, was in motion, he murmured, "Eppur´ si muove!" (But it does move!) We do not know if he really said it, or only that he should have to satisfy the legend of progressive change. We adopted this phrase for the title of our column in *Capitalism, Nature, Socialism* at a low point in the history of our movement for a new form of social life, when the triumph of capitalism seemed irresistible and Margaret Thatcher's cry of "There is no alternative!" seemed to close off all possibilities. Dialecticians know better.

9

Life on Other Worlds

From the earliest years of the American space program, the detection of extra-terrestrial life has been on the agenda. When the *Viking* lander arrived on Mars in 1976 it carried a device for detecting Martian life, an apparatus that was the result of a development program begun with the very first plans for landing an unmanned vehicle on the Red Planet. It was assumed that no little green men would be running around the surface and that life, if any, would be microorgan-ismal. At the beginning of the program there were two competing schemes for detecting life. One consisted of a long sticky tongue that would unroll onto the Martian surface where it would pick up bits of dust. The tongue would then retract, its surface would be passed under a microscope, and the resultant images would be transmitted back to earth-bound microbiologists who would presumably recognize a living organism when they saw it. We may call this the morphological definition of life: if it looks like a cell or wiggles, it's alive. The competing scheme, and the one finally adopted, seemed more objective and more sophisticated. The lander carried a reaction vessel filled with a soluble carbohydrate substrate for metabolism in which the carbon atoms had been radioactively labeled, a kind of radioactive chicken soup. Above the liquid level was a detector that would register the appearance of radioactive carbon diox-ide. Dust was scooped up from the Martian surface and deposited in the soup. If there were living organisms, they would use the carbohydrate as an energy

source, and radioactive carbon dioxide would be released. This is the physiological definition of life: no matter what it looks like, if it metabolizes it's alive.

The reader may imagine the excitement at Mission Control when, indeed,
radioactive counts began to appear and they increased exponentially, as we
expect from a culture of microorganisms dividing in an almost unlimited nutrient. But then things went awry. Suddenly no new radioactive counts were registered, although the apparatus was working. Normally a growing culture of
microorganisms will slow down in its growth and reach stationary phase of
population size, with a steady consumption of nutrients and a steady production of waste products for a long period, but the Martian bugs seemed to have
shut down or disappeared completely—in an instant! After considerable
debate and soul searching it was decided that life had not in fact been detected, and that the carbon dioxide had been produced from a catalytic breakdown
of the carbohydrate on the finely divided clay particles from the Martian surface and that these particles had become saturated. A similar reaction has since
been reproduced in the laboratory on Earth.

The morphological definition of life was regarded as too naive because, as
a century of science fiction has convinced us, Martian life might be very odd-
looking indeed. An extraordinary diversity of forms has arisen in the course of
terrestrial evolution and quite different diversity may have appeared on other
worlds. After all, organismal shapes are just the assemblies of molecules, and
they may take on a bewildering variety, but they are all just different forms of
the same underlying processes and laws. Shape is superficial and subject to the
vagaries of history. It is the molecular processes that are the invariants of life
subject to general physical principles. Molecules are the base; gross forms are
the mere superstructure. So if we wish to search for extraterrestrial life we must
not be led astray by the superficial specificities of living forms that happen to
have occurred on Earth, but search at the molecular level for the constancies
that underlie the variation at the higher levels.

The problem of detecting life on Mars, however, is more profound than the
NASA scientists perceived it to be. The difficulty posed by historical contingency cannot be escaped by concentrating attention on function rather than
form, or molecules rather than gross anatomy, because molecular function also
evolves and betrays the effect of historical contingencies. What the *Viking* lander did was to present Martian life with an "environment" without having ever
seen the life. But, as we have argued in Chapter 5, just as there is no organism

without an environment, there is no environment without an organism. How, among the infinity of possible ways that the physical world can be put together, do we know which represents an environment, except by having seen an organism that lives in it? What the *Viking* experiment showed was that no life on Mars apparently lives in the environment of a restricted range of terrestrial microorganisms. The environment offered to potential Martian life was depauperate, both in what it provided and what it left out. First, it provided only a particular carbohydrate as a nutrient for energy extraction. Even supposing that Martian life is carbon based rather than, say, silicon based, how do we know that it uses carbohydrates rather than, say, hydrocarbons? After all, a bacterium that metabolizes raw petroleum has been produced on Earth. And even if Martian life does metabolize carbohydrate, perhaps it is a sugar that is not fermented by terrestrial bacteria. By mutation and selection experiments, strains of *E. coli* have been made experimentally that will not ferment lactose, their normal energy source, but they will ferment an altered sugar that is not found in nature. Terrestrial organisms have realized historically only a small fraction of the possible basic metabolic patterns.

Second, the Mars lander took no account of most of the complexity that characterizes terrestrial environments. The same experiment done on earth would have failed to detect the presence of most forms of microbial life already known. There is no general microbial culture medium and without a prior knowledge, for example, of the physical substrate specificity, or the inorganic trace elements that are necessary for some species or toxic for others, the search is blind. It would have failed to find sulfur fixing bacteria, nitrogen fixing bacteria that cannot live freely but must be associated with plant roots, fungi and algae that are associated in lichens, extreme thermophiles, halophiles, and so on. For some terrestrial fungi, single spores in isolation will not germinate, but need to be concentrated in a small volume so that their combined low-level metabolism brings the substrate to a critical state, allowing them all to break their dormancy. Is Martian life characterized by dormancy, and if so, what conditions are needed to break it? All of this rich variety of cellular metabolism is the result of historically contingent evolution and none of these forms need ever have existed.

The belief that at the molecular level we will find noncontingent characteristics of life is a consequence of the dominance of a simple model derived from the physical sciences. Biology is seen as a lesser science to the extent that it depends on contingent detail. Perhaps in studying metabolism we have not gone down far enough in the hierarchy of physical nature. What and how organisms eat may indeed be a product of a contingent evolution, but surely there must be some molecular universals that would characterize anything we would want to call "life." Informational molecules? But of course they need not be DNA. Nor does the information need to be concentrated in one sort of molecule. Instead, structures may be self-specifying and may be copied directly by the reproductive machinery, as in the case of cell walls in bacteria which have their own somatic inheritance and which cannot be manufactured without some previous cell wall primer. But why reproduction at all? Like any physical system, living matter necessarily suffers accidents, destructions, and decay, and if there were not some renewal process life would soon end. But why *reproduction*? Why send the old car to the junk heap and buy a new one, if the old one can be repaired indefinitely? All living systems we know of have repair mechanisms, including organ and tissue regeneration, the recovery of damaged cells, and correction of errors in DNA copying. And why individuals? Could not life elsewhere consist of a single physically contiguous object, varying from place to place in its physical extent and from time to time as a consequence of the turnover of its physical constituents? When a tree fell onto the rear part of the four-door sedan of one of our neighbors in Vermont, he converted the car to a pickup truck.

The problem that plagues the investigation of alternative independent lifeforms is the observation that science is necessary because things are different, but that science is only possible because things are the same. The search for life elsewhere that looks simply for a detailed replication of terrestrial life will miss most, if not all, of the events, for it neglects completely the overwhelming importance of historical contingency. However, contingency does not mean that anything goes. The problem cannot be solved by unbounded speculation. There must be something concrete to search for by concrete methods that take into account reasonable physical constraints. NASA does not understand the shape of the problem and is about to repeat the error of the *Viking* lander on a more ambitious scale. It has announced a program in astrobiology, to find life in other planetary systems. But this program is restricted entirely to experimental and engineering projects with no theoretical component. The result will surely be elaborate, with expensive machines designed to detect the simplest terrestrial life somewhere else.

The importance of a correct formulation of the problem is, of course, not in finding life on other planets, a project whose probability of success is exceedingly small. Rather, a proper model for its solution is a model for the management of terrestrial life. Things in the future cannot be exactly as they were in the past. Ecosystems will change and species will go extinct. Life "as we know it" cannot be maintained. But neither is a future possible that is bounded only by imagination and desire. Our methodological problem is to develop an approach to planning and agitation that takes into account both historical contingency and the limits to its possibilities.

10

Are We Programmed?

Living organisms are characterized by two properties that make them different from other physical systems: they are medium in size and functionally heterogeneous internally. Because they are smaller than planets and larger than atomic nuclei, and because there are a large number of interacting processes occurring within them, organisms are at the nexus of a very large number of individually weak determining forces. Their behavior either individually or collectively cannot be described or predicted by reference to a few laws with a few parameters, unlike the laws of motion of the solar system or the laws of quantum physics that apply to very large and very small and rather homogeneous systems. The consequence for science, an enterprise that takes Newtonian mechanics as its model *par excellence,* has been to search for analogies and metaphors for living systems that will somehow reduce their bewildering variety of behaviors to some manageable system of explanation and prediction.

The history of these metaphors mirrors the history of science and technology and the ideologies of successive periods. The founding metaphor of modern biology is Descartes's machine model in which the organism is analogous sometimes to a clock with its gears and levers and sometimes to a mechanical pumping system. Descartes finessed the problem of the unpredictability of human behavior by a neat dualism, putting free will into an entirely nonphysical realm of soul. Problems of faith and morality were assigned to another department, where knowledge was revealed by the Church, leaving science with a free hand to describe the machinery of the body.

Since Descartes, the use of new technology and new ideology in modeling organisms and especially human beings has been the unvarying rule, and in each epoch the metaphor reflects the current state of science, technology, and ideology. The idea that the heart is a pump, that our bones and muscles are levers and pulleys, that our circulatory system is plumbing, and that spinal disks are shock absorbers belongs to the simple technology that dates to the seventeenth and eighteenth centuries. But the development of social ideology also enters. Hanging on the wall of one of our offices is a large educational chart from the late 1920s showing the internal operation of "The Human Factory," with rooms, machinery, and workers reminiscent of the last episode of Woody Allen's *Everything You Always Wanted to Know About Sex*. The gears, pulleys, conveyor belts, and chemical vats, and the workers who operate them, are all being signaled along wires that run through a telephone switchboard tended by women operators. The input to this switchboard comes literally "from the top"—three offices in the skull in which men in suits and ties perform the functions of Intelligence, Judgment, and Will Power.

As technology has changed, so the ruling metaphor has changed. The telephone exchange was clearly too simple to account for the central nervous system, so it became, briefly, a hologram in order to include the new observations that information is stored in a dispersed fashion. But the hologram model didn't do the needed work and we were rescued by the invention of the digital computer. The physical realization of the abstract Turing machine, a digital computer, is an arrangement of electrical and mechanical components, the entire function of which is to serve as the physical host for an abstract set of preexistent directions, the program, that will turn input data about the world into output. The computer itself is the mere electromechanical device, the muscle of the productive enterprise. It is the program, the blueprint, the plan, that is the essence of the productive operation. Nothing better manifests the ideology of the separation of physical and mental labor and the superiority of the mental to the physical than the computer and its program. The immense ideological power of the metaphor of the computer program has resulted in its spread from a model of the central nervous system to a model of the entire organism. The genes contain the program, the essence of the organism, while the cell machinery simply reads the blueprint and executes the directions.

The problem with analogies and metaphors is that we need them in order to understand nature, yet their power to illuminate nature is accompanied by great dangers. Each technological advance reveals a different aspect of our relations with nature, and new domains of technology often imply deeper understanding of nature. The insights can then be applied elsewhere. Nor is it useful to put an analogy under a microscope to see where it does or does not fit. Of course, there will be differences between the model and what is being modeled. As Norbert Wiener wrote, "The best model of a cat is another, or preferably the same, cat." The question is, what does the model do for us to deepen or weaken our understanding?

Let us look at what is implied by the computer analogy.

When somebody says that some behavior is programmed, the implication is that it is inevitable, determined in advance. For scientists, there is the pleasure of puncturing the self-important illusion that we make decisions and choose behaviors freely, with perhaps a touch of an anticlerical poke at the soul. Calling us programmed is a self-deprecatory expression similar to referring to posturing, pompous, and competitive men as "alpha male" or the elementary school claim that the human being is twenty-three cents' worth of chemicals, or that falling in love is a matter of "chemistry."

The technological analogies of the past all served useful purposes. The heart *is* a pump. Its contractions send blood through the body, the strength of the contractions and the amount it fills before contracting tell us how much blood is pumped. The atherosclerotic plaque on the walls of arteries *do* constrict the flow of blood and therefore oxygen to where it is needed. But it also *is not* a pump; plaque is far more dynamic than rust being deposited and removed from the arteries. An artery can be blocked by plaque but also constricted reversibly by stress. The plumbing analogy did not allow for the known possibility of reversal of heart disease or sensitivity to the intricate relation among cardiovascular state, emotional flux, and social location. The analogy of the brain to a circuit network is also helpful: functions are concentrated in specific regions and damage to those regions impairs function. But an activity is carried out in many parts of the brain at once, and when there is damage to one part, the activities may be relocated in other sections. Circuit connections do not guarantee transmission since neurotransmitters are required where the nerve cells meet. Nerves are continually remaking their connections and injured nerve cells can regenerate. Thus, the "hard wiring" of the brain is "soft" (dynamic). It develops during prenatal and post-natal development of the body and depends on the connections being used.

The program model does not mean we always do the same thing. Rather, we have a sophisticated program that can respond differently to different situations, and by comparing the results of a behavior to whether it is good or bad for us, the program can learn. Computers can learn some things well, such as playing chess. In the famous case, Big Blue "learned" by scanning very large numbers of choices and evaluating their outcomes. Increasingly, computer programs are designed to simulate brain behavior. When they do, we are told, "See, the brain is like a computer."

But the notion that we are programmed is misleading in several important ways:

- Brains generate spontaneous activity. When sensory input is reduced, as in sleep or isolation, brain activity gives us dreams, fantasies, or hallucinations. Thus, unlike a computer program, the brain is not at rest when not called upon to act. *Therefore, the brain is never in the same state twice, so that the same stimulus need not evoke the same response.*

- Brain "programs" are influenced not only by the data that can be regarded legitimately as "input" to the program but by processes extraneous to the program that can distract, excite, depress, or otherwise alter the "program" in ways not part of the program. Neurons that are involved in computations may be influenced by hunger, noise, sexual arousal, worries from another sphere of life, exhaustion, or spontaneously generated internal activities. Computers can also do more than one thing at a time. But then it is through time-sharing—essentially having different programs at work that do not influence each other. *The brain is doing many things at once, and these things influence each other.*

- The "program" is not a separate physical entity from the body that is activated by the brain, whereas in a computerized machine or robot the output is conceptually distinct from the sensors and computers and the program itself. In an organism, these are made of the same material as the limbs and eyes. For example, the blood pressure sensors in the kidney can be damaged by high blood pressure and then alter the regulation of blood pressure. As against the hierarchical notion of a programmer aristocracy commanding the peasant body, we have the structures and activities of the body developing and controlling each other.

- The brain has some 10^9 neurons, and these may have hundreds of connections each. Thus the number of circuit arrangements that are possible is vastly greater than the number of subatomic particles in the visible universe. The genome has only some 10^6 to 10^9 genes. Thus there cannot be a different specific genetic blueprint for the construction of each different brain. Rather, there are some more general patterns that are prescribed by the fluxes of proteins: localization of branchedness, probabilities of linkage, proportions of excitatory and inhibitory pathways, synthesis of neurotransmitters, and other very general properties out of which we produce ourselves through interactions with the environments of the uterus and later the wider world. In that interaction, the developing organism selects, transforms, and defines its environment and is transformed by it.

Very little is known about the neurological equivalents of particular behaviors even if we do know the regions of the brain involved. For instance, if people are given arithmetic problems, we can detect heightened activity in some cortical regions, but we have no idea how doing addition is different from long division. We might detect a region of the brain that is especially active as we contemplate works of art, but not separate neurological patterns for looking at expressionist and cubist art. We can identify pathways of neural and chemical activity associated with stress but not why some things are stressful and others not, or how fear of an oncoming automobile differs from fear of losing your job. What we can say is that there are stress responses, coordinated activities of nerves, glands, and muscles that form a more or less coherent cluster of behavior. But these clusters are loosely linked to each other and to cognition, to the processes that evaluate a situation as requiring that mobilization of bodily resources. Our total behavior is therefore a unique combination of more or less stereotyped subunits that makes behaviors look familiar. So, yes, our "printer" may be programmed to print letters as "instructed," but the text is created in a different arena.

How do we interpret the observation that male baboons who have "low social status" in a troop have cardiovascular patterns similar to those of low-status human males? Clearly, both are stressed by their social circumstances. This is not an argument for the universality of hierarchy, but rather a critique of our society that creates a status hierarchy attached to all kinds of privileges, the exclusion from which is stressful. The stress response itself is partly shared with other mammals. (Since we study the patterns in the laboratory, we pick out those aspects that can be compared and remain ignorant of aspects of the stress response that are uniquely human.) But what is stressful is clearly not the same; the stress cluster is linked to quite distinct phenomena. Although the metaphor of the computer program has some use and application when applied at the level of the translation of genes into specific proteins, that use becomes more and more problematic as we move away from that level toward higher and higher levels of organismic function. Genes may be a "program" for protein structure, but protein structure does not contain all the information needed to construct the physical body of an organism at birth, and the physical structure at birth does not predict the course of later development. Most remote of all from a program model is the specific formation, development, and the moment by moment functioning of the brain. To quote Wiener again, "The price of metaphor is eternal vigilance."

11

Evolutionary Psychology

With the waning of religion as the chief source of legitimation of the social order, natural science has become the font of explanation and justification for the inevitability of the social relations in which we are immersed. Biology, in particular, plays a central role in creating an ideology of the inevitability of the structure of society because, after all, that structure is the collective behavior of individuals of a particular species of organism, a manifestation of the biological nature of *Homo sapiens*. Biology has been supposed to provide the answers for two major questions. First, why, despite the ideology of equality that seems an unquestioned fundamental of bourgeois social theory, is there so much inequality of status, wealth, and power? The biologistic answer has been that such inequalities are the consequence of unequal distributions of temperament, skill, and cognitive power, manifestations of genetically determined differences between individuals, races, and the sexes. But this claim leaves untouched the second question. Suppose it were true that there were such genetically determined individual and group differences. Those differences in themselves do not dictate a hierarchical society. Why not "from each according to ability and to each according to need"? What is required is a biologically based framework for human motivations and interactions that will explain, among other things, why unusually skillful basketball players get so rich and famous, but women players less so than men. That is, to complete its program of explaining human society, biology must have a biological theory of human nature.

A biological explanation of human individual behavior and social interactions cannot simply be a story of genetic determination. It must also incorporate an explanation of how the particular genes that are said to be the efficient causes of human behavior came to characterize the species, as opposed to the genes that govern behavior in, say, fish. A modern biological explanation, to be respectable, must be evolutionary. But a plausible evolutionary explanation must be more than a mere narrative, providing a reconstructed historical sequence of characteristics during the evolutionary history of a species. First, it must convince us that characteristic human behaviors, though specific to the human species, are nevertheless detectable alterations of general behavioral properties of other organisms. Somehow what humans do must be special cases of aggression or communication, or sexual competition, or problem solving, or a mechanism of cheating in a cooperative sharing of resources, or any of the other properties that all animals are supposed to exhibit. A unique behavior that cannot be derived from a related one in a related species is a serious embarrassment for the teller of evolutionary stories.

Second, given the ideological function of an evolutionary explanation as providing justification for a behavior, it must be possible to give an explanation of the evolution of the behavior as resulting from natural selection, so that the genes for behavior are not only present but superior to alternatives. It is the selective story that, along with the genetic determination, does the most important ideological work. If a behavior is genetically determined, or at least very strongly influenced by genes, then it will be seen as very difficult to change by merely social arrangements, or, even if it could be changed, the new behavior would be unstable and likely to relapse back to its "natural" state. If the genes for the behavior were established by natural selection, then the welfare of the species is at stake. The most popular view of evolution by natural selection, a direct inheritance from Adam Smith's "invisible hand," is that evolution is an optimizing process in which choosing the most fit individual will maximize a species' efficiency or stability or likelihood of survival. We change what has been established by natural selection at our peril.

Over the last twenty-five years there have been two widely disseminated versions of the evolutionary argument for human social behavior. The first, sociobiology, provided a specific adaptational explanation for every social manifestation that the theory's inventor, E. O. Wilson, could list, including religiosity, entrepreneurship, xenophobia, male dominance, the urge to conform, and ease of indoctrination. Sociobiological theory was an instant success in explanations of animal behavior, but it engendered, from within biology, a strong critical attack on both its pretensions and its status as well-supported natural science. As a consequence, though it remains part of the explanatory structure used by many economists, political scientists, and social psychologists, "sociobiology" has become a term of some opprobrium in biology and even Wilson has gone on to immerse himself in the more acceptable domain of species conservation. In its place there has arisen the subject of "evolutionary psychology," a somewhat more nuanced version of sociobiology that replaces the naive and easily attacked detailed claims of sociobiology with a more general adaptationist theory. The basic assertions of evolutionary psychology are expressed by its best-known proponents, Cosmides and Tooby:

> The brain can process information because it contains complex neural circuits that are functionally organized. The only component of the evolutionary process that can build complex structures that are functionally organized is natural selection. . . . Cognitive scientists need to recognize that while not everything in the design of organisms is the product of selection, all complex functional organization is.[1]

Unfortunately, we are not given helpful directions on how to know a "complex functional organization" when we see it. This general theory is then cashed out, in particular for human social behavior, by claiming that what has been selected are certain specialized mechanisms like "language acquisition device . . . mate preference mechanisms . . . social contract mechanisms, and so on." The list is much less specific than xenophobia and religiosity, but nevertheless covers the same territory. Like its predecessor, sociobiology, evolutionary psychology depends on poorly specified notions of complexity and adaptation and asserts without any hope of proof that traits judged to be adaptive can only have been established by natural selection as opposed to, say, learning by individuals and groups in a social environment. What characterizes evolutionary explanations of human behavior is the lack of any articulated social theory. The closest evolutionary psychology comes to a social theory is to claim that individuals have been selected who have the capacity to enter into "social contracts," that is, the willingness to go along with group norms. How those norms are arrived at, what their historical dynamic is, how individual socialization varies from group to group, between sexes, among individuals, are all outside the theory. It is, in fact, a theory without a social content.

Whereas evolutionary psychology and its parent, sociobiology, derive their appeal outside of science as bases of legitimation for political and economic structures, it should not be supposed that the drive to invent such theories comes from such justificatory needs. There is something else at stake for natural scientists and academic theorists of society. The model of a "real" science is one that is universal in the domain of its explanations. In evolutionary biology, the drive to apply the skeletal structure of evolution by natural selection to every aspect of living organisms is the drive to provide the science with its ultimate legitimation. After all, if the principles of evolution cannot explain the most significant aspects of human existence, our psychic and social lives, then what kind of a science is it? Moreover, the most prestigious domain of modern biology and the one that claims the greatest successful generalization is not evolutionary, but molecular biology. So evolutionary science, if it is to succeed, must not only be universal in its application, but must conform to the extreme reductionism of molecular biology. Social explanation is seen as obfuscatory.

In pursuit of a reductionist explanation some, but not all, of the recent discoveries of neurobiology are used. Gross regions of the brain can be identified that become more metabolically active (that is, consume more sugar or show more electrical activity) when memory, cognitive, or emotional processes occur. Neurotransmitter molecules have been identified that mediate specific kinds of activity such as motor control or memory, and disorders such as Parkinson's and Alzheimer's disease have been associated with their aberrant production. This encourages evolutionary psychologists to believe that the Human Genome Project will reveal genetic determination of neuroanatomy and neurochemistry and, hence, human behavior.

Yet, other discoveries are ignored, such as the ability of nerve cells to develop and reconnect throughout life and the impact of social experience on our whole physiology. The cerebral cortex, acting through its labile connections and by way of neurotransmitters, links social experience to our biology. For instance, the balance of the two branches of the autonomic nervous system, the sympathetic and the parasympathetic, in the regulation of heart function is different in working-class and middle-class teenagers. Thus, causation flows in both directions and a biological difference associated with a behavioral difference is not evidence for internal biological determination, nor do behavioral differences explain social organization.

The drive for intellectual legitimacy also compels psychology, sociology, and anthropology. That search for legitimation has demanded the creation of "social science" out of the "merely" humanistic study of history, anthropology, and sociology. Evolution is a form of history, and nothing is easier than to gain the respectability of a natural science by confounding history and evolution. But because evolutionary biology as the price of its own respectability is driven to an extreme reductionism, evolutionary social theory is no social theory at all.

12

Let the Numbers Speak

After three centuries of reductionist science in Europe and its cultural inheritors, in which the problem of "What is this?" would be answered by "This is what it is made of," modern science increasingly confronts the problems of complexity and dynamics. Whereas the great successes of science have been largely discoveries about isolatable phenomena or small objects in which a small number of determinate causes are operating, the dramatic failures have arisen where attempts are made to solve problems of complex systems and dynamics. It is no exaggeration to claim that complexity is the central scientific problem of our time.

In preceding chapters we have criticized reductionist approaches in various fields, challenging the fundamental assumption of reductionist science that if you can understand the smallest parts of a system in isolation from one another, then all you have to do is to put them together correctly in order to understand the whole. As a research tactic this model certainly works, provided the system being studied is simple enough, and even for very complex systems, many of the bits and pieces are nearly independent of one another and can best be understood by a reductionist research strategy. Descartes's metaphor of the organism as a clocklike machine certainly works for clocks, or for the heart viewed as an isolated pumping machine, but not for whole organisms, or social and economic organization, or communities of species. Our criticism of the simple reductionist machine model has been based in an assertion about the actual nature of things, namely that, in general, their properties do not exist in isolation, but come into being as a function of their context. Thus, it is what philosophers call an *ontological* error to suppose that we can understand composite systems by dividing them into *a priori* parts and then studying the properties of those parts in isolation. But the reductionist research strategy for studying complex systems has also been abandoned for a different reason by many scientists who are ontological reductionists and believe that the world really is a large set of gears and levers with intrinsic, isolatable properties. They have abandoned it because they believe that, in practice, we cannot study all the properties and all the connections of very large systems made up of many different parts with many paths of interaction among them and in which the multiple causal forces are individually weak. These *epistemological* nonreductionists say that it is just too hard, that we do not have world enough and time, or that because of physical, political, or ethical constraints the ultimate power of the reductionist strategy is not available to us. Until relatively recently it was a criminal act to dissect a human corpse.

Laplace is famous for his statement that if he knew the position and velocity of all the particles in the universe, he could predict all future history. This was the strongest claim for reductionism that could be made in a deterministic material universe. But he also knew that the information could not be made available to him and so, using the notions of probability, he treated the effect of all the unexplorable causes as chance. The realization that the world may be too complex to study by dissection, even if in actuality it were machine-like, has given rise to a mode of study that, over the last century and a half, has come to be a major methodology for the analysis of causes in complex physical and social systems. That methodology is *statistics*. In the eighteenth century statistics was a purely descriptive set of techniques for characterizing assemblages of objects, especially human populations, as a political (statist) tool. Beginning in the late nineteenth century, through a union with the theory of probability, statistics became a major mode for the inference of causal relations when, for one reason or another, the preferred reductionist method of dissection and reconstruction is not possible.

Though we ordinarily think of statistics as an analysis of populations, the basis of the statistical approach to inferring causes is a model of the *individual*, and it is an explanation of the properties of the individual that are being sought. The properties of each individual are assumed to be the consequence of a nexus of variable causes whose magnitudes are, relative to one another, insufficient to have an unambiguous effect on each individual object. Every tree that is cut will fall because the single force of gravitation overwhelms all other minor perturbations, but every tree is of a different size and shape because growth is the consequence of the interaction of a very large number of individually weak genetic and environmental causal pathways as well as of microscopic variable molecular events within cells. We do not need statistics to infer gravitation, at least for large objects, but statistical methods are the reigning techniques for inferring the causal relations of genes, environment, and molecular "noise" in nature, because every individual differs in the effects of these variable causes. In order to overcome the difficulty posed by large numbers of causes, each with a weak effect, large numbers of individuals are agglomerated into statistical populations and average values of causes and effects are studied. It is in the formation of these populations and the calculation of the averages that all action occurs.

There are essentially only two techniques of statistical inference. In one, contrast analysis, individuals are sorted into two or more populations based on some *a priori* criteria: males and females, different ethnic groups, age categories, social class. Some kind of average description of some characteristic of interest is then calculated within each group and if these averages are sufficiently different between the groups, then the criterion used for setting up the groups is deemed to be of causal significance. The average that is calculated may be simply the numerical average (mean) of the characteristic, say the mean family income, or it may be the proportion of the population falling into some class, say the proportion of families with incomes above $50,000, or it may be some measure of the variability of the characteristic from individual to individual.

The alternative technique, correlational analysis, is to assemble all the individuals into a single population, to measure two or more characteristics, again chosen *a priori*, and then to look for trends in one or more of these characteristics as other characteristics vary. Does some measure of ill health tend to increase as family income decreases? A commonly chosen variable is time. For all people who have died in the last hundred years, does the proportion of those dying of lung cancer increase as the date of birth is later and later? When some relationship between variables is seen then some inference about causation is made.

Whereas it is often claimed that statistical techniques are ways of letting the objective data speak for itself, in both of these modes of statistical inference all the real work is done by the *a priori* decisions imported into the analysis. What *a priori* categories will be used, in the first mode, to create the contrasting populations? Is gender relevant, or social class, or ethnicity? These decisions must be made before the data are even collected. American sociology is well known for ignoring theory-laden social class as a variable and substituting theory-laden social economic status as a numerical and therefore "objective" measurement. In both contrast and correlational analysis what characteristics are to be measured: mean family income, which is heavily weighted by a small number of very-high-income families, or median family income, which is not biased in this way; days of work lost which, for a given cause of ill health, may be greater for the more affluent than for those who must go to work even sick? Which characteristics should be held constant while others are compared? Do blacks and whites differ in health status if the data is filtered in such a way as to equalize occupational status and income between the two groups? And, finally, which is cause and which is effect? Is low income the cause of ill health, or ill health the cause of low income? At every juncture in the analysis, from the gathering of data to the final analysis, an *a priori* theoretical model of causal relations guides the "objective" statistical methodology. Therefore it is necessary to recognize that causal relations inferred from statistical comparisons may be artifacts of the set of assumptions that enter in the "objective" statistical evaluation of data.

In what remains of this chapter, let's briefly explore the problem of directionality of causation and the relationship between cause and effect, on the one hand, and dependent and independent variables, on the other. A variable that is said to be "independent" is one that is assumed to be determined by conditions outside of, and autonomously from, the effects being studied. The distinction between independent and dependent variables is a fundamental theoretical construct of much correlational statistical work. In environment studies, the level of pesticide treatment may be the "independent" variable and the prevalence of brain cancer the "dependent" variable. In economics the tax rate may be the independent variable and investment the dependent variable. In the new field of "policy," a policy choice such as allocation of resources to health programs can be treated as the independent variable and health outcomes the dependent variable. Then statistical calculations are performed using these *a priori* variables and inferences about causes and effects. One or another statistical rule is used to decide if the putative causal relation is supported by the relation between the independent (causal) and the dependent (effect) variables.

But what happens if cause flows in both directions? What happens if health outcomes of policies result in public action to change policy, if disability affects income? In the last century, Engels wrote of the interchanging of cause and effect, physiologists described self-regulation, and engineers were designing self-correcting industrial processes. In systems of any complexity there are feedbacks, and these affect the relationship between statistical outcomes and causal pathways.

In negative feedback, a change in one element of a system leads to changes in others that eventually negate the original change. The negation may be partial, complete, or even overshoot, so that dumping nitrogen in a pond may reduce the nitrogen level if a radical change in species composition occurs, or applying pesticides may increase pest load by removing more pesticide sensitive competitors of the pest or, frequently, by killing off predators of the pest species. The predators are poisoned directly by the pesticide, but both a negative and positive branch of a feedback loop are involved. Along the positive branch predators are decreased because their food supply, the pest species, is decreased by the pesticide. Along the negative loop the pest carrying insecticide molecules poisons the predator, which results in an *increase* in prey. It is not that predators are more sensitive physiologically to insecticide, but that their location in the loop makes them more vulnerable ecologically. The important point for statistical analysis is that every negative feedback loop has a negative and a positive branch. Along the positive branch, prey increase the predator population, high blood sugar increases insulin, addition of nutrient increases algal growth, high farm prices encourage production. Along this branch both variables increase or decrease together: this is formalized as a positive correlation between the dependent and independent variables. But along the negative branch of the feedback loop, predators decrease their prey, insulin reduces blood sugar, high algal growth creates a mineral shortage, increased production reduces farm prices. Then the two variables move in opposite directions and show a negative correlation.

These feedback loops are embedded in larger contexts and other influences may impinge upon the loop at any point, moving first along the positive or negative branch. Then the same pair of variables, predator/prey, insulin/sugar, production/price, nutrient/algae, may show positive correlations in some situations and negative ones in others. Finally, if influences of other variables percolate along both positive and negative branches there may be no correlation at all, even if the variables are interacting strongly. This may lead to the erroneous conclusion by students that correlation is not the same as causation. Then why do they carry out correlational analyses at all?

13

The Politics of Averages

It is commonplace knowledge that different sorts of averages give very different information about populations and thus can suggest different conclusions from the same basic data. The mean or arithmetic average household income, for example, simply takes the total income of the entire population and divides by the number of households so that one very rich family makes up for a large number of poor ones. If one wants to emphasize how well off people are the mean is the number to use. The median family income, in contrast, is the value below which half of all families fall, thus taking account of the proportion of families in different income categories and providing a more realistic view of the situations in which families find themselves. In the United States the median family income is about two-thirds of the mean. If Bill Gates and other rich entrepreneurs all double their incomes, the average family income in the United States will increase but not the median. Measures such as the income of "the top 10 percent" or "the bottom 20 percent" or their ratios grasp the distributional aspects better while averages are more suitable for how well "we" are doing.

What is not so well known is that all ratios, such as those commonly used in ecology, population studies, and economics, provide the same ambiguity as simple averages and the same opportunity to obscure or reveal the actual situation. This ambiguity arises because the average of a ratio of two variables is not, in general, equal to the ratio of the averages, and this discrepancy is quite large when the variation is large for both the numerator and denominator. There is no application in which this discrepancy is more apparent and more distorting than in the characterization of population and resource density. For example, the population density of a country or region is usually calculated as the total number of individuals divided by the total area. For the United States, according to the 1990 census, the population density was:

$$\frac{248{,}709{,}873 \text{ people}}{3{,}539{,}289 \text{ sq. miles}} = 70.3 \text{ persons/sq. mile.}$$

But this is clearly a gross underestimate of the effective density at which people are living because the estimate takes the large, dense urban populations and treats them as if they were uniformly spread over the vast deserts of the Great Basin. In fact, the effective density at which people are living in the United States turns out to be about 3,000 people per square mile.

The density of a population can be calculated in two ways. In both cases we begin by dividing the entire extent of the population into small areas within which the population is more or less evenly spread, say counties or ponds or patches, depending on the organism. We then measure the area of each patch and count the number of individuals in each to calculate a local density. The question now arises how we are to combine these individual local ratios to characterize the population as a whole. One way is to weight each ratio by the proportion of the entire *area* that is in the local patch to produce a so-called *area-weighted density*. This turns out to be what is actually calculated by the usual ratio of total population to total area. But such an area-weighted density gives great weight to all those areas with few or no individuals in them and thus badly underestimates the real density at which most individuals live. For instance, suppose there are three people living on a one-acre plot and one person on three acres. There are four people on four acres so that the average population density is one per acre. But three people are living at a density of three per acre and one person is living at one-third person per acre. The alternative is to weight each local ratio by the proportion of the entire *population* that is included, and sum these up to produce an *organism-weighted density*, giving a realistic picture of the density at which individuals are actually living. In our simple example, the average effective density is then

$$\frac{3 \times (^3/_1) + 1 \times (^1/_3)}{4} = 2.33 \text{ people/acre}$$

The person-weighted density is always larger, and often many times larger than the area-weighted density, with consequences that are inconvenient for a national government or the World Bank. However, if, as ecologists, we ask the question, "What is the average pressure of human activity experienced by a patch of land?" then the area-weighted measure would be quite appropriate, although it would still leave out the information that some pieces of land are much more highly exploited than others.

As an example we consider the pattern of farm size in Panama in 1973 for which we have a census. Since the total farm population was 575,153 occupying a total farm area of 2,098,062 hectares, it was a mere 0.27 people per hectare by the conventional measure or, inversely, 3.65 hectares per person, not a very high density on a world scale. However, as one might expect, the most crowded 20 percent of the population occupied only 0.2 percent of the farm area and the most crowded one-third of the farmers had only 1 percent of all the land. At the other end of the distribution a tiny 0.1 percent of the farmers occupy a total of 10 percent of the farmland. The effective density of farm occupation, calculated from the person-weighted density, turns out to be 22.07 people per hectare or, inversely, only .045 hectares per person, clearly inadequate to support people even at the highest yields achieved anywhere. The person-weighted density, which is 80 times the conventional calculation, gives a quite different picture of the causes of poverty in the Panamanian countryside.

Just as there are different ways of calculating density, so, reciprocally, there are different ways of calculating average wealth, that is, the average amount of resource available to each individual. The conventional measure, as in per capita income, is to take the aggregate wealth and divide it by the total number of individuals, which is exactly the reciprocal of the conventional measure of density. But again this leaves out of account the effect of the uneven distribution of resources. By analogy with the measure of density we can calculate a *resource-weighted wealth* and an *organism-weighted wealth*. The conventional measure, which is simply the reciprocal of the area-weighted density, turns out to be the organism-weighted wealth and gives an overestimate of how much wealth individuals typically have because it takes no account of the uneven distribution of resources. Once again it is not a complete surprise that this is the measure used in public statistics.

In ecological questions the choice of a measure of density or resource availability depends upon whether one takes the standpoint of the resources or the consumer. Consider, for example, a fairly uniformly spread food plant consumed by an insect with a patchy and clumped distribution, a common situation when there is larval feeding. From the standpoint of the insect as a consumer, most individuals are densely packed on their resource so the organism-weighted density is an appropriate measure for the population. From the standpoint of the food plant, however, most individuals are free of predators, or nearly so, and it is the resource-weighted wealth that counts. Evolutionary arguments about the force of natural selection depend on the organism-weighted density for pressure to adjust the search behavior of the predator, but on the resource-weighted wealth for pressures on the plant to develop secondary poisonous compounds that will resist the insect. Thus the predator and the prey respond to two quite different measures of density arising in the same predator-prey interaction.

There is, then, no single "correct" measure of average density or wealth either in ecology or political economy. The question is: Whose side are you on?

14

Schmalhausen's Law

Ivan Ivanovich Schmalhausen was a Soviet evolutionary biologist working at the Academy of Sciences in Minsk. In the 1940s his book *Factors of Evolution* appeared and was denounced by T. D. Lysenko, whose neo-Lamarckian theories of genetics were then on the ascendancy. At the close of the 1948 Congress of the Lenin Academy of Agricultural Science it was revealed that Stalin had endorsed Lysenko's report in which it was affirmed that the environment can alter the hereditary makeup of organisms in a directed way by altering their development. Schmalhausen was one of the few who affirmed his opposition to Lysenko and spent the rest of his life in his laboratory studying fish evolution and morphology.

In the West, Lysenko's views were simply dismissed. But Schmalhausen could not ignore the Lysenko agenda, which insisted on a more complex interpenetration of heredity and environment than genetics generally recognized. Along with Marxist and progressive scientists in the West, such as C. H. Waddington in the United Kingdom, he accepted the challenge. As a result, he developed a more sophisticated approach for understanding these interactions and helped explain the observations of some of the better studies cited by Lysenkoists.

Schmalhausen argued that natural selection was not only directional, producing new adaptations to new circumstances, but stabilizing. That is, if a characteristic of a species causes it to be well adapted, then random variation in the characteristic caused by external or internal disturbances would reduce the fitness of the organism, so natural selection will operate to prevent such disturbances. The development and physiology of the species will be selected to be *canalized*, that is, insensitive to such random disturbances. These disturbances come not only from the environment but also from genetic variations from individual to individual. Genes are selected which work in such a way that most genetic combinations produce more or less viable and similar offspring. Thus individual genetic variation remains hidden because of the canalization of development.

The selection to produce canalized development and physiology operates over a restricted range of natural conditions that characterize the usual or normal environmental range to which the species is subjected during its evolution. However, under unusual or extreme conditions where selection has not had the opportunity to operate, these genetic differences show up as increased variation. This claim provided an alternative explanation to the observation that populations that are apparently uniform under normal conditions show a wide range of heritable variation under new or extreme conditions. Whereas Lysenko argued that these populations were uniform genetically and that the environment created new genetic variations, Schmalhausen argued that the environment revealed latent genetic differences which could then be selected.

Waddington developed this line of reasoning further with his idea of genetic assimilation. Suppose that there is some threshold condition in the environment for the development of a particular trait. Much below threshold none of the individuals show it, much above threshold they all do. But under some intermediate conditions some will be above and some below threshold. If environmental conditions change so that it is advantageous for all individuals to manifest the trait, then those with the lowest threshold will be favored by natural selection. The average threshold in the population will decrease and eventually produce organisms whose threshold is so low that the trait always appears under any conditions in which the organism can survive. Then the trait has become "assimilated": an environmentally induced condition has become fully genetic.

Schmalhausen's realization that natural selection operates to change the sensitivity of physiology and development to perturbations, but that it only operates under the usual and normal range of environmental and genetic variations experienced by the species in its evolution, leads to a result with wide implications. This result is known as "Schmalhausen's Law." It indicates that when organisms are living within their normal range of environment, perturbations in the conditions of life and most genetic differences between individuals have little or no effect on their manifest physiology and development, but under severe or unusual general stress conditions even small environmental and genetic differences produce major effects.

Two examples of the application of Schmalhausen's Law are in the determination of species distribution and in the effect of toxic substances on population health. Both show the danger of predicting the outcome of perturbations in natural populations using the results of experiments on single factors under controlled conditions.

In Biogeography. At almost any location on the earth, the ecological community is made up of species near the boundary of their distribution and species that are in the middle of their range. When the environment changes, this has a major impact on the species near their boundary. Some may become locally extinct, others may experience great expansions of their abundance and range, and still others will remain more or less as they have been. Further, populations near their boundaries are especially sensitive to changing conditions and are more likely to show big differences from year to year. Thus simple predictions about the effect of climate change are bound to err if they take into account only the direct physiological impact of the environmental change on species one at a time, out of the context of their community interactions. In contrast, species in the middle of their range are likely to show less effect from an environmental change. Therefore, when we ask how a $1°C$ change in temperature will affect the distribution of malaria, we have to ask how close to their boundaries not only the vector mosquito but also its natural enemies and competitors are located. Different localities near the boundary will respond differently for no obvious reason, just because of extreme sensitivity to even undetectable changes of circumstance.

The Thresholds of Toxicity. Tolerable levels of toxic substances are often set on the basis of experiments with animals. Usually the work is done with standardized healthy animals under well-controlled conditions to minimize "error" due to individual differences or variation in the environment. However, this methodology underestimates the impact of a toxin for a number of reasons. If an organism is exposed to a toxic substance of external or internal origin, it has various mechanisms to detoxify that substance. But the toxin is still present. If there is a constant level of exposure, the toxin will reach some level of balance between new absorption of toxin and the rate of removal. This equilibrium depends on the level of exposure and the maximum capacity of the detoxification system to remove the poison.

Of course, we know that the environmental exposure is not constant for all members of a population or even for any one individual over time. And we also know that different members of the population differ in their detoxification capacity and that it may vary over time for the same person. Furthermore, this variability matters and cannot be averaged away.

What good is a model that assumes constant conditions? Here we see one of the powerful ways in which models are both useful and dangerous in science. In physical and engineering sciences it is often possible to isolate a problem sufficiently to ignore external influences, assume that all switches are the same in what is relevant, that all salt molecules are interchangeable, and so on. Then we can measure accurately and get equations that are as exact as we need. But in ecological and social sciences this is not possible—the populations are not uniform, conditions change, and there is always an outside impinging on the system of interest. We cannot even believe the equations too literally. But we can still study these systems. First, we find the consequences of models under unrealistic conditions that are easily studied and give precise results. Then we ask, how do departures from those assumptions affect the expected outcomes? In this case, the standing level of toxicity, a measure of damage done to an organism, is a mathematical function of $d - e$, the maximum detoxification capacity minus the exposure (see Figure 14.1). The maximum removal rate has to be greater than the exposure or else, according to the mathematical model, the toxicity will accumulate without limit. In reality, it will accumulate to the point where other processes, which were negligible in the original model, take over. These might involve any of the consequences of toxicity such as cell deaths. In relatively unstressed conditions, when d is greater than e the graph of toxicity plotted against $d - e$ decreases from zero as capacity exceeds exposure by greater and greater amounts. Furthermore, it is concave upward.

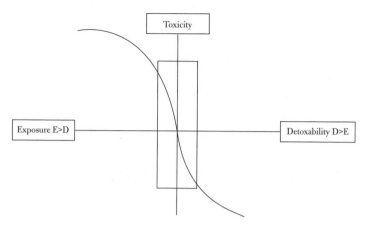

Figure 14.1

That is, it is steeper the closer we are to $d = e$ and flattens out when detoxification capacity is much greater than exposure. If we measure the dose response curve in the range where capacity is much bigger than exposure the results will show little effect of the poison and we will be reassured by claims that there is no detectable effect. Testing is often conducted under optimal conditions on uniform populations of experimental animals in order to get uniform results, reduce the error, and avoid "confounding factors."

If different stressors are confronted by the same detoxification pathways, they can be added at the level of exposure and act synergistically at the level of toxicity. Therefore, if we look at only one insult at a time, the other "confounding factors" increase the damage beyond what we expected.

In the United States, exposure varies with location and occupation. The poor, excluded, and marginalized communities such as inner cities, *colonias,* and reservations are often subject to multiple exposures due to incinerators, *maquiladoras,* poor water quality, malnutrition, and unsafe jobs. Therefore even toxic substances that meet EPA standards will prove more harmful than expected. But these effects will be hard to detect since we will observe an array of health impairments rather than a single harm appearing to different degrees.

Similar arguments hold if the capacity to detoxify varies among individuals: because of the shape of the curve relating toxicity to detoxification capacity, the average toxicity in the population is greater than the toxicity at average detoxification capacity. Once again, if detoxification capacities are reduced each unit of insult has a bigger effect than expected.

We suspect that detoxification capacities are undermined in the course of life for all of us after the first two decades, but that adverse conditions accelerate this erosion so that vulnerability increases more rapidly and life expectancy is reduced, for example, by some five years for African-American women and seven years for African-American men.

The Variability of Results. Under stress, when $d - e$ is small, small differences in either one can have big effects. A population at a disadvantage will show big differences among people for reasons we cannot explain, and different poor communities will differ widely in the rates of adverse outcomes. This can easily be misinterpreted: it appears as if under the "same" conditions some do well and others poorly, and that we can then blame the latter. But what really is happening is that under conditions of any kind of stress, small differences have big effects.

Schmalhausen's Law focuses our attention on the historical relation of a population with its environment, the responsiveness of the physiology to familiar and to new stressors, and the inherent variability of both organisms and environments.

15

A Program for Biology

Recent internal developments in biology and social science urge the necessity to confront the rich complexity of the phenomena of interest at the same time that the large-scale practical problems of greatest concern—eliminating poverty, promoting health, and equity and sustainability—call for more integral, multilevel, and dynamic approaches than those to which we are accustomed. Both areas of knowledge are grappling with ways to escape from the unidirectional causation, *a priori* categories, hierarchies of "fundamentalness," and rigid disciplinary boundaries that have dominated thinking and have led to some of the big mistakes of recent times. Most of these, such as the green revolution, the epidemiological transition, sociobiology, the reification of intelligence testing, and the current fetishism of the genome, err by posing the problems too narrowly, treating what is variable as if it were constant and even universal, and offering answers on a single level only.[1]

Cumulative research in each area points toward a dialectical, dynamically complex alternative, but theoretical and philosophical traditions within the sciences, institutional arrangements of disciplines, and economic interest combine to resist this rather obvious development. Single causes are more readily patented than are complex webs of reciprocal determination and make better headlines. Scientists are rewarded or excluded according to whether their work fits comfortably into the departmental boundaries or definitions of funding programs, since narrower, more conventional projects are more likely to reach publishable conclusions within the time limits of the rehiring and promotion cycle. Therefore, while there exist interdisciplinary, transdisciplinary, or nondisciplinary programs and all institutes are set up to study complexity, the sciences as a whole still stumble over the obstacles that we all recognize.

The major theoretical achievement of the genome project was the refutation of its greatest expectation—that a mapping of the DNA base sequence would also be a map of all the interesting characteristics of the organism, disease vulnerability, individual and group behaviors, and the origin of life. The source of the error lies in the continued repetition of the mantra that "genes determine organisms," because genes "make" proteins and proteins "make" organisms. Even putting aside the overwhelming importance of organisms being the consequence of processes that depend in an interactive and dialectical manner on genes, on environment, and on random events in development, the error begins at the molecular level. No DNA sequence contains all the information necessary for the specification of a protein. A DNA sequence contains a recipe for the sequence of amino acids in a polypeptide. But that polypeptide must fold into a three-dimensional structure, a protein, and there are multiple free-energy minima for such a folding. The particular folding that occurs depends on cellular conditions, on the presence of so-called chaperones and other molecules and cellular structures. But the DNA sequence of a gene does not always have the full information sufficient to determine the sequence of amino acids in the polypeptide. In some organisms, such as flagellates, there is editing of the RNA message that is transcribed from the DNA, a message that does not contain all the information on the amino acid sequence to be assembled. This editing may involve the insertion of many missing nucleotides into the RNA sequence or it may result in a shuffling of blocks of messenger RNA sequence into a final sequence that is then translated into an amino acid sequence. More generally, before an amino acid sequence becomes a protein with a role in cell structure and metabolism, individual amino acids may be chemically modified, the sequence trimmed, or have other amino acid sequences attached to it (post-translational modification).

The alternative to the unilinear sequence is the feedback loop in which all elements on a pathway have equal rank. Even the distinction between the genome and the soma, useful for transmission genetics, is misleading in the discussion of development and evolution. Rather, we have to confront a more complex, multilevel system in which the genome, the proteome, the traitome, the behaviorome, and the societome exist in reciprocal feedbacks of a complex nonlinear dynamics instead of the simple sequence DNA→RNA→protein→trait.

In the context of genetics, the first feedback takes place at the cellular level, where RNA, protein, and metabolites interact with DNA sequences to regulate the timing, rate, and location of the cell's conversion of DNA sequence information into protein. Much of this feedback is a consequence of physical changes in the organism that flow from developmental events themselves. Changes in cell number, shape, and location, and the production of proteins within these cells affect the processes within neighboring cells. But these changes within the organism alter the external environment, which in effect then feeds back into the development and metabolism of the organism itself.

At the same time, experimental results and new technologies of functional mapping of the brain show the surprising plasticity of the central nervous system and the spread of almost all interesting activities throughout the brain. This does not negate gross regional specialization but it casts doubt on the rigidity of regional boundaries and the repeated invocation of "hard-wiring" as explanation and the repeated claims that each newly evolved region of the brain leaves the previous ones intact and limited to their previous functions. The limbic area of humans is not the fossil of the Mesozoic era, the reptilian brain that is supposed to be the deepest part of ourselves in a more than anatomic sense. The modern amygdala receives signals from the cortex, is confronted with new patterns of stimulation, and has evolved in its new context. Nowadays all biologists agree in principle that the organism depends on both the internal processes and the environment. But the distinction between internal and external is permeable. "Environment" reaches inside the organism: each part is environment to all the other parts. Even within the uterus, monozygotic twins sharing the same interior of the same mother may be in importantly different environments if they are attached to separate chorions and are more different than if they shared a chorion. The voice of the mother affects the fetus in ways that make "innate" quite different from "genetic."

C. H. Waddington wrote about genetic assimilation, a process whereby a trait that depends on an external stimulus becomes "genetic" by a lowering of the signal threshold for its appearance to the point where that signal is present in all circumstances compatible with life. Thus calluses appear on the feet of ostriches at birth, before any stress from walking could induce their formation. Natural selection has favored this unconditional appearance of calluses, which then protect the young bird from the damage that would be done during the development of calluses from external stress.

The complementary view of this process is that a part of the "external," which is almost universal, can become incorporated into the developing system as a sort of external yolk. Then its consequences would be seen as "hard-wired." Sensory stimulation is needed to form certain near-universal traits of the brain, such as the organization of the visual cortex, and in the case of social animals the activity of adult caretakers plays a vital role in the development of the young.

Finally, it raises questions about the emergence of the nervous system from apparently passive input-output reflex mechanisms to networks capable of spontaneous activity. The stability and dynamics of such a network depend on relations among long and short positive and negative feedback loops (excitatory and inhibitory pathways). In order for the nervous system to work, the excitatory activity must prevail first and then be damped by the inhibitory slower pathways in order to contain the activity. But a large number of synapses is needed for more complex activity, to make more distinctions in the environment and initiate more differentiated and complex responses. Therefore a system that can do a lot of things in response to external inputs will have many inhibitory pathways. If the connectivity of the network is great enough and the total number of neurons big enough, then the very conditions that maintain boundedness also lead to local instability, that is, spontaneous activity.

In examining the Green Revolution, we see that a view based on unidirectional causation leads to the expectation that since grasses need nitrogen, a genotype that takes up more nitrogen would be more productive; since pesticides kill pests, their wide use would protect crops; and since people eat food, increased yields would alleviate hunger. In each case, the linear inferences were plausible. The counterintuitive outcomes came about because of branching pathways from the starting point: the increase in wheat yield was partly achieved by breeding for dwarf plants that are more vulnerable to weeds and to flooding; the killing of pests was accompanied by the killing of their natural enemies, their replacement by other pests, and the evolution of pesticide resistance. The successful yield increases encouraged the diversion of land from legumes. The technical package of fertilizers, pesticides, irrigation, and mechanization promoted class differentiation in the countryside and displacement of peasants.

When a complex system is perturbed—for instance, by adding a medication to a physiological network—the impact percolates through many pathways. It may be buffered along some pathways, amplified along others, even inverted on some pathways, giving the opposite result from what is expected. (Ritalin is used both to arouse and to calm.) And the stronger the medication, the more effective in its intended effect, the more likely it is also to have big unexpected effects.

A strategy for avoiding the kinds of errors we have been discussing would start from the following propositions:

1. The truth is the whole. Of course, we cannot really see the whole, but it warns us to pose a problem as bigger than we would have, with further reaching ramifications. For instance, we can write an equation for a prey population:

$$\text{Prey population} = a - b \text{ (predator population)}.$$

Here the predator is the independent variable and the prey, the dependent variable. If we measure them carefully, we could find the regression coefficient b and "account for" a large fraction of the variance. But the predator is simply given from outside the model. We always have to ask, where is the rest of the world? In this case we could start by having the predator determined by the prey and have a negative feedback loop. This gives a richer understanding because it shows mutual determination and allows us to see the covariation pattern of predator and prey when the rest of the world enters mostly by way of one of the two species. The first equation is not "wrong." It fits the prey population to the level of the predator as accurately as we possibly can, but it is also an impoverished way of looking at nature.

2. Recognize that everything in the world is not relevant to everything else. The death of a single butterfly does not have palpable effects on the rest of the living world. We must find the boundaries of subsystems within which there are effective interactions and between which there is effective independence. This process of "dissecting nature at its joints" is one of the most difficult tasks in biology because the same material objects, molecules, cells, individuals, and populations belong to multiple functional subsystems, depending on the process being considered.

3. Things are the way they are because they got that way, have not always been that way, need not always be that way. History matters—at the short-term

level of the development of specific individuals, at the medium-term level of the assemblages of individuals into populations and ecosystems, and at the long-term level of evolution.

We ask three questions: Why are things the way they are instead of a little bit different (the question of homeostasis, self-regulation, and stability)? Why are things the way they are instead of very different (the question of evolution, history, and development)? And what is the relevance to the rest of the world?

PART TWO

16

Ten Propositions on Science and Antiscience

Since radicals began to look to science as a force for emancipation, Marxists both as social critics and as participating scientists have grappled with its contradictory nature. Because there is such a rich diversity of Marxist thought about science, I cannot claim that what follows is "the" Marxist position. I only offer in schematic form some propositions about science that have guided the work of at least this Marxist scientist.

1. All knowledge comes from experience and reflection on that experience in the light of previous knowledge. Science is not uniquely different from other modes of learning in this regard.

What is special about our science is that it is a particular moment in the division of labor, in which resources, people, and institutions are set aside in a specific way to organize experience for the purpose of discovery. In this tradition a self-conscious effort has been made to identify sources and kinds of errors and to correct for capricious biases. It has often been successful. We have learned to be alert to the possible roles of confounding factors and to the need for controlled comparison; we have learned that correlation does not mean causation and that the expectations of the experimenter can affect the experiment; we have also learned how to wash laboratory glassware to avoid contaminants and how to extract trends and distinctions from morasses of numbers. Our self-consciousness reduces certain kinds of errors but in no way eliminates them, nor does it protect the scientific enterprise as a whole from the shared biases of its practitioners.

In contrast, so-called traditional knowledge is not static or unthinking. Africans (probably mostly women) brought as slaves to the Americas quickly developed an African-American herbal medicine. It was put together partly from remembered knowledge of plants found both in Africa and in America, partly from borrowed Native American plant lore, and partly from experimenting on the basis of African rules about what medicinal plants should be like. The teaching of traditional medicine always involves experimenting, even when it is presented as the transmission of preexisting knowledge. Finally, the criteria for prescribing various herbal therapies in non-European/North American medicine are probably better grounded than those that guide decisions about cesarean sections, pacemaker implants, or radical mastectomies in U.S. scientific medical practice.

Even what is described as intuitive (as against intellectual) knowledge comes from experience: our nervous/endocrine system is a marvelous integrator of our rich, complex histories into a holistic grasp that is unaware of its origins or constituents. Scientific and intuitive knowledge are not fundamentally different epistemologically; they differ instead in the social processes of their production and are not mutually exclusive. In fact, one of my goals in teaching mathematics to public health scientists is to educate their intuition, so that the arcane becomes obvious and even trivial, and complexity loses its power to intimidate.

2. All modes of discovery approach the new by treating it as if it were like the old. Since it often is like the old, science is possible. But the new is sometimes quite different from the old; when simple reflection on experience is not enough, we need a more self-conscious strategy for discovery. Then creative science becomes necessary. In the long run we are bound to encounter novelty stranger than we can imagine, and previous well-grounded ideas will turn out to be wrong, limited, or irrelevant. This holds true in all cases, in both modern and traditional, class-ridden, and non-class societies. Therefore both modern European/North American science and the knowledges of other cultures are not only fallible but are guaranteed to err eventually.

To call something "scientific" does not mean it is true. Within my lifetime, scientific claims such as the inertness of the "noble gases," the ways in which we divide up living things into major groupings, views as to the antiquity of our species, models of the nervous system as a telephone exchange, expectations as to the long-term outcomes of differential equations, and notions of ecological stability have all been overturned by new discoveries or perspectives. And major technical efforts based on science have been shown to lead to disastrous outcomes: pesticides increase pests; hospitals are foci of infection; antibiotics give rise to new pathogens; flood control increases flood damage; and economic development increases poverty. Nor can we assume that error belongs to the past and that now we've got it right—a kind of "end of history" doctrine for science. Error is intrinsic to actually existing science. The present has no unique epistemological status—we just happen to be living in it.

Therefore, we have to consider the notion of the "half-life" of a theory as a regular descriptor of the scientific process and even be able to ask (but not necessarily answer), "Under what circumstances might the second law of thermodynamics be overthrown?"

3. All modes of knowing presuppose a point of view. This is as true of other species as of our own. Each viewpoint defines what is relevant in the storm of sensory inputs, what to ask about the relevant objects, and how to find answers.

Viewpoint is conditioned by the sensory modalities of the species. For instance, primates and birds depend overwhelmingly on vision. With visual information objects have sharply differentiated boundaries. But that is not the case when odors are the major type of information, as for ants. An anoline lizard sees moving objects as being the right size to eat or as representing danger. A female mosquito perceives an academic conclave as gradients of carbon dioxide, moisture, and ammonia that promise blood meals, while a sea anemone trusts that glutathione in the water is enough reason to thrust out its tentacles in expectation of a meal. The fact that we live on the surface of Earth makes it seem natural to focus our astronomy on planets, stars, and other objects while ignoring the spaces between them. The timescale of our lives makes plants seem unmoving until time-lapse photography makes their changes apparent. We interact most comfortably with objects on the same temporal and size scales as our own and have to invent special methods for dealing with the very small or very large, the very fast or very slow.

4. A point of view is absolutely essential for surviving and making any sense of a world bursting with potential sensory inputs. Much of learning is devoted to defining the relevant and determining what can be ignored. Therefore the appropriate response to the discovery of the universality of viewpoints in science is not the vain attempt to eliminate viewpoint but the responsible acknowledgment of our own viewpoints and the use of that knowledge to look critically at our own and one another's opinions.

5. Science has a dual nature. On the one hand, it enlightens us about our interactions with the rest of the world, producing understanding and guiding our actions. We really have learned a great deal about the circulation of the blood, the geography of species, the folding of proteins, and the folding of the continents. We can read the fossil records of a billion years ago, reconstruct the animals and climates of the past and the chemical compositions of the galaxies, trace the molecular pathways of neurotransmitters and the odor trails of ants. And we can invent tools that will be useful long after the theories that spawned them have become quaint footnotes in the history of knowledge.

On the other hand, as a product of human activity, science reflects the conditions of its production and the viewpoints of its producers or owners. The agenda of science, the recruitment and training of some and the exclusion of others from being scientists, the strategies of research, the physical instruments of investigation, the intellectual framework in which problems are formulated and results interpreted, the criteria for a successful solution to a problem, and the conditions of application of scientific results are all very much a product of the history of the sciences and associated technologies and of the societies that form and own them. The pattern of knowledge and ignorance in science is not dictated by nature but is structured by interest and belief. We easily impose our own social experience on the social lives of baboons, our understanding of orderliness in business, implying a hierarchy of controllers and controlled, on the regulation of ecosystems and nervous systems. Theories, supported by megalibraries of data, often are systematically and dogmatically obfuscating.

Most analyses of science fail to take into account this dual nature. They focus on only one or the other aspect of science. They may emphasize the objectivity of scientific knowledge as representing generic human progress in our understanding. Then they dismiss the obvious social determination and the all-too-familiar antihuman uses of science as "misuses," as "bad" science, while keeping their model of science as the disinterested search for truth intact.

Or else they use the growing awareness of the social determination of science to reject its claims to any validity. They imagine that theories are unrelated to their objects of study and are merely invented whole cloth to serve the venal goals of individual careers or class, gender, and national domination.

In stressing the culture-boundedness of science, these analyses ignore the common features of Babylonian, Mayan, Chinese, and British astronomies and their calendars. Each comes from a different cultural context but looks at (more or less) the same sky. They recognize years of the same length, notice the same moon and planets, and calculate the same astronomical events by very different means.

Social determinists also ignore the parallel uses of medicinal plants in Brazil and Vietnam, the namings of plants and animals that roughly correspond to what we label as distinct species. All peoples seek healing plants and tend to discover similar uses for similar herbs.

Other traditions than our own also have their social contexts. Babylonian priests or Chinese administrators were not bourgeois liberals, but they were not wiser or freer from viewpoint. Nor does the phrase "the ancients say" tell us anything about the validity of what they say. Ancients like moderns belong to genders, sometimes to classes, always to cultures, and they express those positions in their viewpoints. Those ancients whose thought has been preserved in writing were also not a random sample of ancients.

But to be socially determined and conditional on viewpoint does not mean arbitrary. Although all theories are eventually wrong, some are not even temporarily right. The social determination of science does not imply a defense or toleration of the patently false doctrines of racial or gender superiority or even the categories of race themselves, whether in the conventional academic forms or the "Adamic man" and the "mud people" of the Christian identity movement. Racism is a more real object than race and determines the racial categories.

Thus the task of the analyst of science is to trace the interactions and interpenetrations of intellectual labor and the objects of that labor under different conditions of labor and under different social arrangements. The art of research is the sensitivity to decide when a useful and necessary simplification has become an obfuscating oversimplification.

6. Modern European/North American science is a product of the capitalist revolution. It shares with modern capitalism the liberal progressivist ideology that informs its practice and that it helped to mold. Like bourgeois liberalism in general it is both liberated and dehumanized. It proclaimed universal ideals that it did not quite mean, violated them in practice, and sometimes revealed those ideals to be oppressive even in theory.

Therefore, there are several kinds of criticisms of science. A conservative criticism inherits the pre-capitalist critique. It is troubled by the challenge that scientific knowledge poses to traditional religious beliefs and social rules and rulers, does not approve of the independent judgment of ideas and values, does not demand evidence where authority has already pronounced, and is thus disturbed mostly by the radical side of science. Creationists quite accurately identify the ideological content of science, which they label secular humanism, against the liberal formula that science is the neutral opposite of ideology. But no matter how much they search the scientific journals for evidence of conflicts among evolutionists and weak spots in modern evolutionary theory, their challenge is not to make science more "scientific," more democratic, less bound by oppressive ideology, and more open. Rather, they propose to return to faith, to the more obvious kinds of authority, and to anti-intellectual certainties. Their gut-level anti-intellectualism is often expressed in delight at the stupidities of scientists as opposed to the wisdom of the "simple man," a delight that at first seems appealingly democratic. But this is not the assertion that everyone is capable of rigorous and disciplined thinking. Instead, it altogether denies the importance of serious complex thinking in favor of the spontaneous smarts of uneducated certainties. Conservative critics accept the dichotomy of knowledge versus values and opt for their particular values whenever there is conflict.

At the same time, conservative critics reject the fragmented and reductionist aspects of modern science on behalf of a holistic, "organic" view of the world. At an aesthetic and emotional level their holism partly resonates with that of radical criticism, but their holism is hierarchical and static, stressing harmony, balance, law and order, the ontological rightness of the way things are, were, or are imagined to have been.

The most consistent liberal critics of science accept the claims of science as valid goals but criticize the practices that violate them. They approve of science as public knowledge and deplore the secrecy imposed by military and commercial ownership of it. They want democratic access to science determined only by capacity, and they deplore the class, gender, and racial barriers to scientific training, employment, and credibility. They agree that ideas should be judged only on their merits and on the evidence, regardless of where the ideas come from, but they see hierarchies of credibility reinforced by a rich vocabulary for dismissing unorthodox ideas and their advocates as "far out," "quackish," "ideological," "not mainstream," "discredited," "anecdotal," or "unproven." They may be horrified by the uses of science in the production of harmful commodities or vicious weapons, or the similarly vicious justifications of oppression, without relinquishing the belief that thinking and feeling should be kept separate.

Because of the increasingly obvious blindnesses, narrowness, dogmatism, intolerance, and vested interest in official science, alternative movements have sprung up, especially in health and agriculture. They must be examined with the same tools that we use to look at "official" science. Who owns them, where do they come from, what viewpoints do they express, how are they validated, what theoretical biases do they manifest? Embedded as they are in a capitalist context, these alternatives are also a field for exploitation, producing commodities, and often are clothed in shameless commercial hype. They, too, have class roots that lead some to separate individual from social causation (for instance, criticizing the magic bullets of the pharmaceutical industry but peddling their own miraculous "natural" cures, or promoting holistic cancer treatments but ignoring the industrial origins of many cancers). The alternative communities are domains where insightful radical critique mixes with petty and medium-scale entrepreneurship.

Marxist critique attempts to see science, in both its liberating and oppressing aspects and its powerful insights and militant blindnesses, as a commoditized expression of liberal European capitalist masculinist interests and ideologies organized to cope with real natural and social phenomena. Its ideology is both a product of European liberalism and a self-generated contribution to that ideology, not a mere passive reflection of it.

Radical critiques of agriculture, medicine, genetics, economic development, and other areas of applied science point out both the external and internal aspects that limit science's ability to reach its stated goals. The external refers to science's social position as a knowledge industry, owned and directed for purposes of profit and power as guided by shared beliefs, carried out mostly by men. The modes of recruitment into and exclusion from science, the various subdivisions into disciplines, the hidden boundary conditions restraining scientific inquiry become intelligible when we examine its social context. We can approach the dominant modalities of chemical therapy in medicine and farming as expressions of the commodification of knowledge by the chemical industry. But the reliance on molecular magic bullets is also congenial to the reductionist philosophy that has dominated European/North American science since its formation in the seventeenth century, and in turn is supported by the atomistic experience of bourgeois social life. (As we trace the connections, we see that "internal" and "external" are in fact not rigidly alternative explanations, another example of the general principle that there are no nontrivial, complete, and disjunct subdivisions of reality. Yet science is still plagued by the false dichotomies of organism/environment, nature/nurture, deterministic/random, social/individual, psychological/physiological, hard/soft science, dependent/independent variables, and so on.)

The internal refers to the reductionist, fragmented, decontextualized, mechanistic (as against holistic or dialectical) ideologies and liberal-conservative politics of science. Marxist and other radical critics have always called for broadening the scope of investigations, placing them in historical context, recognizing the interconnectedness of phenomena and the priority of processes over things, whereas conservative ideology usually advocates elegant precision about narrowly circumscribed objects and accepting boundary conditions without even acknowledging them.

7. A radical critique of science extends also to the inner workings of the research process. In approaching a new problem, Marxism encourages me to ask two basic questions: Why are things the way they are instead of a little bit different, and why are things the way they are instead of very different? Here "things" has a double meaning, referring both to the objects of study and to the state of the science studying them.

The Newtonian answer to the first question is that things are the way they are because nothing much is happening to them.

But our answer is that things are the way they are because of the actions of opposing processes. This first question is that of the self-regulation of systems, of homeostasis. In the face of constantly displacing influences, how do things remain recognizably what they are? Once posed, it enters the domain of systems theory in the narrow sense, the mathematical modeling of complex systems. That discipline starts with a set of variables and their connections and applies equations to ask: Is the system stable? How quickly does it restore itself after perturbation? How much does it respond to permanent changes in its surroundings? How much change can it tolerate? It asks, when external events impinge on the system, how do they percolate through the whole network, being amplified along some pathways and diminished along others? We work with notions such as positive and negative feedback loops, pathways, connectivity, sinks, delays, reflecting and absorbing barriers. In its own terms, this analysis is "objective." But the variables themselves are social products. For instance, the apparently unproblematic notion of population density has at least four different definitions that lead to different formulas for measurement and different results when the measurements are compared across countries or classes. We could simply divide the total number of people by the total area (or resource):

$$D = \Sigma \text{people} / \Sigma \text{area}.$$

We could ask, what is the average density at which people live? Then we would use

$$D = \Sigma \text{ (people/area) (people in that area)} / \Sigma \text{people};$$

the unevenness of access to resources or land is then included. Or we could do the same but from the perspective of the resource. The total resource per person is

$$D = \Sigma \text{area} / \Sigma \text{people},$$

the average intensity of exploitation of a resource is given by

$$D = \Sigma (\text{area/people}) \text{ (area)} / \Sigma \text{area}.$$

Thus even what seems to be an objectively given measure is laden with viewpoint, and this is either taken into account or hidden. Nancy Krieger, a professor at Harvard University, has used the metaphor of fractal self-similarity to stress that the inseparability of the social and biological occurs at all levels, from the most macro to the fine details of the micro in epidemiology.[1]

The second question is the question of evolution, history, and development. Its basic answer is, things are the way they are because they got that way, not because they have to be that way, or always were that way, or because it's the only way to be. From this perspective we reexamine the first question and ask: What variables belong in the system anyway, and how did they get there? What do we really want to find out about the system? What do you mean "we"? Who says? Do new connections appear and old ones decline? Do variables merge or subdivide? Do the equations themselves change? Should we use equations or other means of description? And since we know that the models we use are not photographically accurate pictures of reality, how would departures from the assumptions affect the outcomes? When does this matter?

What were the givens in the first formulation now become the questions. It is here that the powerful insights of Marxist dialectics, when combined with substantive knowledge of the objects of interest and the manipulative skills of the craft, have been most productive. Here the familiar propositions of the unity and interpenetration of opposites, universal connection, development through contradiction, integrative levels, and so on, so dry in the listings of the formal manuals, burst with rich implications and scintillate with creative potential.

Finally, these same methods are used reflexively to examine the historical constraints that have acted on Marxism itself as a consequence of its own historical circumstances and the composition of Marxist movements. But these methods should not be used in a mechanistic, essentialist way, rejecting notions because they are European and therefore foreign in Latin America, or male and therefore irrelevant to women, or of nineteenth-century origin and therefore inapplicable to the twenty-first. After all, every idea is foreign in most places where it is held, and in all places in the world most of the current ideas are of foreign origin. Rather, the historical context can be used to evaluate the ideas critically, to discover the insights and limitations and the needed transformations. The insights of feminism and the ecology movement, particularly those branches that have already overlapped with Marxism, are especially helpful in gaining the distance needed for this examination. Themes that had been relegated to the periphery of most Marxist vision can now be restored to their rightful places in historical materialism, and societies can be studied more richly as social/ecological modes of production and reproduction.

8. Although different theories use different terms, look at different objects, and have different goals, they are not mutually unintelligible. Linnaeus saw species as fixed at the time of creation, with each particular example being a

corrupted version of the archetypal design. Evolutionary biologists see species as populations that are intrinsically heterogeneous and subject to forces of change. The description of the typical is then seen as an abstraction from the array of real animals or plants. Nevertheless, I still use Linnaean Latin names for genus and species, many of which Linnaeus himself would recognize, and I could talk with Linnaeus about plants, argue about their anatomy or geographic distributions. He would be delighted to learn that our technologies have given us new ways of distinguishing among similar plants. We would disagree about the significance of variation within a species, and I don't know how he would react to the shocking idea that similarity often implies a common origin. But we could talk.

This is even true across larger cultural divides. All peoples name plants and animals. Most peoples assign different names to plants that correspond to different Linnaean species and divide up the botanical world much as we do. They also tend to distinguish more finely among organisms that have to be dealt with differently. And like our own theories, theirs also "work." They guide actions that often enough lead to acceptable results. Whether you are a modern taxonomist who recognizes that half the snakes in Darien are poisonous or a Choco who will tell you that all snakes are poisonous but only kill you half the time, the practical conclusion is similar: when walking in the forest, beware of snakes.

Furthermore, the tools of investigation show a greater continuity than the theories. Galileo would be impressed by our more sophisticated telescopes but would not be completely lost in a modern observatory. Although a Marxist economist might not be interested in the input-output equilibrium models of the neoclassical school or the techniques of cost-benefit analysis so dear to the corporate mind, these would be perfectly comprehensible to her. The claim that different outlooks are incommensurate, speak different languages, and find no points of contact is a gross distortion of the understanding of social viewpoint. Theoretical barriers do not mean the existential aloneness imagined by distant observers.

9. The diversity of nature and society does not preclude scientific understanding. Every place is clearly different and every ecosystem has its unique features. Therefore ecology does not look for universal rules, such as "plant diversity is determined by herbivores," or attempt to predict the flora of a

region by knowing its rainfall. What it can do is look for the patterns of difference, the processes that produce the uniqueness. Thus the number of species
on an island depends on the processes of colonization and speciation increasing numbers and the processes of extinction reducing numbers. We can go
further and relate colonization to distance from a source of migrants, extinction
to habitat diversity and area and community structure, try to explain why the
migrants are of a particular type, and so on. The outcomes will be very different
on tiny islands where populations do not last long enough to give new species
or are so close to the source of migrants as to swamp any local differentiation,
from islands that are very remote, with high habitat diversity.

The use of site specificity to reject broad generalizations is misplaced.
What we look for is the identification of the opposing processes that drive the
dynamics of a kind of system (e.g., rain forest, or island, or capitalist economy) rather than propose a unique and universal outcome.

10. Radical defenders of science cannot defend science as it is. Instead, we
have to come forward as critics both of liberal science and of its reactionary
enemies. The present right-wing attack on science is part of a more general
assault on liberalism, now that the demise of a worldwide socialist challenge
makes liberalism unnecessary and intensified competition during a period of
long-term stagnation makes liberalism seem too costly. Although its opposition to liberalism is opposition to the liberating aspects of that doctrine, the
reactionary attack on liberalism often emphasizes the oppressive or ineffectual sides of liberalism.

We have to call for opening up science to those who have been excluded,
democratizing what is an authoritarian structure modeled on the corporation,
and insist on the goal of a science aimed at the creation of a just society compatible with a rich and diverse nature. We should not hide behind but rather
undermine the cult of expertise in favor of approaches that combine professional and nonprofessional participation. The optimal condition for science is
one foot in the university and one in the community in struggle, so that we have
the richness and complexity of theory coming from the particular and the comparative view, and generalizations that only some distance from the particular
can provide. It also allows us to see the combination of cooperative and conflicting relations we have with our colleagues and ways in which political commitment challenges the shared common sense of professional communities.

We should not pretend or aspire to a bland neutrality but proclaim as our working hypothesis: all theories are wrong that promote, justify, or tolerate injustice.

We should not cover up or lament in private the triviality of so much published research but denounce that triviality as coming from the commodification of careers in scholarship and from the agendas of domination that rule out of order many of the really interesting questions.

We should challenge the competitive individualism of science in favor of a cooperative effort to solve the real problems.

We should reject the reductionist magic bullet strategy that serves commodified science in favor of respect for the complexity, connectedness, dynamism, historicity, and contradictoriness of the world.

We should repudiate the aesthetics of technocratic control in favor of rejoicing in the spontaneity of the world, delighting in the incapacity of indexes to capture life, savoring the unexpected and anomalous, and seeking our success not in dominating what is really indomitable but in far-sighted, humane, and gentle responses to inevitable surprise.

The best defense of science under reactionary attack is to insist on a science for the people.

17

Dialectics and Systems Theory

In a generally sympathetic review of *The Dialectical Biologist*, and in personal conversations, John Maynard Smith argued that the development of a rigorous, quantitative mathematical systems theory makes dialectics obsolete.[1] Engels's awkward "interchange of cause and effect" can be replaced by "feedback"—the mysterious "transformation of quantity into quality" is now the familiar phase transition or threshold effect. He noted that "even in my most convinced Marxist phase, I could never make much sense of the negation of the negation or the interpenetration of opposites." He could have added that hierarchy theory grasps some of the insights of "integrated levels" or "overdetermination."

Mary Boger, a leader of the New York Marxist school, has been urging me for years not to allow dialectics to be subsumed under systems theory. Despite systems theory's concern with complexity, interconnection, and process she has argued that it is still fundamentally reductionist and static, and despite the power of its mathematical apparatus it does not deal at all with the richness of dialectical contingency, contradiction, or historicity. Finally, she added that systems-theoretic "interconnection" does not grasp the subtleties of dialectical "mediation."

Here I attempt to systematize my own views as they have evolved in discussions with Mary Boger, Rosario Morales, Richard Lewontin, and other comrades.

As I entered this exploration I became aware of two opposing temptations. On the one hand, I wanted to emphasize the distinctness of dialectics from contemporary systems theory, to proclaim that our theoretical foundations are not obsolete and continue to have something important to say to the world of science that systems theory has not already adopted. On the other hand, along with Engels I found it gratifying to see science, grudgingly and haltingly and inconsistently but nevertheless inexorably, becoming more dialectical. Both affirmations are true, but their emotional appeal can also lead to errors of one-sidedness. I attempted to use this awareness to question my conclusions as I made one or another claim.

Any description of systems theory and of dialectical materialism is subject to two kinds of problems: in both areas there are many practitioners with quite divergent views. I will not attempt any kind of comprehensive survey of systems theory or "a systems approach," but limit myself to systems theory in the narrow sense as a mathematical approach to "systems" of many parts. And second, systems theory and dialectics are not mutually exclusive. Some systems theorists are also Marxists or have been influenced by Marxism in their research contributions to the development of the theory. Other Marxists have had at least a passing contact with systems theory and have used some of its notions in their Marxist research. For example, Göran Therborn, a Swedish Marxist social scientist influenced by systems theory, approached the nature of the state from two perspectives: the traditional Marxist view of the role of the state as an expression of class rule and a systems theoretic examination of its dynamics as a system with inputs and outputs. The publisher's blurb for his book *What Does the Ruling Class Do When It Rules?* summarizes the work: "Therborn uses the formal categories of systems analysis—input mechanisms, processes of transformation, output flows—to advance a substantive Marxist analysis of state power and state apparatuses."[2]

Nonetheless, the two are different in their origins, objectives, and theoretical underpinnings. In what follows I will discuss several general themes that unite and differentiate dialectics and systems theory: wholeness and interconnection, selection of variables or parts, purposefulness, and the outcomes of processes. Materialist dialectics is not offered as a complete philosophy of nature, a system in the classical sense.[3] Dialecticians are too aware of the historical contingency of our thinking to expect that there will ever be a final worldview. First of all, it is polemical, a critique of the prevailing failings of both the mechanistic reductionist approach and its opposite, the holistic idealist focus. Together these have dominated Euro-North American natural and social science since its emergence in seventeenth-century Britain as a partner in the bourgeois revolution. They have also dominated politics as the broad liberal-conservative consensus that has defined the "mainstream" politics of democratic capitalism.

Secondly, dialectical materialism has focused mostly on some selected aspects of reality while ignoring others. At times we have emphasized the materiality of life against vitalism, as when Engels said that life was the mode of motion of "albuminous bodies" (i.e., proteins; now we might say macromolecules). This seems to be in contradiction to our rejection of molecular reductionism, but simply reflects different moments in an ongoing debate where the main adversaries were first the vitalist emphasis on the discontinuity between the inorganic and the living realms, and then the reductionist erasure of the real leaps of levels. At times we have supported Darwin in emphasizing the continuity of human evolution with the rest of animal life, at other times the uniqueness of socially driven human evolution. We could classify our species as omnivores, along with bears, to emphasize that we are just another animal species that has to get its energy and substance by eating other living things, and are not limited to only one kind of food. Or we could underline our special status as "productivores" who do not merely find our food and our habitat but produce them. Both are true; the relation of continuity and discontinuity in process is an aspect of dialectics that systems theory does not deal with at all.

But critique is not just criticism, and dialectics goes beyond the rejection of reductionist or idealist thinking to offer a coherent alternative, more for the way in which it poses questions than for the specific answers its advocates have proposed at any particular time. Its focus is on wholeness and interpenetration, the structure of process more than of things, integrated levels, historicity, and contradiction. All of this is applied to the objects of the study, to the development of thought about those objects, and self-reflexively to the dialecticians ourselves so as not to lose sight of the contingency and historicity of our own grappling with the problems we study.

Dialectical materialism is unique among the critiques of science in that its roots are outside the academy in political struggle as well as within, that it directs criticism both at reductionism and idealism, that it is consciously self-reflexive, and that it rejects the goal of a final "system." But it is unlike postmodernist criticism of science which uses the contingency of scientific claims to deny the historically bounded but no less real validity of some claims over others in favor of an acritical pluralism.

Systems theory has a dual origin, in engineering and in the philosophical criticism of reductionism. On the one hand it comes out of engineering as cybernetics, the study of self-regulating mechanisms with often rather complex circuitry. Norbert Wiener introduced the term cybernetics in his book of that name.[4] The term became part of common usage in the Soviet Union, but was mostly replaced in the United States by control theory, the theory of servomechanisms, or systems theory. In this form it is the mathematics of feedback, the study of mathematical models. The preface to *The Theory of Servomechanisms*, one of the early classical texts in this field, states:

> The work on servomechanisms in the [Livermore] Radiation Laboratory grew out of its need for automatic radar systems. It was therefore necessary to develop the theory of servomechanisms in a new direction, and to consider the servomechanism as a device intended to deal with an input of known statistical character in the presence of interference of known statistical character. . . .
>
> A servomechanism involves the control of power by some means or other involving a comparison of the output of the controlled power and the actuating device. The comparison is sometimes referred to as feedback.[5]

This form of systems theory is highly mathematical and formal. Its earlier versions assumed systems that were given, the equations known, and measurement precise. But soon systems analysis was taken up by military designers, with the idea of a weapons system replacing the development of particular weapons as the theoretical problem and by management systems as the scientific aspects of directing large enterprises. Here the measurements are fuzzier, the equations not known, and therefore other techniques become necessary. Herbert Simon at Carnegie Mellon University, Mihajlo D. Mesarovic at Case Western Reserve, the International Institute for Applied Systems Analysis in Austria, as well as mathematicians and engineers in the Soviet Union and other centers worked to advance the conceptual frameworks and mathematics of many variables interacting at once and the computing routines for following what happens. More recently, the Santa Fe Institute has made the study of complexity the core intellectual problem.

The major role of engineering and management systems in developing systems theory is reflected in the assumption of goal seeking. Thus Donella H. Meadows, Dennis L. Meadows, and Jorgen Randers define a system as "an interconnected set of elements that is coherently organized around some purpose. A system is more than the sum of its parts. It can exhibit dynamic, adaptive, goal-seeking, self-preserving and evolutionary behavior."[6]

But the "system" of systems theory is not reality itself but a model of reality, an intellectual construct that grasps some aspects of the reality we want to study but also differs from that reality in being more manageable and easier to study and alter. Therefore models are not "true" or "false." They are designed to meet a number of criteria that are in part contradictory, such as realism, generality, and precision.[7] It is the hope of systems analysts that the departures from reality that make them easier to study do not lead to false conclusions about that reality.

The wholeness, interconnectedness of parts, and the purposefulness of systems are emphasized. The first two qualities are inherent in what we mean by a system.

Wholes

The other source of systems theory is found in in critical attempts to counter the prevailing reductionism in science since the last century. Here its boundaries are not well defined but shade off gradually into various holisms.

Holism is not new. The history of science is not the history of its mainstream, the succession of dominant paradigms popularized by Thomas Kuhn. In science, there has always been dissidence, dissatisfaction with dominant ideas, alternative approaches within various disciplines, and divergent "mainstreams" among disciplines. "Holistic" criticism has always coexisted with the dominant reductionism. It was expressed in such currents as vitalism in developmental biology, Bergson's "emergence," in psychology (Bronfenbrenner, Perl, Piaget), ecology (Vernadsky's biosphere, the Soviet "geo-biocoenosis," Clements's and later Odum's ecosystems), anthropology (Kroeber's "superorganic"), and other fields as a grasping for wholeness and interconnection. In this aspect it is usually referred to in the United States as a "systems approach" or "systems thinking." Some authors engage in systems theory in both the narrow and the broad meanings. Especially ambitious and central was Ludwig von Bertalanffy's General Systems Theory starting in the 1930s.[8] Biological complexity was usually a central challenge. William Ross Ashby's *Design for a Brain* poses the problem as one of reconciling mechanistic structure and seemingly purposeful behavior:

> We take as basic the assumptions that the organism is mechanistic in nature, that it is composed of parts, that the behavior of the whole is the outcome of the compounded actions of the parts, that organisms change their behavior by learning, and that they change it so that the later behavior is better adapted to their environment than the earlier. Our problem is, first, *to identify the nature of the change which shows as learning*, and secondly, *to find why such changes should tend to cause better adaptation for the whole organism.* [Emphasis in original][9]

Ecology also has brought to public consciousness the rich interconnectedness of the world. Examples are regularly put forth of the unexpected, often counterproductive effects of interventions directed at solving a particular problem. Pesticides increase pest problems, draining a wetland can increase pollution, antibiotics provoke antibiotic resistance, clearing forests to increase food production may lead to hunger. Barry Commoner's dicta that everything is connected to everything else and that everything goes somewhere have become part of the common sense of at least a part of the public.

The powerful impact of the realization that things are connected sometimes leads to claims that "you cannot separate" body from mind, economics from culture, the physical from the biological, or the biological from the social. Much creative research has gone into showing the connectedness of phenomena that are usually treated as separate. It is even said that because of their interconnectedness they are all "One," an important element of mystical sensibility that asserts our "Oneness" with the Universe.

Of course, you *can* separate the intellectual constructs "body" from "mind," "physical" from "biological," "biological" from "social." We do it all the time, as soon as we label them. We have to in order to recognize and investigate them. That analytical step is a necessary moment in understanding the world. But it is not sufficient. After separating, we have to join them again, show their interpenetration, their mutual determination, their entwined evolution, and yet also their distinctness. They are not "One." The pairs of mutualist species or predator and prey are certainly linked in their population dynamics. Sometimes the linkage is loose, as when each affects the life of the other but the effect is not necessary. Sometimes very tightly, as in the symbiosis of algae and fungi in lichens. Snowy owls and Arctic hares drive each other's population cycles in a defining feedback loop. Mutualists may evolve to become "one," as Lynn Margulis has pioneered in arguing for the origins of cellular structures. But predator and prey are not "one" until the last stages of digestion. Psychotherapists work both with asserting connection in examining family systems and with criticizing "codependence," the pathological loss of boundaries and autonomy. There is a one-sidedness in the holism that stresses the connectedness of the world but ignores the relative autonomy of parts.

As against the atomistic and absolutized separations of reductionism, holists counterpose the unity of the world. That is, they align themselves at the "Oneness" end of a spectrum that ranges from isolated to "one." They look for some organizing principle behind the wholeness, some "harmony" or "balance" or purpose that gives the wholes their unity and persistence. In technological systems, there is a goal designed by the engineers that is the criterion for evaluating the behavior of the system and for modifying the design. To the extent that the development of systems theory has been dominated by designed systems, goal-seeking behavior appears as an obvious property of systems as such, and therefore it is sought in the study of natural systems.

In the study of society, this may lead to a functionalism that assumes a common interest driving the society. But a society is not a servomechanism; its component classes pursue different, both shared and conflicting goals. It is not a "goal-oriented" system, even when many of its components are separately goal seeking.

Within the framework of static holism it is difficult to accommodate change as other than destructive, so that conservation biology often emphasizes preservation of a particular species or ecological formation, rather than conditions that permit continued evolution.

Dialecticians value the holistic critique of reductionism. But we reject the sharp dichotomy of separation/connection or autonomy/wholeness and an absolute subordination of one to the other. This is not a complaint about being "extreme." "Extreme" is a favorite reproach by liberals, for whom the desired condition is moderation, a middle ground, "somewhere in between" mainstream compromise. Their favorite colors are "not black or white but shades of gray." In contrast the dialectical criticism is "one-sidedness," the seizing upon one side of a dichotomous pair or a contradiction as if it were the whole thing. *Our spectrum is not a gradient from black through all the grays to white but is a fractal rainbow.*

Of course, despite Hegel's dictum that "the truth is the whole" we cannot study "the whole." The practical value of Hegel's affirmation is twofold:

First, problems are larger than we have imagined and we should extend the boundaries of a question beyond its original limits. Even systems theory construes problems too small, either because the domain is assigned to the analyst as a given "system" or because additional variables known to interact with the initial system are not measurable or do not have known equations, or because of traditional boundaries of disciplines. Thus a systems analysis of the regulation of blood sugar may include the interactions among sugar itself, insulin, adrenaline, cortisol, and other molecules but is unlikely to include anxiety, or the conditions that produce the anxiety such as the intensity of labor and the rate of using up of sugar reserves, whether or not the job allows a tired worker to rest or take a snack. Models of heart disease are likely to include cholesterol and the fats that are turned into cholesterol but not the social classes of the people in whom the cholesterol is formed and breaks down. Systems analysis would not know how to deal with the pancreas under capitalism or the adrenals in a racist workplace. Models of epidemics may include rates of reproduction of viruses and their transmission but not the social creation of a sense of agency that may allow people to take charge of their exposure and treatment.

The second application of the understanding that the truth is the whole is that after we have defined a system in the broadest terms we can at the time, there is always something more out there that might intrude to change our conclusions.

Dialectics appreciates the pre-reductionist kind of holism, but not its static quality, its hierarchical structure with a place for everything and everything in its place, nor the *a priori* imposition of a purposefulness that may or may not be there. Thus it "negates" materialist reductionism's negation of the earlier holism, an example of the negation of the negation that John Maynard Smith found so opaque but could have recognized as the non-linearity of change.

What Are Parts?

Wholes are thought of as made out of parts. Systems theory likes to take as its elements unitary variables that are the "atoms" of the system, prior to it, and qualitatively unchanging as they ebb and flow. Their relations are then "interactions," as a result of which the variables increase or decrease, emit "outputs" and thus produce the properties of the wholes. But the wholes are not allowed to transform the parts, except quantitatively. The long-distance conversation does not transform the telephone, the market does not change the buyer or seller, and power does not affect the powerful nor love the lover. It is the priority of the elements and along with it the separation of the structure of a system from its behavior—rational assumptions for designed and manufactured systems—that keeps systems theory still vulnerable to the reproach of being large-scale reductionism.

The parts of dialectical wholes are not chosen to be as independent as possible of the wholes but rather points where properties of the whole are concentrated. Their relation is not mere "interconnection" or "interaction" but a deeper interpenetration that transforms them so that the "same" variable may have a very different significance in different contexts and the behavior of the system can alter its structure. For instance, temperature is important in the lives of most species, but temperature has many different meanings. It acts on the rate of development of organisms and therefore their generation time and also on the size of individuals; it limits the suitable locations for nesting or reproduction; it may determine the boundaries of foraging or the time available for searching for food. It influences the available array of potential food species and the synchrony between the appearance of parasites and their hosts. It modifies the outcomes of species encounters.

But temperature is not simply given to the organisms. The organisms change the temperature around them: there is a layer of warmer air at the surfaces of mammals; the shade of trees makes forests cooler than the surrounding grassland; the construction of tunnels in the soil regulates the temperatures at which ground-nesting ants raise their brood; the color of leaf litter and humus determines the reflection and absorption of solar radiation. Through the physiology and demography of the organism, *effective* temperature, its range, and its predictability are quite different from the weather box temperature of a place. On another timescale, temperature acts through various pathways as pressures of natural selection, changing the species, which again changes its effective temperature. Thus "temperature" as a biological variable within an ecosystem is quite different from the more easily measured physical temperature that can be seen in the weather box as prior to the organisms.

Although systems theory is comfortable with the idea that a certain equation is valid only within some limits, it does not deal explicitly with the interpenetrations of variables in its models, their transformations of one another. In a sense, Marx's *Capital* was the first attempt to treat a whole system rather than merely criticize the failings of reductionism. His initial objects of investigation in volume 1, commodities, are not autonomous building blocks or atoms of economic life that are then inserted into capitalism, but rather they are "cells" of capitalism chosen for study precisely because they reveal the workings of the whole. They can be separated for inspection only as aspects of the whole that called them forth. To Marx, this was an advantage because the whole is reflected in the workings of all the parts. But for large-scale reductionists the relationship goes from given, fixed parts to the wholes that are their product. The priority and autonomy of the part is essential to systems analysis. "Autonomy" does not of course mean they have no influence on one another. The variables of a system may increase and decrease but remain what they are.

Parts of a system may themselves be systems with their own structure and dynamics. This approach is taken by hierarchy theory in which nested systems each contribute as parts of higher-level systems.[10] This allows us to separate domains for analysis. However, the reverse process, the defining and transforming of the subsystems by the higher level, is rarely examined.

Much statistical analysis, for instance in epidemiology, separates the independent variables that are determined outside the system from the dependent variables that are determined by them. The independent variables might be rainfall or family income; the dependent variable might be the prevalence of malaria or the suicide rate. In contrast, systems approaches recognize the feedbacks that give mutual determination: predators eat their prey, prey feed their predators; prices increase production, production leads to surpluses that lower prices; snow cools the earth by reflecting away more sunlight, and then a cooler earth has more snow. In feedback loops, changes in each variable are in a sense the causes of the changes in the others. What then happens to causation? What makes one "cause" more fundamental than another?

We can attempt to answer this question in two ways. First, we may ask where a particular pattern of change was initiated at a particular time. For instance, we might ask of a predator-prey system, why does the abundance of both predator and prey vary over a five-hundred-mile gradient? We can analyze the feedback relationship to show that if the environmental differences along the gradient enter the system by way of the prey, say through temperature increases that increase its growth rate, this will increase the predator population so that the two variables are positively correlated. But if the environmental differences enter by way of the predator, perhaps because the predator is itself hunted more in some places than others, then increases in hunting reduce the predator and therefore increase the prey. This gives us a negative correlation between them. Therefore if we observe a positive correlation we can say that the variation is driven from the prey end and if a negative correlation then the variation is driven from the predator end. The prey mediates the action of the environment and is the "cause" of the observed pattern in the one system, the predator in the other. Similarly, in a study of the capitalist world economy in which I examined production and prices during the 1960s and 1970s, I found that the major agricultural commodities exhibited a positive correlation between production or yield per acre and prices on the world market. This supports the view that price fluctuations arise mostly in the larger economy and affect production decisions rather than appear as responses to fluctuations in production—and this despite obvious and dramatic changes of production due to the weather or pests.

Whether this is generally true is an empirical question. In a complex network of variables the driving forces for change might originate anywhere. When we attempt to ask "Does economics or geopolitics determine foreign policy?" or "Is the content of TV driven by sales or ideology?" the question is unanswerable in general. The complex network of mutual determinations requires a complex answer that is hinted at in the awkward term "overdetermination," which recognizes causal processes as operating simultaneously on different levels and through different pathways. Or it brings us back to Hegel: The truth is the whole.

Then where is the locus of historical materialism? Doesn't it require that the economy determine society?

No! "The economy" as a set of factors in social life has no inherent priority over any of the other myriad interpenetrating processes. Sometimes it is determinant of particular events, sometimes not. As long as we remain within the domain of a systems network tracing pathways, everything influences everything else by some pathway or other. Changes in the productive technology change economic organization and class relations and beliefs about the world, but changes in the technology arise through the implementation of ideas, and exist in thought before they are made flesh. Or, as the founding document of UNESCO stated, "Wars are made in the minds of men." Then is social life a product of intellect? Or is intellect an expression of class and gender? Approached in this way, all is mediation, and the assignment of absolute priority is dogmatism.

But this is quite different from identifying *the mode of production and reproduction,* which is present not as a "factor" in the network but as the network itself. It is the structure of that network, that mode, that defines workers and capitalists as the actors or "variables" in the network, makes it possible for sexism to have commercial value, makes legislation a political activity, and allows major events to be initiated by the caprices of monarchs. It is the context within which the various mediations play themselves out and transform each other rather than being a factor among factors.

Goal Seeking

The third quality of systems, purposefulness, also betrays the origin of systems theory. The outcomes are evaluated for their correspondence to the built-in purpose, while deviations from that purpose are seen as nonadaptive, contradictory, and self-destructive behaviors. These appear as system failures. The engineer can discard or a manager can reorganize the structures that lead to them. But in reality only some systems are purposeful, even when they are constructed to satisfy some purpose. In others, while the "elements" are actors each with their own purposes and may be said to seek goals, the system as a whole does not.

Dialectical "wholes" are not defined by some organizing principle such as harmony or balance or maximization of efficiency. In my view, a system is characterized by its structured set of contradictory processes that gives meaning to its elements, maintains the temporary coherence of the whole, and also eventually transforms it into something else, dissolves it into another system, or leads to its disintegration.

Outcomes

Once mathematical systems theory defines a set of variables and interrelations it then asks the simple mathematical question, what is the future trajectory of those variables starting from such and such initial conditions? From then on, all depends on the mathematical agility of the analyst or the computer program to come up with "solutions" of the equations. A solution is the path of the variables. The desired result is prediction, the correspondence between the theoretical and observed values of the variables.

There are only a few possible outcomes of equations:

1. The variables may increase or decrease out of bounds. This may mean a real explosion, disrupting the system. But it can also mean that past a certain point the equations are not valid.

2. The variables may reach a stable equilibrium. It then remains there unless perturbed and returns toward equilibrium after a perturbation. If the processes include randomness, then a solution may be a stable probability distribution.

3. There may be more than one equilibrium, in which case not all of the equilibria are stable. Each stable equilibrium is the end result for the variables that start out "near" that equilibrium, within some range called its "basin of attraction." The basins of attraction around the equilibria are separated by boundaries where there are unstable equilibria. The outcome then depends on the starting place, and the variables move toward the equilibrium in whose basin of attraction they start out.

4. The variables may show or approach cyclic behavior, in which case how quickly the variables cycle and the magnitude of the fluctuations describe the solution. A cyclical pattern also has its basin of attraction, the range of initial conditions from which the variables approach that cycle.

5. The trajectories may remain bounded but instead of approaching an equilibrium or a regular periodicity show seemingly erratic pathways, sometimes looking periodic for a while and then abruptly moving away, and different initial conditions no matter how similar may give quite different trajectories. This is referred to as "chaos," although in fact it has its own regularities.

The behavior of a system will depend on the equations themselves, the parameters, and the initial conditions. Much of the content of systems theory is the description of the relations between the assumptions of the model and the outcomes for the variables, or identifying the procedures for validating the models.

The outcomes are expressed as quantitative changes in the variables. This is an extremely useful activity for making predictions or deciding upon interventions in the system or system design. But it is also limiting, and it imposes constraints on the models. Most models require specifying the equations and estimating the parameters and variables. Therefore those that are not readily measurable are likely to be omitted. For instance, we can write compartment models for epidemics that take as variables the numbers of individuals in each compartment—those who are susceptible, infected but not infective yet, infective, or recovered and immune. We make some plausible assumptions about the disease (rates of contagion, duration of latent and infective periods, rate of loss of immunity) and turn the crank, watching as numbers shift from one compartment to another. Then we can ask questions such as, will the disease persist, how long will it take to pass the peak, how many people will die before it is over, what would be the effect of immunizing x percent of the children? We could add complications of differences due to age and even subdivide the population into classes with different parameters.

Contagion also depends on people's behavior, the level of panic in the population. This changes in the course of the epidemic as people observe acquaintances getting sick and dying, and may take protective action. But how much experience is needed to change behavior? How much panic before they will lose their jobs rather than face infection? What degrees of freedom do people have? How long will an altered behavior last? Do people really believe that what they do will affect what happens to them? Will they remember for the next time? Since we have neither the equations for describing these aspects nor measurements of panic or historical horizon or economic vulnerability, such considerations will not usually appear in the models but at best only in the footnotes. In recent years, modeling has become a recognized major research activity. But it has had the effect of reducing modeling to the quantitative models described above.

Most systems modelers take it for granted that quantitative information ("hard" data) is preferable to qualitative ("soft") information and prefer prediction or fitting of data to understanding. In their view of science, progress goes simply from the vague, intuitive, qualitative to the precise, rigorous, and quantitative. The highest achievement is the algorithm, the rule of procedure that can be applied automatically by anyone to a whole class of situations, untouched by human minds. That is the rationale behind Maynard Smith's suggestion that systems theory replaces dialectics. Marxists argue for a more complex and non-hierarchical relation between quantitative and qualitative approaches to the world.

A much smaller effort goes into qualitative systems modeling, which would allow us to deal with these "soft" questions. Instead of the goal of describing a system fully in order to predict its future completely or to "optimize" its behavior, we ask how much we can get away with not knowing and still understand the system. Whereas the engineering systems presume rather complete control over the parameters so that we can talk about optimizing the parameters, the systems we are most concerned with in nature and in society are not under our control. We try to understand them in order to identify the directions in which to push but do not trust our models to be more than useful insights into the structure or process.

Dialecticians take as the objects of our interest the processes in complex systems. Our primary concern is understanding them in order to know what to do. We ask two fundamental questions about the systems: Why are things the way they are instead of a little bit different, and why are things the way they are instead of very different? And from these follow the practical questions of how to intervene in these complex processes to make things better for us. That is, we seek practical and theoretical understanding rather than a good fit. Precision and prediction may or may not be useful in this process, but they are not the goals of it.

The Newtonian answer to the first question is that things remain the way they are because nothing much is happening to them. Stasis is the normal state of affairs, and change must be accounted for. Order is the desired state, and disruption is treated as disaster. A dialectical view begins from the opposite end: change is universal and much is happening to change everything. Therefore equilibrium and stasis are special situations that have to be explained. All "things" (objects or patterns of objects or processes) are constantly subject to outside influences that will change them. They are also all heterogeneous internally, and the internal dynamics is a continuing source of change. Yet "things" do retain their identities long enough to be named and sometimes persist for very long times indeed. Some of them, much too long.

The dynamic answer to the first question is homeostasis, the self-regulation observed in physiology, ecology, climatology, the economy, and indeed in all systems that show any persistence. Homeostasis takes place through the actions of positive and negative feedback loops. If an initial impact sets processes in motion that diminish that initial impact, we refer to it as negative feedback, whereas if the processes magnify the original change the feedback is positive. Thus positive and negative applied to feedback have nothing to do with whether we like them or not. When positive feedbacks have undesirable results that increase out of bounds, we refer to them as vicious circles.

It is often said that negative feedback stabilizes and positive feedback destabilizes a system. But this is not always the case. If positive feedback exceeds the negative then the system is unstable in the technical sense—it will move away from equilibrium. In that case an increase of negative feedback is stabilizing. But if the indirect negative feedbacks by way of long loops of causation are too strong compared to the shorter negative feedbacks the system is also unstable and will oscillate. Then positive feedback loops can have a stabilizing effect by offsetting the excessive long negative feedbacks. Long loops behave like delays in the system. The significance of a feedback loop depends on its context in the whole. The complex systems of concern to us usually have both negative and positive feedbacks.

Homeostasis does not imply benevolence. A negative feedback loop should not be seen as the elementary unit of analysis or of design. A simple equation may give the appearance of "self-regulation" in the sense that when a variable gets too big it is reduced and when it gets too small it is increased. But the reduction and the increase may have quite different causes. An increase in wages may lead to employers cutting the labor force, increasing unemployment, and thus making it easier to reduce wages. A decrease in wages may lead to labor militancy that restores some of the cuts. The outcome (if nothing else happens) is a partial restoration of the original situation. Neither party is seeking homeostasis, and the wage-employment feedback is not designed or pursued by anyone to maintain economic stability. It is simply one possible manifestation of class struggle. Thus homeostasis does not imply functionalism, a view that assigns purpose to the feedback loop as such.

This distinction is important, especially when we examine apparently unsuccessful attempts to achieve socially recognized goals. Meadows, Meadows, and Randers present the problem as follows:

> This book is about overshoot. Human society has overshot its limits, for the same reason that other overshoots occur. Changes are too fast. Signals are late, incomplete, distorted, ignored or denied. Momentum is great. Responses are slow.[11]

From this systems-theoretic point of view, the socialized Earth's error-correcting feedbacks are inadequate. And if you assume that social processes are aimed at sustainable, healthful, equitable relations among people and with the rest of nature, then the defect is in the feedback loops, the mechanisms for achieving these goals. But if agriculture fails to eliminate hunger, if resource use is not modulated to protect people's health and long-term survival, it is not because of the failings of a mechanism aimed at these goals. Rather, most of world agriculture is aimed at producing marketable commodities, resources are used to make profits, and the welfare effects are side effects of the economy. It is the contradictions among opposing forces (and between those of the ecology and the economy) rather than the failure of a good try by inadequate information systems and deficient homeostatic loops that are responsible for much of the present suffering and the threat of more.

When a change occurs in a component (or variable) of a system, that initial change percolates through a network of interacting variables. It is amplified along some pathways and buffered along others. In the end, some of the variables (not necessarily the ones that received the initial change or those nearest the point of impact) have been altered while others remain pretty much the way they were. Therefore we identify "sinks" in the system, variables that absorb a large part of the impact of the external shock, and other aspects of the system that remain unchanged, protected by the sinks. We can even have situations where things change in ways that contradict our common sense; for example, when adding nitrogen to a pond can lower the nitrogen level or an inflated military budget undermines national security. (This outcome depends on the location of positive feedbacks within a system.)

But "unchanged" requires some further examination. The "variable" is not a thing but some aspect of a thing, perhaps the numbers of individuals in a population, not "the population."

One simple system consists of a predator that feeds on a single prey. All else is treated as "external." It is sometimes the case that the predator is regulated only by the prey. Then a change in conditions that acts directly on the reproduction, development rate, or mortality of the prey—which is not due to the predator—will be passed along to the predator. Increased prey leads to increased predators and this reduces the prey back to its original value. The "prey" variable may remain unchanged while the predator population either increases in response to increased availability of prey or diminishes if fewer prey are produced. The predator variable acts as a sink in this system. Tracing the ups and downs of predator and prey finishes the tasks of the systems analysis.

But what I referred to as "prey" is only the numbers of prey. If prey reproduction has increased with more food but the population of prey has not changed, it is because the prey are being produced faster and consumed faster. That is, the prey population is younger. Individuals may be smaller and therefore more vulnerable to heat stress. They may be more mobile, migrating to find unoccupied sites. If the prey are mosquitoes, a shorter life span may mean that they do not spread as much disease even if there are more of them. They may spend more time in cool moist shelters where they meet additional predators and the model has to be changed. Natural selection in a younger population might focus more on those qualities that affect the survival and early reproduction of the young. Thus the variable "prey," which was unchanged in the model, can be actively transformed in many directions not dealt with in the model.

The particulars of the dynamics, the relations among the positive and negative feedbacks in a system—sources and sinks, connectivity among variables, delays along pathways and their effects—are all in the domain of systems theory in the narrow sense. The parts of the system become the variables of models, and equations are proposed for their dynamics. Systems theory studies these equations. Mathematical rules have been discovered for determining when the system will approach some equilibrium condition or oscillate "permanently," that is, as long as the assumptions still hold.

Modern computational methods allow for the numerical solutions of large numbers of simultaneous equations. The parameters are measured, the initial conditions of the variables are estimated or assumed. (The distinction between parameters and variables is that the parameters are assumed to be determined outside the boundaries of the "system" and are only inputs while the variables change each other within the "system.") The computer then calculates successive steps in the process and comes up with numbers, the predicted states of the variables at different times. The numerical results are compared to observations. If the correspondence is good enough, it is assumed that the model is valid, that it "accounts for" the behavior of the system being studied, or 90 percent of the behavior, or whatever level we decide is acceptable. If not, more data may be collected to get better estimates of parameters or the equations may be modified.

However, systems theory starts with the variables as givens. It deals with the problems of selecting variables only in a very limited way. When we approach any real system of any complexity, the question of the right variables to include in the model is itself complex. It is the classical Marxist problem of abstraction.[12] Some practical systems modeling criteria are reciprocal interaction, commensurate timescales, measurability, and variables that belong to the same discipline and can be represented by equations of change. The system should be large enough to include the major pathways of interaction, with identification of where external influences enter the network. Systems theory makes use of growing computing capacity to give numerical solutions to the differential or difference equations that describe the dynamics. In order to have precise outcomes it is necessary to have good estimates of the parameters, things like the reproductive rate of a population, the intensity of predation, the half-life of a molecule, or the cost-price ratio in an economic production function. The gathering of these measurements is difficult, so that estimates are often taken from the published literature rather than made afresh. Parameters that cannot be measured readily cannot be used.

Once variables are selected, they are then treated as unitary "things" whose only property is quantity. The mathematics will tell us which quantities increase, which decrease, which fluctuate or remain unchanging. The source of change is either in the dynamics of the variables in interaction or in perturbation from outside the system. ("Outside the system" means outside the model. In a model of species interactions a genetic change within a species is regarded as an external event, since it is external to the demographic dynamics although it is located inside the cells of the bodies of individual members of a population.) But all variables are themselves "systems" with internal heterogeneity and structure, with an internal dynamics that is influenced by events on the system scale and also change the behavior of the variables. Thus dialectics emphasizes the provisional nature of the system and the transitory nature of the systems model.

The variables of a system change at different rates, so that some are indicators of long-term history and others are more responsive to the most recent conditions. Thus in nutritional surveys we use the height of children as an indicator of long-term nutritional status, growth over a lifetime, while weight for height indicates food intake over recent months or weeks and therefore measures acute malnutrition. Because each variable reflects its history on its own timescale, they are generally not in "balance" or harmony. Ideology need not "correspond" to class position, political power to economic power, or forests to climate. Rather, the links between variables in a system identify processes: ideology responding, not corresponding, to class position, economic power enhancing political power, political power being used to consolidate economic power, colder climate trees such as spruce and hemlock gradually displacing the oak and beech of a warmer period. But all of these processes take time, so that a system does not show a passive correlation among its parts but rather a network of processes constantly transforming each other. In Darwinian evolutionary theory both the adaptedness of a species to its surroundings and its nonadaptedness are required, the former showing the outcomes of natural selection and the latter identifying it as a process that is never complete and showing the history of the species. Complete adaptedness would have been an argument for special creation, not evolution, proclaiming a harmony that manifests the benevolent wisdom of the Creator.

The second question, why things are the way they are instead of very different, is a matter of history, evolution, and development that is concerned with the long-term processes that change the character of systems. The variables involved in long-term change may overlap with the short-range ones, but are not in general the same. Many of the short-term processes are reversible, oscillating according to conditions without accumulating to contribute in the long run.

At any one moment the short-term events are strong processes, temporarily overwhelming some of the long-term directional changes that are imperceptible in the short run. Yet the two scales are not independent. The reversible short-term oscillations through which a system confronts changing circumstances have themselves evolved and continue to evolve as a result of their functioning in the long run. And they leave long-term residues: the breathing in and breathing out of ordinary respiration may also result in the accumulation of toxic or abrasive materials in the lung; the repetitive cycles of agricultural production can exhaust the soil; the periodicity of the tides also has its long-term effect of lengthening the day through tidal friction; the buying and selling of commodities can result in the concentration of capital. Long-term changes alter the circumstances to which the short-term system responds as well as the means available for that response.

Here mathematical systems theory is less useful, since the mathematics is much better developed for studying steady-state systems than evolving ones. (The work of Ilya Prigogine on dissipative systems is only a partial exception to this limitation.)

Conclusion

Systems analysis is one of the techniques for policymaking. As its technical side becomes more sophisticated it also is usually less accessible to the non-specialist. Therefore it often reinforces a technocratic approach to public policy, and does that in the service of those who can afford to contract its services. The ruling class and its representatives are referred to in the trade by the more neutral term "decision makers." This is not unique to applied systems theory but is a common correlate of its increasing use within a managerial framework. A special effort has to be made to counteract this tendency, to demystify the study of complexity, and to democratize even complex decision making. The Soviet author Viktor G. Afanasev, before he embraced the free market, wrote an interesting book, *The Scientific Management of Society*, which emphasizes the systems-theoretic aspects of planning as a technocratic procedure with only perfunctory nods in the direction of popular control of the planning process as a whole.[13]

Systems theory can be understood as a "moment" in the investigation of scientific problems within complex systems by means of mathematical models. Its value depends in large measure on the context of its use, and here dialectics has a broader role that can inform that use:

1. The posing of the problem, the domain to be explored, what is taken as the "fundamental elements," what is taken as the givens of the problem, and the boundaries that are not questioned. To do this well requires not only a substantive knowledge of the objects of interest, their dynamics and history, but also an understanding of process. There is also frank partisanship, since what is taken as given and what is assumed to be "fundamental" is a political as much as a technical problem. For instance, a model of a society that consists of atomic individuals making decisions in the void cannot escape the dead end of bourgeois individualist reductionism no matter how elegantly the mathematics is developed. An economic model that consists of prices and production and profits and such can give projections of trajectories of prices and production and profits and such (at best; in reality they do this very badly). But it will never lead to an understanding of economics as social relations.

Sometimes the variables are given to the systems analyst: the species in a forest, the network of production and prices, the gizmos in a radio, the molecules in an organism. That is, the "system" is presented to us as a problem to be solved rather than as an objective entity to be understood. But often it is presented more vaguely: how do we understand a rain forest or the health of a nation? The way in which a problem is framed, the selection of the system and subsystem, is prior to systems theory but crucial to dialectics. A dialectical approach recognizes that the "system" is an intellectual construct designed to elucidate some aspects of reality but necessarily ignoring and even distorting others. We ask what the consequences would be of different ways of formulating a problem and of bounding an object of interest.

2. Selection of the appropriate mathematical formalisms (equations, graph diagrams, random or deterministic models, and so on). Whereas technical criteria influence these choices there are also issues of the purposes of the model, the partially conflicting goals of precision, generality, realism, manageability, and understanding. The important thing here is not to be limited by the technical traditions of a field but to examine all these choices not only for hidden assumptions but also for their implications.

3. Interpretation of results. Here qualitative understanding is an important supplement to numerical results. In the course of an investigation we may

go from vague qualitative notions through quantitative explorations to more precise qualitative understanding. This is only one example of non-progressivist, nonlinear thinking that is captured in our "mysterious" negation of the negation.

Progress is not from qualitative to quantitative. Quantitative description of a system is not superior to qualitative understanding. When approaching complexity, it is not possible to measure "everything," plug it all into a model, and retrieve intelligible results. For one thing, "everything" is too big. Qualitative understanding is essential in establishing quantitative models. It intrudes into the interpretation of the results. The task of mathematics is to make the arcane obvious and even trivial. That is, it must educate the intuition so that confronted with a daunting complexity we can grasp the crucial features that determine its dynamics, know where to look for the features that make it what it is, suspect mainstream questions as well as answers.

A dialectical understanding of process in general looks at the opposing forces acting on the state of a system. This is now accepted more or less in ordinary scientific practice. Excitatory and inhibitory neurons, sympathetic and parasympathetic stimulation, opposing selection forces or an opposition between selective and random processes are all part of the tool kit of modern science. However, this has still not been generalized to thinking of process as contradiction.

4. When does the system itself change and invalidate the model? We need a permanent awareness of the model as a human intellectual construct that is more or less useful within certain bounds and then can become nonsense. The internal workings of the variables in a model, the dynamics of the model itself, or the development of the science eventually reveals all models as inaccurate, limited, and misleading. But this does not destroy the distinction between models that are terribly wrong from the start and those that have relative validity.

5. Doubt is an essential part of the search for understanding. Some areas of science have been consolidated to the point of near certainty. Others are border regions of our knowledge where there is a plurality of insights and opinions and conflicting evidence. Here doubt and criticism are essential. And beyond that is the unknown, where we have divergent intuitions and

where our biases can roam freely. But when we have the same doubts persisting for long periods it is not a sign of a postmodern pluralist democracy but a sign of stagnation. Useful doubt is not the expression of an esthetic of indecision or a response to the petulant reproach of "you're so damn sure of yourself" or an acknowledgment that truth is "relative," but rather a historical perspective on error, bias, and limitation.

The art of modeling requires the sensitivity to decide when in the development of a science a previously necessary simplification has become a gross oversimplification and a brake to further progress. This sensitivity depends on an understanding of science as a social process and of each moment as an episode in its history, a dialectical sensitivity that is not taught in the "objectivist" traditions of mechanistic systems analysis.

Thus systems theory is best understood as reflecting the dual nature of science: part of the generic evolution of humanity's understanding of the world and a product of a specific social structure that supports and constrains science and directs it toward the goals of its owners. On the one hand it is a "moment" in the investigation of complex systems, the place between the formulation of a problem and the interpretation of its solution where mathematical modeling can make the obscure obvious. On the other hand it is the attempt of a reductionist scientific tradition to come to terms with complexity, nonlinearity, and change through sophisticated mathematical and computational techniques, a groping toward a more dialectical understanding that is held back by its philosophical biases and the institutional and economic contexts of its development.

18

Aspects of Wholes and Parts
in Population Biology

Since the seventeenth century, the mechanistic reductionist worldview associated with Descartes has dominated European and American thought about nature and society. According to this view, the world is made up of separate objects, things. These things are essentially passive; they normally remain the way they are but can be set in motion by external causes. They can be examined in isolation from one another and their properties measured. The resulting quantitative differences are the most important things about them. Finally, once we have measured and described them, we can combine them into structures that will behave according to the properties analyzed in isolation.

This conceptualization of the world was successful because it corresponded to the daily experience of capitalist life, making its precepts seem self-evident. It provided the guidelines that enabled science to answer the questions posed to it by that society and then guaranteed further success by defining as legitimate questions only those that could be answered within its framework.

Although it has evolved in diverse and complex ways over the next three centuries and has been forced to confront more dynamic and complex systems than it arose to study, its outlook has remained intact and dominant.

However, in recent decades new holistic challenges to the mechanistic approach have arisen in many specific fields and as a philosophy as well. This new holism has grown up partly outside and partly within the existing scientific institutions. Some forms of it have been able to draw on diverse traditions that either survived from pre-capitalist organism or arose directly as challenges to early capitalism. These include religious, antiscientific, feudal, holisms organized around a tightly integrated Great Chain of Being, where connection was tightly fixed and rigidly if benevolently hierarchical, but also include variants of a communal heretical type that sought egalitarian connectedness.[1] Some also incorporate the insights of pre-capitalist Asia, where Chinese medicine and the Buddhist and Taoist schools emphasized wholeness, connection, and balance, and the philosophies of Native American and other indigenous peoples.

The new holism also makes use of the Marxist criticisms of mechanism and of the research of some unconventional groups within established science.

But the present popularity of holism as a growing opposition comes less from comprehensive philosophical dissent and more from the criticism of the consequences of mechanisms in different fields of applied science, often giving rise to movements labeled *alternative*. It has been promoted by feminism, the ecology movements, alternative agriculture, alternative health, and various schools of psychological-social counseling.

Often we observe political conflicts around how broadly a problem is to be defined, with the liberal-conservative forces usually insisting on the narrow isolation of a problem and the radicals generally urging a broader context, a concern with long-term and indirect effects, and the linking of natural and social pressures in the same system. Is hunger caused by insufficient production of food or by social relations that guarantee food's insufficiency and inequitable distribution? Do poor people get tuberculosis because of the Koch bacillus, or is the Koch bacillus one of the ways that poverty kills? Is the cause of a plant epidemic some fungus or the monoculture—in part required by the dominant economic relations—that allows its rapid spread? Although holistic critics often pose these rhetorical questions in the form of mutually exclusive alternatives, the major thrust is toward making the issues more inclusive and complex and focusing attention on the higher levels of organization.

Holistic and Alternative Criticism of Health and Agriculture

Critics of the existing health system have emphasized its failure to look at the broader contexts of health for the following reasons:

1. One-to-one clinical medicine is beside the point for people who do not have access to health care. The most elementary demand of oppressed groups is for available health service, and popular revolutions (such as in Cuba) take enormous pride in bringing health service, even of the prevailing kind, to the whole country. Demands for national health insurance, socialized medicine, community and alternative clinics are other expressions of this criticism.

2. The physical availability of health service is not sufficient. Not only the cost but the social content of the doctor-patient interaction and the perceived effectiveness of treatment will determine whether people will use the established services. Issues of medical condescension, sexism, and racism are part of "accessibility," and patient advocacy becomes a political demand to deal with the patient as an active whole.

3. The pattern of health and disease in a population is a much broader issue than the availability of health service. Physicians cannot prescribe food for the hungry, rest for the overworked, or clean air for the miners and textile workers. Critics insist that whereas at a clinical level poverty may help the pneumococcus kill people, at a population level pneumococcus is the way that poverty kills people. The conflict between the approaches of Koch (microbiology) and Virchow (social epidemiology) is how holism is pressed as the social causation of disease.[2]

4. The health care provided by modern medicine is itself flawed and, as Ivan Illich has emphasized, is often the cause of illness.[3] This comes about because of a mechanistic reductionist model and the fragmentation of health issues into narrow subfields.

Cartesian dualism still separates mind and body, even when as in psychosomatic medicine attempts are made to build links between them as separate entities. Recent discoveries showing how conscious activity in the cerebral cortex affects the action of the autonomic nervous system and the whole physiology have created new specialities such as biofeedback and introduced techniques such as meditation or visualization in an attempt to use that influence therapeutically.

The criticism of agriculture is often strikingly similar to the criticism of medicine. It emphasizes the persistence of hunger; the development of technology without considering its impact on different classes and on women; and how modern agriculture undermines its own productive base by highly mechanized systems that increase soil erosion, salinization, and compaction. This type of agriculture destroys the complex microbial and invertebrate communities and increases vulnerability to new pests. (The secondary pest is the agricultural equivalent of iatrogenesis.) Both high-tech medicine and agriculture dismiss the previously accumulated folk knowledge as superstition and make recipients of the new technology powerless. Finally, both work from a narrow intellectual base that exacerbates the contradiction between the increasing scientific rationality in the small and the irrationality at the level of the whole enterprise, a contradiction that guarantees unpleasant surprises and "side effects."

But the eclectic theoretical foundations of the new holism is also unsatisfactory. There is an emphasis on the whole that subordinates and even obliterates the parts. The notions of balance, harmony, and stability as the organizing principles of wholes make it difficult to cope with the dynamic aspects of natural processes and with conflict.

In contrast, a more dialectical view of complexity stresses: (1) the historically contingent nature of wholes; (2) the qualitative differences among kinds of wholes such as organisms, ecosystems, and societies, each with its own origins and dynamics; (3) the ontological equality of part and whole, and their reciprocal determination; (4) the absence of any universal organizing principle. Rather, the way to understand systems is to identify the opposing processes that allow its persistence and those that eventually transform it.

Classically, the problem of parts and wholes has been seen as the question of emergence. Do wholes have properties that are, in some sense, "more than the sum of the parts"? The meaning of "sum" is taken in many different ways and more or less sets the terms of the problem. The geneticist or ecologist concerned with the numerical prediction of changes in populations or communities often takes "sum" quite literally, so that any non-additivity in, say, the fitnesses of genotypes when regarded as composed of individual genes at a locus is regarded as an evidence of emergence. A somewhat more sophisticated view is that deviations from an additive scale are evidence of "interaction" rather than emergence; that is, a sensible person will recognize that particular combinations of cases will deviate from the simplest additive scheme because of special interactions, and we can estimate the importance of these interactions by techniques like the analysis of variance that isolate the interaction variations from the additive main effects. Thus the presence of interactions are not taken to negate the underlying additivity of phenomena, but to add a complication that causes deviation from the simplest additive scale in particular cases.

To those who protest that dominance and epistasis are evidence that an additive metric is not the "natural" one for fitness, which ought to be multiplicative even on the simplest biological hypothesis, the emergentist displays the case of overdominance. Here the transformation between the scale of gene dose and the scale of fitness is not simply non-additive, it is non-topological, because heterozygotes are between the two homozygotes on the scale of gene dose but not between them on the fitness scale. But this claim to emergence of fitness is easily evaded by extreme compositionists who argue that an adequate explanation at a lower level of phenomena will show the relationship to be a simple metric one. For example, on the one hand, the overdominance in fitness of sickle cell anemia heterozygotes is the consequence of two quite different selective processes superimposed on one another. Homozygotes for sickling homoglobin die from anemia, and heterozygotes are slightly less fit than homozygous normals in this respect. On the other hand, homozygous normals may die of malaria, whereas heterozygotes are the equivalent of homozygous sicklers in this component. Overall fitness, being the projection on a single metric axis of the two independent fitness components, is "artificially" non-metrically related to the two "real" underlying physiological properties that are not emergent. Thus reductionism comes to the rescue of anti-emergentism. An alternative claim is that emergence disappears if the correct characteristic of the genotypes is chosen as a scale. A common explanation of heterosis is that the correct genetic scale is not the dose of A or *a* alleles, but the number of different alleles present in an individual. Each allele codes for a protein that has its own optimal operating range for, say, temperature. Moreover, simple dominance provides that one dose of an allele will result in an adequate supply of the protein. So the possession of two different alleles, each with a slightly different temperature optimum, will, in a fluctuating environment, provide a greater range of function than available to a homozygote. By scaling genotypes on an axis of diversity rather than dose of one of the alleles, the topology of fitness is preserved and the claim of emergence evaporates. To this ploy emergentists respond that the baby has been thrown out with the bath water. Fitness is not some "artificial" mathematical construct produced for the calculating convenience of population geneticists, but the "real" property on which all of evolution rests. Moreover, the direction of future genetic evolution of a population depends critically on whether fitnesses are or are not topologically ordered with allele dose and only allele dose, because the existence of a stable polymorphism as opposed to the elimination of one allele or another depends critically on this relationship. Thus, we are not free to choose any scale. The relation of net fitness to allele dose is the "natural" scale forced on us by the actual dynamics of the evolutionary process.

In ecology the same struggle goes on. In the famous study of Vandermeer on the dynamics of a community of four competing ciliates, the values of r and K were estimated separately for each species in isolation and the $a(ij)$ interaction coefficients among pairs of species were estimated in isolated pairwise interactions.[4] When all four species were put together in the same universe, their dynamics agreed qualitatively with the predictions from the parameters estimated in isolation, but not quantitatively; that is, the order of abundances and stability of the species populations were as predicted, although not their actual numbers. At most, this is a victory for the proponents of interactionism. However, competition experiments between genotypes within a species have often shown a lack of transitivity. Thus, type I may outcompete type II in a pairwise competition, and type II may be superior to III when tested pairwise, yet III may outcompete either in pairwise comparisons or when all three are tested together.[5] Yet the lack of transitivity of competitive rank can be explained, at least in principle, by the claim that different resources are limiting for the different genotypes or species, and that competition is a one-dimensional projection from these many independent dimensions.

And so the struggle among compositionists, interactionists, and emergentists continues and repeats itself in every branch of synthetic biology. Certainly, the study of social behavior is permeated by it, adding another layer to the problem. What are the appropriate elements of explanation of human social structure? Perhaps the most extreme compositionist view is that taken by Lumsden and Wilson, who regard the structure of culture as the collection of individual preferences and behaviors of the individual human beings making up a society.[6] Lumsden and Wilson, however, are interactionists; that is, they do not propose that the individual preferences and behaviors are uniquely coded genetically within the individuals but rather are the consequence of individual biologies finding themselves in particular environmental contexts. What characterizes an individual, then, is not a unique behavior but a norm of reaction of possible behaviors, each invoked by a particular environment. There is no strong claim about the shape of these norms of reaction, nor any necessary assumption of additivity between genes and environment. The extreme compositionism of these authors comes at the level of social organization itself. Whatever the origin of individual behaviors may be, culture for them is the collection of those behaviors both across individuals and across elementary units of behavior called *culturgens,* each with its own etiology in the separate interactions of the genes and environment. The arrows of causation are from the individuals to the social organization, not the reverse. It is important to realize in social behavior that whatever biological theory may be held about the causes of individual behavior, including the theory that individual behavior is itself influenced by the collectivity, a separate social theory is needed that is not in any way biological to make claims about the way individual manifestations will be reflected in the collectivity. So-called *biologistic* theories of social structure contain such a social theory implicitly. It is a compositional theory that places the individual as ontologically prior to the social, although there may be feedback from the social to the individual so that individuals accommodate themselves to social structure.

In our view, the struggle over whether wholes are "more than the sum of their parts" or the precise sense in which this is thought to be true is the wrong issue, because it already accepts an incorrect view of parts and wholes. In brief, the standard view takes parts as given prior entities that can be defined in isolation and can have their properties considered in some ideal isolated prior state before these units become articulated into wholes.

This has both logical and contingent difficulties. First, nothing can be a "part" unless there is a "whole" for it to be a part of. Units may exist in isolation from each other, but these units are not "parts" until they are brought together in a "whole." Conversely, wholes imply parts of which they are made. A thing is not a whole in any meaningful sense of that word unless there are parts that make it up; that is, like the concepts of "good" and "bad," or "large" and "small," the concepts of "part" and "whole" are dialectically related and reciprocally determine each other's status. This logical problem is of real consequence because the question about properties of parts and wholes must be related, and in a more revealing way, as we see later.

Second, nothing exists in isolation. Everything exists in the world in some context even if, rarely, that context is the nearly total lack of interaction with other parts of the world. In the case of hydrogen, oxygen, and water that are so often cited in discussions of parts and wholes, the properties of hydrogen and oxygen that are said to be properties of the parts are, of course, properties of biatomic gaseous hydrogen and oxygen, not of isolated hydrogen and oxygen cations and anions. Moreover, if we were concerned with the properties of these ions, would we mean ions in solution (of water!) or in their extremely unstable monoatomic state at extremely low concentrations so that interactions between them are rare?

Because parts do not come together to make wholes but come into being in them only as the whole comes into being, the real questions about parts and wholes are:

1. What is the relation between the units described as "parts" in one whole to the units described as "parts" in some other whole?
2. What are the properties of units within their respective wholes, that is, in their respective contexts?
3. What are the similarities of contextual properties of units identified as the"same" units in different contexts?
4. What is the causal relation between properties of contextually defined "parts" and the contextual "whole"" of which they are parts?

Notice that none of these questions, not even number 4, can properly be posed as whether wholes are more than the sum of the isolated properties of their parts. In fact, the first three deal with the way in which parts have their properties determined, and it is fundamental to our dialectical view that parts acquire their properties as they are parts of wholes, rather than bringing prior properties to those wholes. No individual human being can fly by flapping his or her arms and legs, and this is true whether that human being is stranded on a desert island ("isolated") or standing at the corner of 42nd Street and Broadway. Yet human beings do fly as a consequence of social interaction and culture that have created airplanes, pilots, fuel, airports, and so on. It is not society that flies, however, but individuals in society. "Parts" have acquired properties contextually. In like manner, no historian can begin to remember unaided a tiny fraction of the facts needed to carry out his or her profession. Yet historians do "remember" a virtual infinity of facts by recourse to books, newspapers, libraries, all social phenomena.

Sometimes the attempt to define properties of parts in isolation is given up in favor of a kind of abstraction of properties by averaging over context. This is the underlying theory of the analysis of variance that seeks to associate main effects of factors with some context-independent property. If there is any non-additivity at all between factors, however, no such isolation of effects is possible, and effects of one factor are always dependent on the context of the other factors.[7] The questions, as we have posed them, begin with an epistemological rather than an ontological issue, the identification of units as parts, and in particular whether the parts of one whole are to be identified with the parts of some other whole. In anatomy, the identification is sometimes so obvious as to require no comment. In every sense that seems to be interesting, the wing of a wren and the wing of an eagle are the same functional and developmental unit. But the wing of a wren and the wing of a fly are not obviously the same unit, even putting aside the much overrated issue of developmental and genetic homology. Perturbation to the development of a fly's wing has a very different consequence for the development of the rest of the organism than a similar perturbation in a vertebrate. Moreover, a fly's wing is an essential part of its courting behavior and a butterfly's wing is a thermoregulator, whereas for wrens it serves neither function. Even for the thermoregulatory function of wings in butterflies, these "parts" are reflectors in some species and absorbers in other closely related forms. Because thermoregulation is carried on by a complex suite of characters including wing color, wing position, wing shape, body orientation, and time of activity, the melanic regions on the wing of one species are not parts of the same system as the melanic spots of another species, for in one case they are components of a heat absorption system and in another of a heat reflection system.[8] In neuroanatomy and behavior, the identification of parts across species is fraught with dangers. The identification of vocalization in primates with speech in humans is tempting, but the region of the brain that is the vocalization center in apes is not the speech center in humans. Stimulation of this region causes meaningless grunts in people as it does in chimpanzees. The speech area in humans maps topographically to a region of the ape brain concerned with tongue and lip movements, but at the same time this speech area has strong commissures to the region of vocalization. So speech is not simply a hypertrophy of grunts but a novel function involving the juxtaposition of bits and pieces of anatomy with relevant properties. Nor is speech simply the combination of tongue and lip movements with grunting because the destruction of Broca's area does not prevent either of those activities.

The heart of the problem is the confusion between the creation of parts by anatomization of wholes, a process carried out in analysis by the observer, and the ontological claim that wholes are actually created by prior existing parts. The world is not, of course, a seamless web. It is broken down into systems that inter-act weakly with each other and within which there are stronger interactions of subsystems. Within those subsystems are sub-subsystems interacting even more strongly. But the identification of these subsystems does not come from some prior existence of independent parts, but from the actual structure of interactions within the whole. Whether Broca's area is or is not a sensible "part" of the brain, whose evolution as a unit is to be studied, whether defense of the group is a legit-imate unit of activity to be compared between species or cultures cannot be decided on contextless grounds. The functioning of the organism, the colony, the community, the culture will redefine its own appropriate units and confer on those units' relevant properties. The evolution of the mammalian ear ossicles from the reptilian jaw suspension has not simply changed the function of various parts, it has redefined the relevant parts into which a sensible functional descrip-tion of both jaw suspension and auditory apparatus must be made.

The contrast between preexistent parts with preexistent functions and consequent parts with contextually created functions is nowhere clearer or more relevant to practice than in the difference between "interest groups" and "social classes" as units of social analysis. The analysis by interest groups assumes that there are roles in society that transcend actual history and that provide the causal force for the construction of social orders. So, as in ants, there are the tasks of defense, food gathering or production, leadership, and reproduction and the differentiation of individuals into these roles creates, at least among humans, interest groups with competing demands. The farmers, the military, the owners of factories, the political leaders, the mothers, the con-sumers, each make different demands on resources. Social structure is a way of mediating among the competing interests in the service of stability. If one interest group, say the military, achieves a temporary excess of power over resources, the society becomes less desirable for the others and ultimately less stable. Social class is a radically different analytic concept, for social classes are seen as the product of the social structure of interactions rather than as their determinants. Classes are created and defined by the act of social production. The primary struggle, in this view, is not over the distribution of limited resources, but over the form of production of resource and how it is to be controlled. As the organization of production changes, so the relations between classes change with the possibility of the disappearance of the classes themselves. Interest groups, however, are seen as eternal.

As for the part-whole dilemma, the problem of reductionism arises from a confusion between the process of knowing and the process of physical determination. By reductionism, we mean a commitment to the view that more complex phenomena are, in fact, the consequence of determination by processes at "lower" levels; that is, the properties of societies are determined by the properties of the individuals whose properties, in turn, are determined by the interaction of their genes and an autonomous environment, whereas the properties of the genes are determined by the properties of DNA, and so on down to quarks. So the action of natural selection is "nothing but" the differential survival of genes and can be reduced to the relative fitness of single alleles on the average, and culture is nothing but the coming together in culturgens expressed as the preferences of individuals. Clearly, reductionism takes parts to be ontologically prior to wholes and would generally reject an emergentist view of the properties of wholes.

There is a program of the study of nature, which we may call *reduction*, that asserts that the truth about nature can be uncovered only by studying the details of processes. The program for study and the ontological commitment should not be confused. It is entirely possible to hold an anti-reductionist view of nature while insisting on the importance of details at lower levels for an understanding of nature. The three-dimensional structure of a folded protein is largely determined by its amino acid sequence, although there may be a few alternative stable folding states for a given sequence. However, many different amino acid sequences may give rise to the same three-dimensional structure. If one compares the three-dimensional structure of lysozyme from avian eggs with that from the bacteriophage T4, the three-dimensional structures are essentially identical. An examination of the amino acid sequences, however, shows absolutely no homology between these widely divergent organisms. The determination of the three-dimensional structure has clearly been a consequence of a long history of natural selection either maintaining or producing by convergence a molecule of special function. The knowledge of the three-dimensional structure alone would not allow us to distinguish this very strong force of natural selection from a simple similarity because of similar underlying amino acid composition. Knowledge of the detailed sequence showed its causal irrelevance. The same possibility for distinguishing between historical similarity and similarity enforced by natural selection exists when amino acid sequences or proteins are compared with their DNA sequences, because of the many–one relationship that exists between the lower and the higher level; that is, the situation at a lower level can be a symptom of the forces acting at higher levels even when it is not their cause. *Reduction* looks to lower levels of analysis for differentiating symptoms of forces at higher levels, whereas *reductionism* claims that forces at lower levels are the actual causes of the phenomena higher up. Modern biology has made immense progress in understanding through the process of reduction, but at the same time the evidence has accumulated that structures at one level do not bear a one-to-one relationship to structures at other levels, and forces must be understood at their appropriate level. Natural selection does not occur at the level of the gene, although its effects can sometimes be calculated at that level, and hypotheses about the action of natural selection can often be tested by observing the changes in gene frequency.[9] The fate of individuals is often the consequence of social forces. It is virtually never their cause.

Once we have established the relative autonomy of the different levels of organization, it becomes necessary to stress their interconnection as well. Variables that we might assign to distinct domains such as physiology, behavior, population dynamics, and community structure come together in particular systems in ways that depend on the system's history.

In order to include behavioral variables with physiological, social, and demographic variables in complex models, we must observe some aspects of behavior that unite it with these processes and also identify some special features:

1. Behavior is similar to other responses to external or internal environment such as shivering, dormancy, or phototropism and partakes with them of the interpenetration of organism and environment. Organisms select, transform, and define their environments through their own activity.

2. Any action by the organism has some impact on its surroundings. This impact can be perceived and responded to by other organisms. When the major significance of an organism's response to its environment is the response of other organisms, we are on the way to communication.

3. Organisms respond to the environment both as a particular physical impact and as information. High temperature as a physical factor accelerates the rates of chemical processes. As information, it may be a predictor of summer beginning. The role of light is more completely that of information, and in organized communication the information content of a signal is its main characteristic. Nevertheless, all information transfer has a particular physical form and takes place in physical structures that are not merely processors of information. Although automata theory can talk about conceptually separate input, state of system, processors, and outputs, these are not really all that separate in living systems. The brain respires, consumes nutrients, receives mechanical pressures: its enzymes have temperature characteristics, rates of synthesis, and breakdown that all depend on the rest of the organism; the controllers are themselves controlled.

4. Responses to complex environments already involve a kind of proto-abstraction. The desert harvester ant stops foraging and returns to her nest when the temperature rises. But she can be lured back above ground by the odor of bait, the duration of her venture into the heat depending on temperature. The opposing push and pull of food and temperature stress becomes qualitatively the same, but quantitatively opposing influences excite and inhibit foraging behavior; that is, incommensurables can in fact be compared.

5. In the vertebrates, the cerebral cortex is not only concerned with higher functions. It is the link between the social and the physiological, transforming the activity of older parts of the organism and making human physiology a socialized physiology.

6. Each behavior is unstable. It docs not continue indefinitely but is normally completed in some sense and replaced by other activities. Therefore, in mathematical models of networks that include behavioral components, stability is no virtue and the demonstration of stability would usually suggest an inadequate model. Even in the absence of external stimuli, a behavioral system generates its own spontaneous activity so that an external stimulus does not encounter the same black box each time, and the variability of response is not an indicator of faulty experimental technique, but an essential property.

In what follows, we look briefly at three models of complex processes that involve behavioral components that are integrated into networks with variables that would usually be assigned to different levels. The models are abstractions derived from, but not fully faithful to, real systems and are intended primarily to illustrate two points: the necessity for inclusion of variables from different domains in the same model and the diffuse, reciprocal nature of control in complex systems.

The first model looks at the regulation of blood sugar and its relation to psychological states. The variables are as follows: E is epinephrin (adrenaline); G is glucose level in the blood; I is insulin level in the blood; A is anxiety symptom. Although the term is vague and not easily measured, we can usually recognize increased and decreased anxiety. Anxiety brings out more adrenaline; adrenaline increases the symptoms of anxiety, and people with low or rapidly falling blood sugar experience anxiety that is relieved by rising glucose. Figure 18.1a shows the assumed relations among these variables. In this and subsequent figures, → indicates a positive effect in the direction of the arrow, —○ a negative effect.[10] For our purposes, it is important to note that there is positive feedback in the relation between epinephrin and anxiety, negative feedback between glucose and insulin, and all the variables are self-damped; that is, each of the substances is removed from the system and would diminish unless restored, and that bouts of anxiety eventually abate. There is also the longer negative feedback, glucose-anxiety-epinephrin-glucose. (The sign of a feedback loop is the algebraic product of the signs of the links around the loop.) Such systems normally behave as you would expect—an increase in glucose brings out insulin, which then reduces the glucose and reduces both anxiety and epinephrin. Insulin reduces glucose and increases anxiety and epinephrin, etc. But, if the self-inhibiting loop of anxiety on itself is weak compared to the E,A positive loop, then in the epinephrin-anxiety subsystem as a whole the positive feedback may outweigh the negative. In that case, we would have an anomalous situation in which an increase in ingested glucose would result in a fall in blood glucose and in average insulin level; an increased dosage of insulin would increase average blood sugar levels and reduce average insulin. All these effects would be the result of an overshoot process: glucose reduces anxiety, which reduces adrenaline so much that less glucose is released from the liver, more than making up for the original increase.

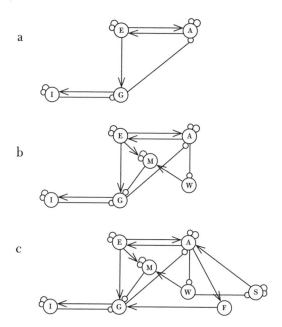

Figure 18.1

The clinical picture would be confusing and might easily be misinterpreted as a genetic condition in which cellular response to insulin is altered. A course of therapy that strengthens the self-damping of the anxiety would also correct the anomalous physiology. (It might be objected that a purely subjective condition such as anxiety cannot be physically efficacious. However, "anxiety" in our model stands for the unspecified set of neural and chemical conditions that are the counterpart of the subjective perception.)

Now consider a person employed in strenuous physical work in a factory or construction site or at home (Figure 18.1b). The physical exertion increases the metabolic rate and uses up blood sugar. The person experiences the subjective impact of reduced sugar and may take protective action by resting (a negative link from "anxiety" to work, W). Some people may also eat a snack, so that there is a positive link from A to F (food) and from F to G. However, these options may not be available. If there is close supervision of labor, resting or eating may bring down the wrath of the supervisor, increasing the anxiety and introducing a new positive feedback into the system (Figure 18.1c). At this point, we may have to break down "A" into several differentiable psychological components. A sufficient drop in blood sugar can lead to a bewilderment that prevents the protective action so that the negative loop G-A-F-G is replaced by

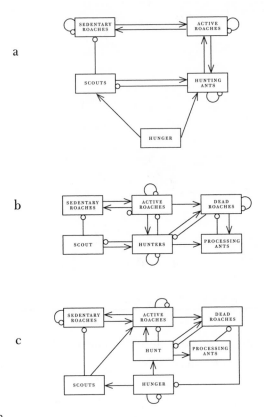

Figure 18.2

a positive loop G-A$_1$-F-G. Finally, coworkers may intervene both to provide a snack and to prevent further harassment by the supervisor.

The main points here are: (1) Each person has her/his own network of interactions, some of which are shared by everyone (e.g., the insulin-glucose link) and some are quite individual. The network is simultaneously physiological, psychological, and social without it being possible to isolate these as separate domains; (2) an event impinging directly on any one of the variables percolates throughout the whole system, being damped along some routes, amplified along others, and sometimes even reversed; (3) what happens depends on the structure of the network, the pattern of positive and negative feedbacks, paths, and sinks; (4) therefore the diagnosis of a health problem should include the identification of the network, and the appropriate loci of intervention may be anywhere in the system.

The second model involves a predator-prey relationship between laboratory populations of ants *(Pheidole dentata)* and cockroaches (Figure 18.2a).
A colony of large cockroaches lives under an egg carton in an aquarium
that also contains a nest of predatory ants. The roaches are gregarious and
usually inactive. The ants can eat roaches, but it is not their favorite food.

An occasional ant scout encounters the roaches. It will tend to back off if
the ants are well fed, but if the nest is hungry, the scout will grab the roach.
The roach can easily twitch and throw off the ant. However, if a roach is seized
by an ant more than about three times within a minute, it becomes restless and
moves out from under the carton. It may also indicate its distress and activate
other roaches.

If the roaches are not further molested, they calm down. However, if there
are enough ants about, the active roach keeps encountering and stirring up
more ants. The initial dynamics are shown in Figure 18.2a.

The positive feedback between active roaches and ant-hunting activity can
make the system unstable. After a threshold is crossed both roach activity and
ant-hunting activity increase sharply. Hunting ants activate other ants.

As soon as more ants grab a roach than it can shake off, it is stopped,
killed, and ants begin to process it. Then ant activity results in removing active
roaches more rapidly than it stirs up more roaches (Figure 18.2b).

One positive feedback loop becomes negative, but two new positive loops
are created. The system is still unstable. Finally and slowly, the dead roaches
are consumed in the nest and hunger is diminished, restoring the original situation. The hunger, previously a parameter of the system, has now entered
into reciprocal interaction and becomes a covariable, introducing a long positive loop (Figure 18.2c).

This mechanism resulted in bursts of hunting and slaughter of roaches every
5 to 10 days in one nest of *Pheidole dentata.* But if the ant colony is too small to
mobilize a sufficient number of hunters or if a steady supply of preferred food
controls hunger, the cyclic behavior could be suppressed. High temperature
could also reduce both scouting and hunting so that the roaches are mainly

sedentary and the   positive paths are weak.

If the roach colony is too small or its reproduction too slow, it may be wiped out by this interaction. If the ants mobilized only very weakly (as most ponerines do) or if high temperature inhibits aboveground activity or if the ant colony is too small, the positive feedback of roach activity on itself by way of the ants can be broken.

Thus the same general qualitative structure can give rise to rather different outcomes. The system's behavior here is inseparable from demographic, physiological, and interspecific processes. And in the course of the cyclic interactions, the structure of the network changes.

In the third case we consider the Hessian fly, a major pest of wheat in North America, with periodic outbreaks causing extensive damage. Varieties of wheat exist that are resistant to the fly but have lower yields than susceptible varieties. During an outbreak, it is in the interest of all farmers to switch to resistant varieties. Once they have done so, the population of Hessian fly declines. Now it is in the interest of each farmer individually to switch back to the susceptible varieties, and for his/her neighbors to continue with the resistant ones. The farmers return quickly to the susceptible varieties and the conditions are established for the next cycle. Here we have a cyclic fluctuation of a fly population, agricultural yield, and wheat genotypes driven by alternating cooperative and competitive behavior of wheat farmers determined by a private property economy.

Although critics from outside of established institutional science propose holistic approaches,[11] scientists charged with planning or directing major practical projects have confronted the necessity to expand the scope of their problems through bitter experiences such as the Green Revolution. It is now a commonplace that life is complex, at least in the prefaces to the studies. Two main approaches have been developed to confront complexity with a kind of reductionist holism: the democratic and the corporate models.

The statistical, democratic model asserts that there are many "factors" or independent variables that are equivalent qualitatively as "factors" and differ only in magnitude. Therefore, the task of the sciences of the complex is to assign relative weights to these factors through analysis of variance and multivariate techniques, and without theoretical preconceptions. New variables are defined by statistical associations of old variables as "principal components."[12]

The corporate view is that every system has a boss, a dominant or controlling factor analogized with the executive office. Then the study of complex systems becomes the search for who's in charge here. In the frequent cartoons of the early space age, the alien space ship opens and we hear "take me to your leader," never "take us to your collective."

In contrast, we argue for the notion of reciprocal control and diffuse, fluctuating hierarchy among components of a system.

Reciprocal Determination of Differential and Long-Term Behavior

Ecologists have long been aware of a peculiarity of the classical predator-prey equations in which, for prey X and predator Y,[13]

$$dX/dt = X(a - bY)$$

and

$$dY/dt = Y(bX - c).$$

Here a can be interpreted as the birth minus death rate of the prey in the absence of predation, b is the predation rate, and by way of food supply also determines the birth rate of the predator, and c is the death rate of the predator. Additional constraints would make the equation more realistic but are unnecessary for our present purpose. There is an equilibrium point given by $Y = a/b$, $X = c/b$, around which the populations cycle. The average values of X and Y are those same equilibrium values.

The peculiarity is that the equilibrium or average value of X is determined from the differential equation for Y, and the equilibrium or average value of Y is determined from the differential equation for X. However, this peculiarity is more general: the equal sign in a first-order differential equation joins two unequals. The left side defines a property, the capacity to change X, which brings together a number of variables that influence X in some function $f(X,Y,Z. . .)$. But if the system reaches an equilibrium, then $dX/dt = 0$ and therefore $f(X,Y,Z) = 0$. If the system does not reach equilibrium but is bounded, then the average value of $dX/dt = 0$ and the average value of $f(X,Y,Z) = 0$. Consider, for example, the number, N, of caterpillars that emerge and feed in the spring. Suppose that their emergence is speeded up by temperature but they are preyed upon and removed by foraging ants, A. Suppose, for example,

$$dN/dt = N(aT - bA).$$

The average rate of change over the season is zero (if we start and finish with the same number), and this average temperature equals average number of foraging ants (with appropriate scaling factors a and b). The ants are not related to the temperature directly, but only by virtue of jointly affecting caterpillar abundance.

Further, if it takes two ants to subdue a caterpillar, the probability of two ants being within reach of the same caterpillar at the same time is proportional to A^2. Then

$$dN/dt = N(aT - bA^2).$$

But the average value of A^2 is the square of the average of A plus the variance of A. Thus the variance in the number of foraging ants is set by the average temperature!

This seems magical at first, because we have not shown any causal link between the temperature and ants or, indeed, have not said anything about how ants are determined. However, this relation only holds if the system as a whole reaches an equilibrium or is bounded and if there is no unaccounted-for autonomous factor that is determining the equilibrium. That in turn requires that there is some pathway from N to A that permits coexistence. But nothing more is required.

The generalization is as follows: In a bounded system of variables whose dynamics are given by first-order differential equations, each variable X specifies some function f of one or more variables in the system and associated parameters. This f gives the rate of change of X. But X has determined that the average value of f is zero; that is, X has established a long-term relation among the variables joined together in f. There is thus a reciprocal determination of short-term (differential) and long-term (average) behavior of variables in a persistent system.

Diffuse Control in a Simple Model Ecosystem

Here in Figure 18.3 we look at the equilibrium or average behavior of a model ecosystem of four variables: a resource R; two consumers A_1 and A_2; and one predator P, which feeds only on A_2. Only R is assumed to be self-damped. The representation in Figure 18.3a and the mathematics argument follows Puccia and Levins. The directions of direct impact of one variable on another are shown by positive $\rightarrow$ and negative $\multimap$ links and correspond to our common sense about the biology.

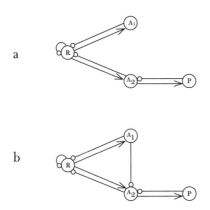

a

b

Figure 18.3

This system is stable under constant conditions. If a change occurs either in the external environment or in the biology of one of the species, it will appear in the model as a change in some parameter(s) directly increasing or decreasing the rates of change of one or more variables. We represent this by a positive or negative input to the node in the graph representing that variable.

The altered parameter has an initial effect on the variable through which it enters the system. The effect then percolates through the network and eventually a new equilibrium is reached. The question we ask is: How does a changed parameter entering the system at any variable affect the equilibrium levels of all of them? In the simplified model studied here, the sign of the effect is the sign of the initial impact multiplied by the signs of the links on a path to the variable of interest, multiplied by the feedback of the rest of the system (the complement) not included in the path, and then divided by the feedback of the whole system.

Because the system is stable, the feedback of the whole in the denominator is negative. This feedback is an inverse measure of the sensitivity of the system as a whole to parameter change. The feedback of R alone is negative, but of the other isolated variables is zero. The trophic relations (R,A_1), (R,A_2), (A_2,P) give negative feedback loops of length two. And combinations of disjunct negative feedback loops such as (R) and (A_2,P) are negative.

In Table 18.1, we show the direction of change of the equilibrium values of each variable when a parameter change enters at a given variable and produces a direct, immediate increase in that variable.

EFFECT ON

INPUT TO	R	A_1	A_2	P
R	0	+	0	0
A_1	-	+	0	-
A_2	0	0	0	+
P	0	+	-	0

Table 18.1

More than half of the entries are zeroes. These arise because the complementary subsystems have zero feedback. For instance, input to R has no effect on the level of R because its complement is

which has no feedback involving all three elements. Similarly, input to P has no effect on the level of P because its complement is

which has no loop or combination of disjunct loops that includes all three elements. In the absence of P, the remaining system is unstable; this means that A_1 and A_2 cannot coexist at an equilibrium.

Note the following:

1. The notion of "controlling factor" is ambiguous. R is the essential resource that makes the community possible. But changes in the parameters of R (inputs to R) affect the equilibrium level only of A_1. A change in R is associated with changes in A_2 and P, but that change arises from A_1. A change in A_1 is associated with a change in A_2 only if they both arise from P. Therefore we do not talk about which factor controls which others, but rather which point of entry to the system affects which variables.

2. A_1 and A_2 are consumers of R and therefore competitors. But changes entering at neither one affect the abundance of the other. And their roles

in the community are different. A_1 responds to changes in all variables except A_2, whereas A_2 responds only to inputs entering at P. This has nothing to do with A_2 being less sensitive physiologically but only to its structural position: P is subsystem with zero feedback so it acts as a sink for all influences that reach A_2 from anywhere in the system except P itself, absorbing their impacts and buffering A_2. Similarly, P absorbs all impacts entering at A_2 so that the rest of the system docs not respond.

3. If we examine a large number of sample communities with this same make-up but different parameters, we will observe a statistical pattern of correlations among variables. This pattern depends on a variable's position in the structure and on which parameters are responsible for the differences from place to place (where they enter the system). Because inputs to R affect only A_1 they give rise to no correlations among variables. Input to A_1 affects all variables except A_2. It generates negative correlations between A_1 and R and between A_1 and P, although P does not consume A_1 but no correlation between P and its own prey or between the competitors A_1 and A_2. In fact, no single parameter change entering at a single node produces any correlation between A_2 and P that might suggest their relationship.

We have claimed that some variables have no effect on others. This is true only of equilibrium levels. For instance, an input to A_1 does not change the abundance of A_2, but it decreases both R and P. Therefore, A_2 has less food than before (and presumably a lower birth rate) and less predation (and lower death rate). The result is the same numbers, but of older individuals. The age distribution has been altered. On the other hand, inputs to P leave R and P unchanged. A_2 declines in numbers but has the same age distribution as before.

Suppose now that we introduce a negative link from A_1 to A_2 (Figure 18.3b). This requires only two changes in the graph: now input to R decreases P, and input to P decreases P. The former result comes from the negative path R-A_1-A_2-P. Thus, R can determine P, but the $A_1 A_2$ relation determines that R can determine P. The second, counterintuitive result comes from the new positive feedback loop R, A_1, A_2, R, which reverses the sign of the path (in this case, from outside to P). An initial increase in P reduces A_2, increasing R which increases A_1 and decreases A_2 even further. The additional reduction A_2 beyond the predation effect reduces P. Thus, A_1 can determine P in three ways: It prevents a direct positive impact of R on P by acting as a sink; it provides a negative path from R to P; and by entering into the positive feedback loop with A_2 and R it causes P to respond anomalously to its own input; A_1 is a good candidate for a "controlling factor." But we see from the table that A_1 itself is determined by inputs to three of the four variables in the system.

The structure of the community also influences the course of evolution of the component species. Genotypes that increase the survival or birth rates of any of the species will be selected over alternative genotypes. But only in the case of A_1 will the species increase in population as a result. And with the A_1, A_2 negative link, Mendelian selection in P will actually result in a decrease in P.

But genes do not do only one thing. Suppose that genotypes exist in A_2 that either increase the sensitivity to the toxin of A_1, but help evade predation by P, or conversely reduce sensitivity to A_1, but increase vulnerability to P. The course or selection will depend on the relative abundance of A_1 and P as threats to individuals of A_2. Suppose that P is sufficiently abundant that avoidance of predation is the stronger force. Selection for this genotype will appear as a positive input to A_2 and a negative input to P. The positive input to A_2 has no effect on the abundance of A_2 but increases P. The negative input to P increases P (because of the A_1, A_2, R positive feedback), increases A_2, and decreases A_1. Therefore, the conditions that favor this pathway of evolution are reinforced.

However, if we begin initially with abundant A_1 and rare P, selection in A_2 will increase vulnerability to predation but weaken the A_1, A_2 link. This behaves as a positive input to A_2 and P. The positive input to P increases A_1, strengthening selection against the toxicity. The positive input to A_2 increases P, but the positive input to P reduces P. Therefore, P may increase or decrease. With a very strong A_1, A_2 link, P will decrease. Thus initial conditions may set the evolution on one of two alternative paths. Finally, we note that strong inputs to R produce large A_1 and small P. Therefore, in a sense R can determine the evolution of the A_1, A_2 relation and the intensity of predation.

The preceding argument supports our general conclusion: Control in complex systems resides in the structure of the network rather than in individual variables. Each variable controls some aspect of the system, but what it controls and how it affects it is, in turn, dependent on other components. Yet having assigned control to the whole is not sufficient.

The next step is to study concretely how which aspects of the system determine the pattern of long- and short-term control.

19

Strategies of Abstraction

> In the analysis of economic forms, moreover, neither microscopes nor chemical reagents are of use. The force of abstraction must replace both.
>
> —Karl Marx, preface, *Das Kapital*

Complexity is now in fashion. Books, meetings, even whole institutes are devoted to complexity. It is a recognition that the long traditions of reductionist science, so successful in the past, are increasingly inadequate to cope with the systems we are now trying to understand and influence. The great errors and failings of attempts to apply science to matters of urgent concern have come from posing problems too narrowly, too linearly, too statically. Infectious disease did not disappear as was predicted thirty or forty years ago. Pesticides increase pest problems, antibiotics create new pathogens, hospitals are foci of infection. Food aid may increase hunger. The straightening and "taming" of rivers increase floods. Economic development does not necessarily lead to equitable, just societies.

It is therefore intensely practical and even permissible to assert some principles of a more dialectical view of things:

- The truth is the whole (Hegel).
- Parts are conditioned and even created by their wholes.
- Things are more richly connected than is obvious.
- No one level of phenomena is more "fundamental" than any other. Each has a relative autonomy and its own dynamics but is also linked to the other levels.
- Things are the way they are because they got that way.

- Things are snapshots of processes. They remain the way they are long enough to be recognized and named because of opposing processes that perturb and restore them.

- The dichotomies into which we split the world—physiological/psychological, biological/social, genetic/environmental, random/deterministic, intelligible/chaotic—are misleading and eventually obfuscating.

We can ask: Why are things the way they are instead of a little bit different? Why are things the way they are instead of very different? The first is the question of self-regulation and homeostasis. The second is the question of evolution, development and history. And then we have to ask, what are the relations between the stabilizing and destabilizing processes? How do the reversible short-term processes of restoration and maintenance that can buffer against long-term forces for a while also give rise to directional changes that alter the stabilizing processes and eventually overwhelm them?

Of course, we cannot really look at the "whole," but Hegel's injunction has two kinds of practical value. First, a problem should be posed large enough for a solution to fit. It is usually better to present a problem that is too big and then reduce it than to start with the problem too small. For in that case, we may never be able to expand it enough. If we fail to do so, we are either condemned to ingenious solutions to trivial questions or to explanations that are mostly external: some external influence caused what we observe, but we have no explanation for that external influence. It is merely given, perhaps observed and measured.

It takes some imagination and experience to know how to pose a question big enough, because this goes against all our training. Second, even after we have posed the problem as broadly as we know how, we always have to be aware that there is more out there that might overwhelm our theories and thwart our best intentions.

Once we accept both the need for wholeness and also its impossibility, we have to resort to processes of abstraction that can give rise to useful models. In 1965, I urged that since each model is partly false we need independent models to converge on the truths we are looking for. But I did not deal with the question of how to choose these models. Now I want to focus more explicitly on the processes of abstraction. In this effort, I have been influenced by Bertel Ollman's perceptive work *The Dance of the Dialectic*, which discusses several kinds of abstraction.[1] Different abstractions from the same wholes capture different aspects of the reality but also leave us with different blindnesses. Thus it is always necessary to recognize that our abstractions are intellectual constructs, that an "object" kicks and screams when it is abstracted from its context and may take its revenge in leading us astray. A particular tree species in a catalog of the trees of the West Indies is not the same tree that we saw on the windswept beach, purple fruit fragrant with volatile compounds and flavored with salt spray, leaves showing the zigzag trails of *lepidopteron* larvae. It is merely Linnean binomial abstracted into "typicalness."

We Choose Our Abstractions

Our abstractions always reflect choices. Bertolt Brecht warned that we live in a terrible time when "to talk about trees is a kind of silence about injustice."[2] He was wrong about the trees—they now figure prominently in the study of justice. But the point is well taken. Much abstraction is evasive of what matters, chosen for reasons of safety or convenience. The preferred objects of neoclassical economics, individuals making choices in ahistoric markets, can lead to elegant theorems about rational choice, but hide exploitation, monopoly, class conflict, and the evolution of capitalism. They even abstract away the specific qualities of the four major market types under capitalism: markets for commodities, labor markets, capital markets, and financial markets, each with their own histories and patterns of ownership, power, and conflicts. Without a historical view it is possible to work with the abstraction of a perfect market. That it is unrealistic is not in itself a devastating criticism—we also abstract away friction from perfect gas models. But if markets are never perfect, and if furthermore they deviate from "perfection" in ways that serve their owners, and become less "perfect" as the power of corporations increases, then the abstraction is not only unrealistic but also actively obfuscating.

We are of course free to abstract as we please. The test of the usefulness of an abstraction is whether it captures what we want of reality, is encumbered with a minimum of scars from the process, and leads somewhere. Abstractions that are full of definitions and axioms but give no theorems are not productive abstractions.

Descriptive abstractions are attempts to turn heuristic notions into quantifiable measures. We use indices of biodiversity or resemblance, population density, nutritional status, efficiency. But once we have defined an index it has a life of its own and might not capture what we are looking for. Consider, for instance, population density. If a population is spread out over districts or farms of different size, an obvious definition of density would be

$$D_1 = \Sigma p_i / \Sigma A_i$$

where p_i is the population in district i and A_i is its area. But if we are interested in the question, "How crowded are people?" we might ask how many people live at each density. Then a more suitable measure would be

$$D_2 = \Sigma (p_i / A_i) p_i / \Sigma p_i$$

where p_i/A_i is the density in the i district, the second p_i is the number of people living in that density and Σp_i normalizes the measure to preserve the dimensionality of people over the area.

We have found more than a hundredfold difference between D_1 and D_2 in some cases.[3] Efficiency is another index that seems more "natural" than it really is. In agricultural production, the biblical measure of efficiency is seeds harvested per seed sown. In land-scarce Europe it is more likely to be measured by yield/unit area, in the land-rich and labor-poor United States we boast of yield per labor day, whereas ecologists are interested in measuring energy harvested compared to energy invested. We could even invent nonsense indices such as the number of endemic beetles in a country divided by the number of deputies in the national assembly. Once created, it acquires the objective existence of other constructs. It can be measured, compared across countries, traced historically, and so on. What makes it a nonsense index is that it does not help us answer any questions other than about itself.

Perspective, Extent, and Level

The abstractions of greatest interest are the variables and parameters of dynamic systems we are interested in. Ollman distinguished abstractions of perspective, extent, and level. A sample ecological abstraction is shown in Table 19.1. We start with the perspective of the effects of temperature on insects. At the biophysical-biochemical level we know that an increase of temperature increases the rate of chemical processes. The muscular activity of insects is close enough to that level that the Harvard astronomer Harlow Shapley could estimate the temperature from the rate of movement of ants on his observatory floor.

In the next row we choose the level of the individual, its "horizontal" extension limited to replicates of the same insect treated as samples from a

PERSPECTIVE	HORIZONTAL SCALE	TEMPORAL SCALE	DYNAMICS	CONSTANTS
Temperature tolerance	Individual fly	Minutes to hours	Mortality	Fly biology, temperature
Adaptation to temperature	Individual fly	Days to a week	Growth and development, acclimation	Fly biology, temperature regimes
Behavior in relation to temperature	Population of flies of one species	Minutes	Attraction to food versus heat stress	Habitat pattern of temperature, food resources
Demography	Population of flies of one species	Seasonal	Reproduction versus mortality	Habitat, community of species
Community	Ecosystem of interacting species	Months to years	Competition, predation	Habitat, community of species
Micro-evolutionary	Single species	Years	Natural selection versus migration and drift	Habitat, community of species

Table 19.1

Abstraction of perspective, extent, and level in fruit-fly ecology

population, the particular temperature chosen to permit observation, and the time frame in minutes. In a bottle, the only dynamic is mortality caused by desiccation or denaturation of proteins. The only variable is the number of insects alive, so that we may have an equation for that variable

$$dx/dt = -mx$$

where x is the number still alive and m is the death rate. After all my brave talk about complexity and wholeness, I have come up with a single equation with one variable and one parameter: Where is the rest of the world? This is the question we must always ask about any model: *Where is the rest of the world?*

The parameter m depends on the physiological state of the insect. This is partly determined genetically but in *Drosophila melanogaster* it can change with exposure to different temperatures for two to three days. In the lab I could control the temperature they are exposed to, but in nature it will depend on the habitat and the behavior of the flies.

Survival also depends on the size of the fly, since the surface-volume relation makes small insects lose a greater fraction of their water per second than larger individuals. When half the flies have died, the survivors are on the average bigger than those that died. Size is also dependent on temperature since development speeds up with moderately higher temperatures and growth is less accelerated. The result is that at higher temperatures smaller individuals are produced, but they are produced sooner. But size also depends on the genotype.

When flies are subject to frequent temperature stress, genotypes that produce larger flies at those temperatures may be selected for. On a timescale of generations, say months or years, higher temperature improves survival by selection for larger size. This is observed in that flies collected in hotter, drier climates in Puerto Rico are the same size as those from the rain forest, but at the same temperatures in the laboratory they are larger than their rain forest conspecifics. Their size has been increased by the selection effects of temperature and reduced by the direct impact of temperature on their development. Thus temperature increases survival by selecting for size, reduces survival by accelerating development, increases survival through physiological adaptation, and reduces survival by desiccating the flies.

Now shift the level to populations of flies in their habitat. I observe the numbers of flies around my traps of fermenting fruit. The timescale is still minutes, the level is now the population of actively foraging flies, the dynamic is the movement of flies attracted to the fruit but repelled once they feel desiccation stress. Thus we may produce a model of the dynamics:

$$dx/dt = A - rx$$

where A depends on the total population of $D.$ $melanogaster$ in the foraging range and the abundance of fruit that might compete with my traps, which in turn depends on the vegetation and season but is considered constant for the duration of my one-day study. The parameter r depends on the temperature effects discussed at the individual level. The rest of the world enters through A, the total local population.

Population depends on the balance of birth and death rates. In this sense all organisms follow the same law of population. Temperature enters the birth rate by way of generation time. Among mosquitoes, a rise in temperature within a moderate range shortens the generation and therefore results in larger populations of smaller individuals who cannot fly as far or remain active as long, and who have lower fecundity and shorter lives. At some point above their optimum temperature the increase in mortality outweighs the shorter generations and populations decrease.

The same approach can be taken to examine the number of ants foraging in a given area. But it would be a mistake simply to transfer the categories and methods adopted for fruit flies. Ants are social; we can observe the numbers coming and going from each nest. Species competition is directly visible and influences the impact of temperature. At the level of the colony on a timescale of minutes to hours, the numbers of foragers leaving the nest is the result of the push out, the success in finding food (signaled to mobilize more foragers), and the pressures to return. Thus we can start with an equation

$$dx/dt = p(T - x) - rx.$$

The push out, p, is related to the need for food in the nest, the number of immature ants to be fed, the good news that successful foragers bring home and signal chemically, and the total number of foragers available in the colony, T. In our time frame, p, T, and r could be treated as constants. The return depends on foraging, heat stress, and species interactions. For instance, we found that on one Caribbean island the ant *Brachymyrmex heeri* would come to tuna bait and surround it completely. If the bait was in the shade, the lion ant *Phediole megacephala* mobilized soon after (its nest was farther away) and could displace the brachys in about twenty minutes. But if the site were then in direct sunlight lion ants were soon stressed and left, and brachy returned.

We could alternate sunlight and shadow experimentally by the appropriate placing of opaque students. If the alternation were rapid, say every ten minutes, the competition was fierce and brachy got most of the food. But if sunlight and shadow alternated over hours, the food was divided between them more or less equally. If we move to the timescale of weeks, the numbers of foragers in the nest and the demand for protein from the larvae changed. These changes depended on the success of the foraging, whereas the hourly flux was taken as given.

Feeding success depends on the total available food in the foraging area, the distance from the nests of all foraging species, the kinds of interactions among them, and the species-specific responses to the weather. The foraging area itself depends on the density of nests and on the scale of months we have a dynamic of production of queens and their loss to predation before they are able to dig a nest. The variable is the number of nests. The colony formation rate depends on the accumulation of nutrients in the form of reproductives, which in turn depends on the foraging success at the shorter-term level. Thus we can have a hierarchy of models in successively larger timescales in which the constants of one level become the variables of interest on another. The species interactions also affect evolution: on an island where the fire ant *Solenopsis geminata* and its smaller, less aggressive relative *S. globularia* coexist, the fire ant expels its cousin from the cooler sites into the exposed beaches and bare rocks. Therefore they are exposed to a different environment from what their own preferences would produce. Populations of *S. globularia* are exposed to more selection for heat tolerance and are more tolerant than populations of the same species on islands without the fire ant.

Comparing the cases of flies and ants, we see that the theoretical approach, nested sets of abstractions, is applicable to both. But the kinds of observations and the specific questions we can ask are different. Our work depends both on generalization and respect for specificity.

But it is now artificial to continue making temperature the point of view. If nobody had ever thought about temperature before in ecology, I might be straining to prove its relevance. I could write a series of papers on "the role of temperature in fly development," "the role of temperature in fly foraging," "the role of temperature in fly communities," and so on. If my concern were to illustrate the abstraction we call "abstraction," I could continue tracing the role of temperature across levels and extensions. It makes more sense to change our point of view and ask what determines fly communities and the abundance and diversity of *Drosophila*. We will not exclude any role for temperature that might become relevant, but it is no longer our perspective. The total population depends on the long-term food supply, competition from other species, and predators. This leads us to the point of view of species-interaction dynamics.

The elementary pairwise interactions between species have been studied extensively. But whatever the model, the core relation is the feedback loop, negative for predator-prey relations and positive for competition and mutualism. It can be a direct two-species loop or much longer and indirect. The negative feedback loop is shown in Figure 19.1. It has some immediate consequences. For instance, it explains why the use of pesticides is often counterproductive. Suppose that a pesticide kills both predator and prey. Its effect in the community is found by tracing the direct negative impact of the pesticide and the indirect effect through species interactions. The predator is harmed both directly, as the toxic impact of the pesticide, and indirectly, by killing its food. The prey is also harmed directly, but the pathway by way of the predator is positive (negative impact on predator times negative link from predator to prey): it is poisoned, but so is its enemy. Thus while the predator is always harmed the prey may increase or decrease.

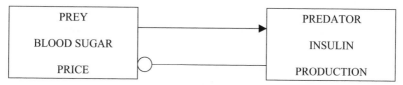

Figure 19.1

A simple negative feedback loop. Positive links are shown by a sharp arrowhead, negative links by circles. Predator/prey, insulin/blood sugar and price/production have the same dynamic structure.

The predator-prey loop tells us about the statistical correlation between the two species. If the rest of the world enters by way of the prey, an increase in the prey is transmitted to the predator as an increased food supply and the correlation between them is positive. But if the environment enters directly through the predator, any impact is transmitted in the opposite direction to the prey, generating a negative correlation. When we abstract a single species pair, the most important thing about the rest of the world is whether it impacts the loop from the prey end or the predator end.

Suppose instead that we had ignored the effect of the prey population on its predator. Then we would model with unidirectional causation. The predator is the independent variable and the prey the dependent variable. If we measure the predator population diligently we might propose a regression model in which:

$$\text{Prey} = a + b \ (x \text{ predators}).$$

We could estimate a and b with great precision and get a good fit and conclude that predators "account for" 60 percent of the variance. Such a procedure is not wrong. It is a legitimate procedure in the sense that it answers the question it asks. We might even arrange to hold temperature and other "confounders" constant so that the "error" variance is as small as possible. But the parameter b may be quite different in different field situations, even of opposite sign. And if environmental variation acts directly on both species the regression may be zero.

The regression approach is not wrong. But we criticize it because what it leaves out is crucial to answering the question "What determines the abundance of prey?" and therefore it offers a superficial answer, in part because it does not account for the "independent" variable and because of this it will give inconsistent results from place to place or time to time, though all of them statistically valid.

The abstraction we used to study species' abundance ignores everything else about a population. But individuals differ in their nutritional status, age, sex, genotype, and so on. We can restrict the horizontal scale to a species pair but consider effects within each species. If the environmental change enters the system as more food for the prey, fecundity increases. The predator also increases so that birth and death rates for the prey are increased whether or not there is an increase in numbers (which depends on the presence or absence of self-damping of the predator). The positive correlation between predator and prey means that when the prey are abundant they are also young and well fed, and that when they are rare they are also older and scrawnier. The predator is well fed when abundant and poorly fed and older when rare, the opposite of Malthusian expectations. But if the system is driven from the predator end, the correlation is negative and the predators are poorly fed when they are most abundant and well fed when rare.

Once we understand the simple negative feedback loop, it can be applied to situations that are physically different with the same dynamics. Thus the insulin-glucose loop or the relation between prices and production in a capitalist economy have the same dynamic properties.

We can expand the abstraction "horizontally" to include more species and consider the self-damping loops. Now the impact of some environmental change depends on the whole network of feedbacks. It can be expressed formally as the derivative of the equilibrium level with respect to any parameter change. (A parameter change, which might come from nutritional state of the organism or from genetic substitution, is "external" to the model even if inside the bug.)

When we look at the whole ecosystem (only relatively "whole," of course), we create a new abstraction. We work with a network in which the vertices are species population sizes and their links are the direct interactions. This network can be described in terms of pathways between variables, feedback loops, the stability and resistance of the whole and of the subsystems. It can then be used to find the direction of change of the variables when external events impinge on one of the species and the effects percolate through the whole. It allows us to understand why sometimes the obvious effect of a pathway is reversed so that adding nitrogen to a pond reduces nitrogen levels, pesticides increase pests, some species remain the same despite environmental change, and what properties of the system lead to oscillations or abrupt transitions.

We can now change the point of view to that of evolution. Then the horizontal scale is one species, the temporal scale is long, the "population" consists of a set of genotypes influencing temperature tolerance, and the environment is represented by selection coefficients.

Each of these abstractions is both legitimate and incomplete, whereas the set of all of them together is a closer approximation to the reality. For all of them, it is necessary to recognize their status as abstractions, intellectual constructs.

Pluralism

The view of theory that depends on a diversity of perspectives is quite different from the fashionable "postmodernist" advocacy of pluralism. The divergent abstractions of perspectives have to be loosely consistent with each other and validated within their limitations. We demand only loose consistency. In the history of genetics, the linear array of genes on the chromosome seemed to contradict the cytological observations of bumpy, branching "lamp brush" chromosomes. This was a tolerable contradiction, eventually resolved by recognizing that the "lamp brush" referred to gyres in the chromosome that could be m flux during development.[4] Unlike formal logic, where a contradiction makes all propositions provable and demolishes the whole edifice, in science a certain level of contradiction is almost always present and is a motor for more research. Its influence is usually limited to a domain of nearby propositions. What makes these contradictions benign is the belief that they will eventually be resolved. Postmodernist pluralism grants equal validity to all viewpoints and sees their discord as a virtue.

If we examine the development of our knowledge, we recognize some things we can be pretty sure of. These ideas have a long-term stability and have frequent verification with numerous cross-links to other information that is reliable. But even their status as certainties is not absolute. We can look at the history of science from the viewpoint that theories have a half-life. To stress this point, I ask students to imagine under what circumstances the Second Law of Thermodynamics might be overthrown.

Then there are claims that are terribly wrong from the start and do not contribute at all to advancing our understanding. Creationism, Holocaust denial, and doctrines of racial or gender inferiority and similar "perspectives" are of this type. They are perspectives that obscure the realities and are introduced into the scientific agenda from nonscientific, even antiscientific, concerns. We can confront them with the *postulate of partisanship*: all theories are wrong that promote, justify, or tolerate injustice.

This postulate of partisanship does not refute them. It does not tell us how they may be wrong: errors of conceptualization, of observation, of validation, of interpretation or application. But it is a powerful working rule that can guide our research.

At the advancing front of our sciences are unresolved questions in which diversity and controversy are part of the process of finding out, and different disciplines with their own perspectives enrich the process. This is the domain of constructive pluralism. Beyond this frontier there are questions about which we have no means of resolution and speculation has full freedom. And finally there are the questions it has not yet occurred to us to ask. But if the diversity persists about the same questions for long periods of time, this is not evidence of the health of our science but of its stagnation or of the scientific dispute being a surrogate for clashing interests.

The processes by which we arrive at our consensus in science is very different when there are interests at stake. What follows is a first attempt to formalize this process of adversary science.

In this abstraction the view is that of the observer of science watching a problem as it changes. The first term in the equation is the survival of evidence from one period to the next. The parameter a_1 is the erosion rate. The second term is the creation of new evidence. It is produced more rapidly when the other side is more threatening (larger $y/(x+y)$) but more slowly if the total mass of evidence is large ($e-(x+y)$). Finally, c_1 is the rate of production of evidence from other fields independent of the dispute. The second equation is similar. In this model, for one parameter set we obtained the process shown in Figure 19.2. In this case, the relative evidence $x/(x+y)$ shows a complex pat-

tern over time. There is nothing sacred about these equations. Anyone following the development of these ideas could ask, "But haven't you failed to take
into account x?" Or, "That isn't necessarily so. In our field . . . " or "What is
evidence for an ecologist may not carry much weight for a pharmacologist." It
is easy to propose other models in which the impact of evidence inhibits further research sympathetic to that evidence, discourages further research
aimed at refuting it, or may reflect other relations. At this stage of the inquiry
the important thing is to recognize the dispute as an object of study and "evidence" for and against a proposition as dynamic variables. This makes it possible to ask when a dispute will lead to resolution, when it can stalemate with
a fixed level of conviction, when it will fluctuate over time, with some conclusion seeming to be obvious at one time and absurd at another.

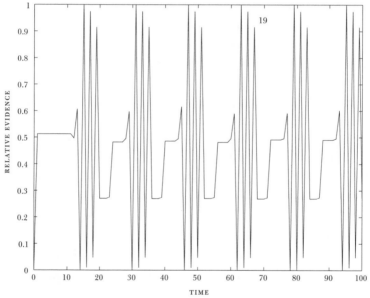

Figure 19.2

A model of consensus formation displaying complex patterns.

We have abstracted away the substance of the dispute and worked from the
viewpoint of sociology of science. Therefore, argument about the specific
model is extraneous, because it is meeting the model on a different level of
abstraction. The validation of this work is not prediction about any particular
dispute but rather the verification that studying the dynamics of dispute in
this way leads to worthwhile insights into controversy.

How Does It Apply to Us?

Human ecology is the most complex ecosystem of all. It is a convergence of biological and social processes in which our biology has become socialized, but for that is no less biological.

Our species is obviously not in equilibrium with its environment. We are a young species, a scant 5,000 generations out from the savannas where we took shape, some 500 generations into agriculture and a mere twenty generations or so afflicted by capitalism. (*Drosophila* can have twenty generations in a year. What have they accomplished in 2007?) By our own actions, factors that could previously be treated as external, independent variables, are now also affected by human action. The pathways of causation have been closed to become loops with reciprocal effects and an altered dynamics. Furthermore its non-equilibrium status is not an inconvenience that we could abstract away in order to seek an invariant "human nature" but the central problem. The rates of change and the extent of interconnectedness are accelerating. Consciousness or volition is an emergent fact of our evolution and an ecological force. This reality is often invoked to deny the possibility of scientific study of human affairs. Our attention is called variously to "human nature," "the human condition," or "the human factor," or the importance of irrationality in human affairs to suggest that we cannot understand enough about our world to make any helpful decisions about society.

The student of human affairs is also a part of that system, with perspectives that are formed in the networks of which she or he is a part. This leads to the biases that are most difficult to detect because they are shared in the scholarly community and serve to determine respectability of ideas and define the common sense. A base of support outside that community, often rooted in grassroots activism, is a powerful antidote to the consensus bias. Each type of society has its own ecology, its own relations with the rest of nature. This includes patterns of land use, resource extraction and rehabilitation, population dynamics, social inequalities, and relations with microorganisms and with the chemosphere and climate.

Back to Earth

Abstracting is only one part of the process of seeking understanding. The inverse process is the return to the world in which the abstractions were made. I am not concerned here with the familiar statistical hypothesis testing or Popperian falsification that results in a hypothesis being accepted or rejected. We may get good statistical fit without knowing the system any better if too much has been left out, or we may get good answers to the wrong questions. Rather, the inquiry is whether the processes of abstraction as a whole and the observations they lead to have increased our understanding of the world and offered some guide to action for deepening our understanding or making the world a better place. For this purpose, several kinds of questions may be directed at our conclusions.

The Bayesian Question

Do the results make sense? We always have some prior knowledge and expectations that come from our previous knowledge or "intuition." Intuition is an integral of diverse knowledge, experience, impressions, and preferences that may often give insights whose source we cannot explain, and sometimes lead us terribly astray.

But priors should not be ignored. If our conclusions are inconsistent with our prior expectation we should investigate why. We might then follow two pathways: we might assume our priors to be correct and look for reasons why the theory led us to a result that contradicts that expectation. This is the more common approach when the priors represent a whole history of a science and the research encompasses only a small area. For instance, the failure of a natural selection model to account for observed characteristics in a group of populations does not refute natural selection as an evolutionary explanation, but we might question whether other evolutionary forces account for our particular observations.

The second pathway examines our expectations to ask why a wrong conclusion seemed so plausible. This is the more radical response since it can challenge fundamental assumptions of a field. If done with care it can show us new directions. But if every table in our notebook leads us to proclaim a new paradigm there would be total stasis.

If our conclusions are consistent with our priors, we bring out the champagne. But then, being aware of the inevitability of surprise in science, we can still ask why they are consistent. Is there something in the research project, oriented by our priors, that forced confirmation of our expectations?

What If We're Wrong?

This is especially important if the research leads to policy decisions that can affect people's lives. The inevitability of surprise makes it necessary to consider how to deal with the intrinsic uncertainty of the world. One approach we have taken is to ask: How do other species, with a billion years of evolutionary experience behind them, deal with uncertainty? We have found four major modes for coping with surprise, which are not mutually exclusive: detection and rapid response; prediction; broad tolerance of whatever might happen; and prevention. In practice, a mixed strategy enhances the survival of species. In science it leads to a research strategy that combines prioritizing from our best judgment of a situation with a secondary line of work, perhaps much less promising, but with a potential for important consequences.

The Polya Question

Were the different perspectives that were in agreement really different enough or were they slightly different repetitions of the same evidence and argument? In George Polya's monumental work on *Mathematics and Plausible Reasoning*, he discusses testing the hypothesis that in a plane geometric figure the number of vertices minus the number of lines plus the number of enclosed areas equals 1.[5] We might test branching figures, triangles, rectangles, pentagons, hexagons, and so on, and they would all support the conclusion. But each additional polygon adds less and less support to the assumption. We need to do something different, such as looking at a disconnected figure like two triangles. Then the conclusion is obviously false. Or we can expand the problem to include solid figures, or figures on a sphere or torus. Eventually it leads to the notion of Euler's number and the robust conclusion that vertices minus lines plus areas minus volumes plus . . . equals Euler's number for that space.

In the dispute about climate change, a rising temperature in several cities is suggestive. Adding more cities to the list gives a diminishing return. But independent lines of evidence—ocean temperatures, cores from glaciers, decline of coral reefs, spread of species into places that had been too cold for them, accumulation of greenhouse gasses—each may have some separate idiosyncratic explanation or source of error but jointly converge on an unavoidable conclusion. We have to seek lines of evidence as independently as we can in order to support a large-scale conclusion.

The Search for Anomaly

Our approach to cancer epidemiology is based on the idea that though there are specific mutagenic agents that induce particular cancer types there is also a more generic vulnerability that allows the mutations to spread. This is related to the environment and to the way of life. Therefore, we expect death rates from different types of cancer to show correlated distributions. In the United States the age-adjusted correlation among all types of cancer over states is .20 and in South Korean cities (not age adjusted) it is .33. But for non-Hodgkin's lymphoma the corresponding figures are .12 and .08. The rates for men and women for each cancer type are highly correlated, but less so for a few types, such as lung cancer where smoking exposures may be different for men and women, and leukemia, where we do not have an explanation as yet. That is, non-Hodgkin's lymphoma differs more from the other cancers than they do from each other.

Further, we find that adjacent states or provinces are very similar in their rates for most cancers but not for non-Hodgkin's lymphoma. Nebraska has a high rate and Kansas low, Wisconsin high and Minnesota low, Seoul high and the surrounding Gyeonggi province low. The non-Hodgkin's anomaly directs our attention to look for specific conditions that separate adjacent, largely similar areas. But unlike formal logic, where a false proposition brings down the whole edifice, anomaly does not destroy the approach of looking for correlation patterns among cancer rates as support for the idea of environmental causation of cancer. Anomaly enriches the study and serves as a guide toward finding the idiosyncratic factors.

The Ethical Question

When they encountered an unfamiliar animal in a picture book, my children would ask, "What does it do to children?" Although philosophers go through great contortions to separate questions of reality from questions of ethics, the historic process unites them. Theories support practices that serve some and harm others. Ethicists may debate, over dinner, the rational reasons for feeding the hungry, but for people in poverty food is not a philosophical problem. Any theory of society has to undergo the test, What does it do to children?

What Is Left Out?

A theory may answer its own questions more or less adequately, but our intellectual landscape is filled with herds of 800-pound elephants. For instance, until recently historical studies usually left out women and workingmen, and in the United States also most African Americans. Social changes were attributed to noble individuals or legal decisions that ratified processes at work in the larger society but did not create them. The inclusion of the excluded is not only a question of justice, it is crucial for understanding U.S. history as a struggle to win those rights that were proclaimed as universal principles, intended for the few, but taken seriously by the excluded. For instance, the allocation of women's labor between production and reproduction has been a major factor in the economy, demography, and intellectual life of the country.

Conclusion: Mathematics and Philosophy

All of these aspects converge to demand our engagement with dynamic complexity not only in dealing with each problem but also as an object of study. The two kinds of tools we have available are mathematics and philosophy.

Mathematics is used mostly in modeling in order to predict the outcomes of systems of equations. But it also has another use: educating the intuition so that the obscure becomes obvious. When we abstract from the reality of interest to create mathematical objects, we do this because some questions that would seem intractable can now be grasped immediately. We can look at the fluctuating abundance of insects and conclude, "Since these bugs vary by several orders of magnitude during the year and yet remain within bounds from year to year, there must be some density dependence operating (a negative feedback)," or see that in a particular patient insulin seems to increase blood sugar and have to ask, "Where is there a positive feedback loop at work?" This kind of qualitative mathematics is essential so that we are not overwhelmed by the sheer numbers of equations and variables of predictive models. The teaching of mathematics to scientists must include the mathematics that aims at understanding rather than solving equations or projecting numbers.

The philosophical tools provided by dialectics abstract the general properties of dynamic complex systems. They therefore permit us to see how different approaches fit together or conflict, and help us ask the critical questions about our systems: Where is the rest of the world? How did things get this way? What can we do about it?

20

The Butterfly ex Machina

In 1963, MIT professor Edward N. Lorenz published an article with a set of three differential equations meant to describe atmospheric conditions.[1] The solutions to these equations did not do what was thought to be the only decent thing for variables whose motion was described by equations, namely, either approach an equilibrium or a permanent repetitive oscillation. Instead, Lorenz's variables ended up with trajectories more like a tangled skein of yarn, and every time the equation was solved numerically on a computer the results were different. Then others began to look for similar behavior elsewhere. Robert May showed that even the simple and familiar logistic equation of discrete population growth models can show this kind of aberrant behavior for some values of the initial growth rate.[2]

It was no wonder that these new unexpected behaviors were labeled chaotic, and that chaos has caught the public imagination in a world that seems so unpredictable and in which people are so helpless. The popularization of chaos is having an impact comparable to that of the discoveries of quantum mechanics in the 1920s and 1930s, when the uncertainty principle and probabilistic transitions of atomic states became a metaphor for the uncertainty, randomness, and ultimate irrationality of life in a Europe still reeling from the unexpected horrors of the First World War. (It is noteworthy that the other conspicuous aspect of quantum theory—that change occurs in leaps rather than by slow, continuous increments—was not incorporated into the popular consciousness. It seemed to come down on the wrong side of the "evolution versus revolution" debate.)

The same thing is happening now with chaos. Some authors counterpose "chaos" to "order" as if chaos showed no orderliness. Others have decided that complexity implies chaos. They use the terms almost interchangeably and conclude that the goal of prediction, and with it the possibility of having any program of change certain enough to make commitments and sacrifice for, is illusory. The physicist Peter Carruthers, on the National Public Radio program *Talk of the Nation* on January 17, 1994, said that chaos overturned the whole basis of science.[3]

Deepak Chopra, in arguing the case for an alternative, holistic medicine based on the Ayurvedic tradition, counterposed the simple linear processes that occur out in the open on the "Newtonian table" to the mysterious world "under the table" of nonlinear, quantal, and chaotic motion.[4]

German socialist Peter Kruger, in an interview with Michael Hilliard stated:

> And currently you can observe a very interesting development in physics, that is the theory of chaos. I think that it's incredibly important in understanding the behavior of humankind and seeing its future. And it would be very, very fruitful for all Marxists who define themselves in the narrower sense to study the theory of chaos. They will see from this that the idea of designing an ideal society can only be a grand failure.[5]

Chaos is very appealing in the postmodernist mood that would deny any lawfulness in the world, reject theory as "grand designs," and see all theories as merely matters of discourse.

The claim that Earth's orbit is chaotic suggested to some that we may fly into the sun or off into cold space at any moment. But others have seen chaos as beneficial (e.g., in the rhythms of heart contractions) and see regularity as a risk factor. Throughout these discussions there is frequent casual reference to "the science of chaos." Perhaps the most dramatic expression of pop chaos is Lorenz's provocative "Predictability: Does the Flap of a Butterfly's Wings in Brazil Set Off a Tornado in Texas?" (the title of an address given to the American Association for the Advancement of Science in 1979).

There is of course no "science of chaos." Chaos refers to a class of mathematical phenomena within the general subdiscipline of nonlinear dynamical systems comparable in scope to time series or local equilibria in linear analysis.

How common chaos is in nature or in social life still remains to be determined. Not all nonlinear equations are chaotic. In fact, population dynamics cannot be truly chaotic because the size of a population is always an integer, whereas chaos requires a continuum of possible values so that different initial conditions can be arbitrarily close to each other. Not all nonlinear equations can be made chaotic by an appropriate choice of parameters. The implications of particular kinds of chaos are still to be worked out. Furthermore, there are very few mathematical proofs of anything having to do with chaos, and thus most of the research still consists of setting up equations, computing numerically the trajectories of solutions starting with different initial conditions, displaying them on a computer monitor, and saying, "Doesn't this look like chaos?"

But chaotic dynamics does represent a radical departure from previously familiar behaviors and from the basic Laplacian approach implicit in the scientific agenda: if I know exactly the initial conditions and laws of motion of all the variables in a system, then I can predict the whole future course of that system.

The first modification of Laplace's notion was the recognition that we cannot know the initial conditions "exactly." Every measurement has its own confidence interval, and in many systems the act of measuring changes the system. Secondly, the laws of motion are only approximate descriptions of what really happens, because no system of variables is really isolated. There is always an "outside" that imparts an additional push. This external push may be extremely small, but that small push may be enough to alter the outcome. Third, the models always include simplifying assumptions, such as the lumping of variables as if they were identical except for the property of interest, or ignore friction, or treat the external as acting uniformly. Models of population genetics treat individuals as interchangeable except for their genes, whereas population ecology distinguishes ages and nutritional states but ignores genetic differences. Epidemiological models separate infective and uninfected individuals but do not usually deal with a range of susceptibilities in populations.

Therefore, the Laplacian expectation was modified: if I know the initial conditions and laws of motion "approximately," then I can know the future of the system "approximately." Of course, exceptions to this were recognized early. Suppose that we are studying the trajectories of marbles rolling off a peaked roof. If two marbles start out on the same side of the peak near each other, then they usually will end up near each other. But if they are on opposite sides of the peak, then no matter how close together they start they will diverge to different end states. And if we make even the smallest error in locating the starting position of a marble near the peak, we may be completely wrong in predicting the outcome.

The peak of the roof is a boundary separating two domains of behavior, two basins of attraction. Whenever a system has more than one possible final outcome, depending on the initial conditions, there is correspondingly more than one basin of attraction separated by boundaries. But "most" points lie comfortably within their basins of attraction, and accurate measurement leads to accurate prediction.

But suppose that there are the equivalent of peaked roofs everywhere, that most values of the variables are near boundaries. Then, no matter how accurate our measurements, our predictions may be far from accurate, and two examples of the same model with only slightly different starting points may give quite different trajectories. This is one of the properties of chaos.

But chaos is also in the eye of the beholder. Once the initial bewilderment passes, patterns become discernible. Different kinds of chaos can be distinguished: regularities in chaotic dynamics, bounds to chaotic trajectories, correlation patterns among chaotic variables, prescriptions for detecting chaos-prone or chaos-resistant systems, and ways of intervening that suppress or enhance chaotic properties.

At first glance chaotic trajectories look like random numbers. And indeed chaotic equations can be used to generate the "pseudorandom" numbers for studying random processes. However, the randomness is only apparent, and with the right viewpoint patterns become obvious.

One task of mathematics is to make the arcane obvious and even trivial. Throughout history, changes of perspective have made it possible for quite sophisticated ideas to become part of the common sense of the public and used in everyday discourse. In medieval Europe, literate monks could add, subtract, and even multiply, but they had to resort to specialists for division. The shift from Roman to Arabic numbers was decisive for making long division a part of the culture of the educated.

The pendulum became widely used in Renaissance Europe, but the "swings of the pendulum" had already become a common metaphor for describing changes in politics or fashion. And long before systems theory, positive feedback was part of common sense as "a vicious circle."

Or consider the graphs of price trends that often appear on the front pages of newspapers. Readers now perceive at a glance and without special effort that upward on the page means greater, farther to the right means more recent, and steep slopes mean rapid change. But the idea that non-spatial variables such as prices and time can be represented by spatial arrangements of points and lines on a plane is not "natural." It carries behind it a history of abstraction embodied in measure theory and the general notion of mapping.

The same applies to mathematical chaos. The unfamiliarity of these new kinds of dynamics and their recalcitrance when we apply methods appropriate to older mathematical systems encourage philosophies of despair. But with a change of viewpoint and a bit of practice their properties become obvious. In what follows we explore one kind of chaos to show how a change of perspective makes the dynamics intelligible.

1. Simple Discrete Chaos

Chaos can occur in continuous or discrete equations. However, much more is known for the discrete case. If the dynamics of a variable are described by an equation of the form

$$x_{n+1} = g(x_n), \tag{1.1}$$

then there are several properties that can be demonstrated rigorously and used as working hypotheses for other cases. Li and Yorke's famous article "Period Three Implies Chaos" showed that if equation 1.1 has a solution of period three, then it also has solutions of every other period, that there are also non-periodic solutions that pass close to the periodic ones, and that there is "extreme sensitivity to intitial conditions."[6] These three properties constitute a definition of chaos for the discrete equation without delays. In other situations the last property is usually the focus of attention. Li and Yorke also offered a method for demonstrating that there is a solution of period three: if you can find a sequence of consecutive points in a trajectory such that

$$x_3 < x_0 < x_1 < x_2, \tag{1.2}$$

then there is a solution of period three and hence chaos.

In our own research, modeling the dynamics of the growth of grass in a savanna derived from the research of Tilman and Wedin, we wanted a simple qualitative approach for understanding the dynamics from the shape of the curve $g(x)$.[7] The shape had to reflect that the quantity of grass present affects growth in opposing ways. By way of reproduction, the more grass now the more grass later, but through litter accumulation on the ground, old grass inhibits new growth. Therefore, the curve $g(x)$ would start at zero, rise to some peak level, and taper off asymptotically toward zero when litter completely covers the ground and suppresses growth. (The full model also took into account the decay of litter and its release of nutrients.) We considered using either of the two equations

$$x_{n+1} = Ax_n / [1 + (A - 1)x_n^2] \qquad (1.3)$$

and

$$x_{n+1} = x_n \exp[b (1 - x_n)]. \qquad (1.4)$$

They have roughly the same shape. In both, $g(0) = 0$, and $g(x)$ rises to a peak and then decreases asymptotically toward 0. Both have equilibrium points at $x = 1$. Yet the first equation always has a stable equilibrium and is never chaotic, whereas the second equation may have oscillatory solutions and even chaos when b is large enough. Therefore, the notion of the "shape" of the function needed refinement. In what follows we show how to find the solutions of the difference equation 1.1 graphically and then introduce a series of landmarks, the tools for understanding "shape."

2. Dynamics of the Interval Map

Figure 20.1 shows an example of one kind of equation that may be chaotic, the interval map with the curve $g(x)$. Draw the 45° diagonal where it intersects the curve,

$$g(x) = x, \qquad (2.1)$$

and thus this is the positive equilibrium point. If the slope of $g(x)$ is less steep than -1 at the equilibrium, then the equilibrium is stable, whereas if the slope is steeper than -1, it is unstable. The stability of the equilibrium depends on the derivative $dg(x)/dx$ at the equilibrium only. It is a local property compatible with any "shape" in the global sense.

Start at any point x_n along the x-axis and go vertically to the curve $g(x)$. Then from the intersection with $g(x)$ draw a horizontal line until it intersects the diagonal. The x-value at that intersection is x_{n+1}. Now repeat the process: first draw a vertical line to $g(x)$ and then a horizontal line to the diagonal. This gives the next value, and so on. Depending on the shape of the curve, the sequence of steps may approach an equilibrium, enter into a periodic oscillation, or become aperiodic. If you draw curves $g(x)$ derived from any equations or from a data set, or even freehand, you can repeat the steps and get a feel for the process.

The equilibrium value x^* is important not because all processes go to equilibrium but because it is a landmark in the interval along with other landmarks that give the shape. The other landmarks are found as follows:

Find the peak of $g(x)$. Draw a horizontal line to the diagonal. This gives the maximum value M. Now draw the vertical down from here to $g(x)$ and once again horizontally to the diagonal. This identifies the lower bound m. The interval $[m, M]$ is the region of permanence: all trajectories eventually fall within this interval.

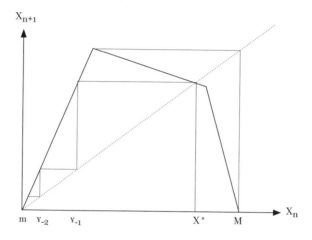

Figure 20.1

The curve $x_{n+1} = g(x_n)$, the generation of its solution, and the landmarks of $g(x)$. The procedure is always to go vertically to the curve and then horizontally to the 45° bisector to find the next x. The intersection of the bisector with $g(x)$ locates the equilibrium. Start from the peak and move horizontally to the bisector to get the maximum M. Then moving vertically to $g(m)$ and horizontally to the bisector locates the minimum m. The inverse process, horizontal from equilibrium to $g(x)$, gives y_{-1}, and then moving vertically to the bisector and horizontally to $g(x)$ locates y_{-2}, and so on.

Next, from the equilibrium point draw a horizontal line to $g(x)$. This identifies the pre-image y_{-1} of the equilibrium value, the point from which the trajectory would get to equilibrium in a single step. Now go vertically to the diagonal and horizontally to the left to the curve $g(x)$. This locates the pre-image of y_{-1}, which is labeled y_{-2}. Repeat the process, horizontal to $g(x)$ and vertical to the diagonal. This is the inverse process of generating the trajectory and gives the pre-image set $y_{-1}, y_{-2}, y_{-3}. \ldots$

If y_{-2} lies within the permanent region (that is, if $y_{-2} > m$), then inequality 1.2 is satisfied and the equation is chaotic.

Semicycles: The negative and positive semicycles S_- and S_+ are the number of consecutive steps during which the variable is below (S_-) or above (S_+) equilibrium. Semi cycles are easier to identity than periods because they do not require that a variable return to exactly the same previous value.

It can easily be proven that a trajectory starting at any initial condition gets within the permanent region in at most three semicycles. Further, the variable x_n crosses over a member of the pre-image set in each step. Therefore, the maximum length of a semicycle is the number of members of the pre-image set on that side of equilibrium that are within the permanent region.

Note that although chaos implies periodic solutions of every period, the lengths of the semicycles can still be bounded. The longer periods are then formed from many semicycles.

These results can be used to identity chaotic equations in several ways. If we have the functional relation $g(x)$, then a simple calculation or plotting of the landmarks suffices to determine whether y_{-2} lies within or outside the permanent region. If we have only data points from which to plot $g(x)$, we can still make the same determination although with a margin for error. If we are working from data rather than the equation we may not know the equilibrium point and therefore cannot be sure of the semicycles. But if any ascending sequence is more than three times as long as the shortest ascending sequence then the equation is chaotic. Thus even short sequences of observation may be sufficient to make a determination.

3. The Logistic Equation

The logistic equation has a special importance in the study of discrete chaos because of its widespread use in population genetics and ecology. Yet until Robert May's observations, it was used mostly for studying equilibrium (stable or unstable genetic polymorphism or species coexistence).

Because the logistic equation

$$x_{n+1} = rx_n(1 - x_n) \qquad (3.1)$$

is quadratic in x_n, it can be solved for the landmarks.

The landmarks are the following:

$x^* = (r - 1)/r$ equilibrium

$M = r/4$ maximum x in permanent region

$m = r^2(4 - r)/16$ minimum x in permanent region

The positive pre-image set is empty. Therefore, when $x_n > x^*$, $x_{n+1} < x^*$. The trajectory can only be greater than equilibrium for one consecutive time interval. The first three members of the negative pre-image set are given by

$$y_0 = x^*, \text{ which equals } (r - 1)/r. \tag{3.2}$$

$$ry - 1 (1 - y_{-1}) = x^* \tag{3.3}$$

so that

$$y_{-1} = 1/r$$

and

$$ry_{-2}(1 - y_{-2}) = y_{-1} \tag{3.4}$$

so that

$$y_{-2} = \tfrac{1}{2}(1 - \tilde{A} [(r_4^2)/r^2].$$

The same procedure can be used to find the other members of the pre-image set. The number of these that are greater than m, and therefore within the permanent region, gives the length of the longest semicycle in any long-term solution (after the first three semicycles in order to be sure the trajectory in inside the permanent region).

. Finally, we can ask, at what value of r does m cross above y_{-1} to preclude chaos or below y_{-2} to ensure chaos? To do this, solve numerically for y_{-1} or $y_{-2} = m$ (or alternatively, $g(m) = x^*$ and $g[g(m)] = x^*$). Thus,

$$r^2(4 - r)/16 = 1/r \tag{3.5}$$

and

$$r^2(4-r)/16 = \tfrac{1}{2}[1 - \tilde{A}\{(r_4^2)/r^2\}]. \tag{3.6}$$

The roots are approximately 3.67 and 3.94. The equilibrium becomes unstable at $r = 3$ and oscillations appear. For $r < 3.67$ there cannot be chaos, whereas for $r > 3.94$ there is necessarily chaos. (Actually we can get a better limit by finding the value of r for which m equals the pre-image of the maximum. This is approximately 3.83.)

The negative pre-image set is a sequence $y_{-1}, y_{-2}, y_{-3}, y_{-k} \ldots$ that converges toward zero, a point of accumulation. As r increases there are more and more members of the pre-image set within the permanent region, and they are closer and closer together. If zero is in the permanent region, then there are infinitely many members of P_-, and therefore there is no limit to the length of the negative semicycle. If the variable gets close to zero it will stay small for a very long time. In reality, a population with such dynamics would become extinct. This occurs at $r = 4$. But if $3.83 < r < 4$, then even though there is chaos, the semicycles cannot become infinite and x cannot become trapped indefinitely near zero.

Because there was only one solution to equation 3.6, as r moves across this critical value, there is a single transition from periodic to chaotic motion. But in other models it is possible for r not to have any real roots, and the transition to chaos may be impossible. This may happen because a parameter analogous to r will in general affect all the landmarks. In equation 1.3, as A increases, m and the y_{-k} all get smaller, but the y's decrease faster than m and are always outside the region of permanence. Then, the equation is immune to chaos.

Or it may be that the equation equivalent to equation 3.6,

$$y_{-2}(r) = m(r), \qquad (3.7)$$

may have several real roots. Then, as r increases, the equations may move in and out of chaos. If there is a double root, then the equation may be chaotic only for a single point value of r. Because the same parameters usually affect several landmarks, not all nonlinear equations can become chaotic just by changing the parameters, and we have no reason to make assumptions about how common chaos is in natural or social life. However, if $g(x)$ consists of two or three straight line segments, then we can manipulate the landmarks independently and design chaotic and nonchaotic systems at will.

4. Discussion

Some other properties of equation 1.1 can be deduced from the shape of $g(x)$. The local stability of an equilibrium depends on the slope $g'(x^*)$ at equilibrium, $-1 < g(x^*) < 1$ giving stability. But the chaotic properties depend on the relations among the landmarks. Therefore, it is possible to have a curve $g(x)$ that gives a locally stable equilibrium and yet is chaotic. This is shown in Figure 20.2a. We can also check the local stability of periodic solutions of equation 1.1.

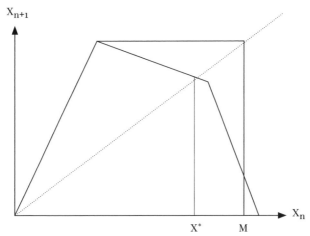

Figure 20.2a

The behaviors of solutions for 2- and 3-segment $g(x)$. (a) The slope at equilibrium is flatter than −1. Therefore, the equilibrium is locally stable. But segment 3 is steep enough to ensure that m is less than y_{-2} and the equation is chaotic. We would observe erratic oscillations until the process is trapped by the equilibrium. (b) Both slopes are steeper than 1 or −1. Therefore, all periodic solutions are unstable, and we would observe a typical nonperiodic chaotic pattern. (c) The equilibrium is unstable because the slope is too steep at the equilibrium. Segment 2 has a flat slope, and thus periodic solutions that have points in segment 2 will be stable. These will be of cycle length 2. (d) The equation is chaotic with m close to 0. Orbits that include points on segment 3 move from that segment to segment 1 and will need several steps to cross equilibrium again. The product of the slopes around an orbit that has k steps in segment 1 and one step in segment 3 will be $s_1^k s_3$. Because segment 3 is very flat, this product can lie between −1 and zero if k is not too big. Thus periodic orbits of intermediate length may be stable, but very short ones miss segment 3, and very long ones have too many steep slopes. If segment 3 is horizontal, any long orbit is stable. (e) In this chaotic equation, only orbits that include a point on segment 2 can be stable. But these correspond to orbits of cycle length 2. They will be of small amplitude compared with the unstable orbits and irregular oscillations.

The requirement for a stable periodic solution is that $-1 < \prod g'(x_i) < 1$, where the product is taken over all points on a periodic orbit. In Figure 20.2b, we show an example of a function whose periodic orbits are all unstable; thus, we would only observe aperiodic trajectories. However, in Figure 20.2c the curve $g(x)$ is flat near the peak value but steep at equilibrium. Here any periodic solution that is stable must certainly have one point in the flat segment 2. The next point is near the maximum, and thus the stable period 2 orbits have maximum amplitude. In Figure 20.2d, any stable periodic solu-

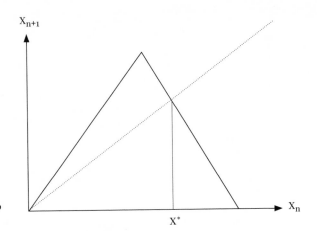

Figure 20.2b

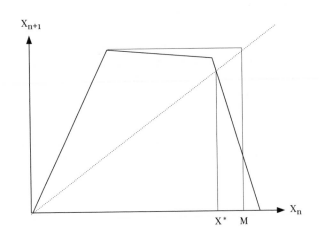

Figure 20.2c

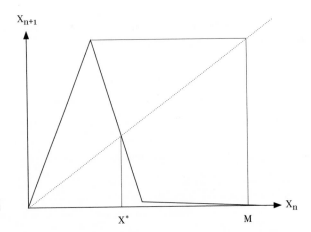

Figure 20.2d

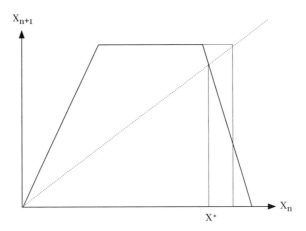

Figure 20.2e   X*

tion must have a point on the flat segment 3 and therefore a moderately long period. But it cannot have too many points on segment 1, or the product will still be below –1, and the orbit will be unstable. Finally, in Figure 20.2e, the flat segment, segment 2, is followed by a point not too much greater than equilibrium. Therefore, there will be stable low-amplitude oscillations, whereas longer cycles will be of greater amplitude and will be unstable.

The analysis of dynamic properties from the "shape" of the curve $g(x)$ can be applied to situations in which we have reason to believe that the real situation deviates from the model in particular ways. For instance, suppose that the curve in Figure 20.3 represents the dynamics of an epidemic when it is assumed that all susceptible individuals have equal susceptibility. We know that this is not true, although we do not know what the shape of $g(x)$ should be. The greater susceptibility of part of the population would alter $g(x)$ by making it steeper at low prevalences, where each case transmits the infection to more people than in the model, and flatter at high values of x when most of the uninfected people are more resistant. Therefore, the pre-image set y_{-i} would be displaced to the left, whereas the flattening of $g(x)$ would later on move m to the right. Thus, heterogeneity of susceptibility shortens the semicycles and makes chaos less likely. Or consider a $g(x)$ for the growth of a population of herbivorous insects in a cultivated field. Suppose now that we intervene with some pesticide whenever the population exceeds some economic threshold. If that threshold is above the equilibrium, then the only effect of the intervention is to make $g(x)$ steeper to the right of the equilibrium, which itself does not change. The lower bound m is decreased, and thus the semicycle may become longer

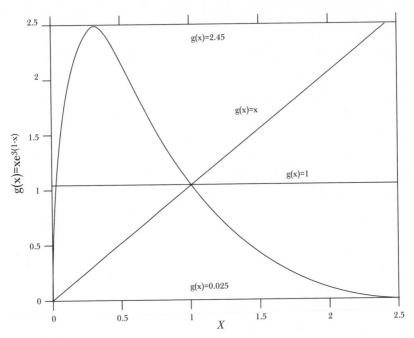

Figure 20.3
An idealized population model from which to show realistic departure.

and we may be provoking chaos. An intervention when populations are small reduces the slope of $g(x)$ and may therefore make periodic solutions stable.

Or suppose that $g(x)$ applies to a pest population in a crop of beans. In the off-season the insects decline slowly in the wild vegetation. Then the next year $x_{n+1} = sg(x_n)$ where s is survival in the wild. The curve is lowered uniformly by the same fraction s. This reduces all slopes and favors stability. Now suppose that a loan from the World Bank encourages farmers to plant several crops of beans per year with less time between crops. Then s increases, steepening all slopes and making periodic and chaotic solutions more likely.

This kind of qualitative analysis is more robust than the more precise models that give equations for $g(x)$ because it makes fewer restrictive and usually unrealistic assumptions about the shape of that function.

5. Conclusions

The popular hyperbole about chaos and the claims that it is ubiquitous, that it is the antithesis of order, that it overthrows science, that it makes the world unintelligible and unpredictable, and that it guarantees that programs for change will be ineffective are all unjustified.

We do not know how common chaos is in the world. It is difficult to detect chaos for two opposite reasons: on the one hand, obviously irregular trajectories may be produced by external perturbations as well as by chaotic dynamics, and telling them apart is quite difficult. On the other hand, nonperiodic oscillations with constrained semicycles may look periodic. Nor does the mathematics provide an answer: adding complexity to a dynamical system does not necessarily give chaos because the parameters change the landmarks in constrained ways. Finally, in nature the dynamics of physiological and demographic processes are influenced by natural selection. The parameters evolve and may evolve toward or away from producing chaotic behavior according to the fitness consequences.

It is clear that chaotic systems are not without order. There are restrictions on the possible solutions to some range of values, there are bounds on semicycle lengths, there are indications about the stability and instability of long or short and small or large amplitude trajectories, and we can understand which features of the curve $g(x)$ give or prevent chaos. In systems of differential equations, which have to have three or more variables or delays in order for chaos to appear, the shape of the trajectories in the three- or higher-dimensional space is also not arbitrary. We can make predictions about the correlations among the variables or the correlations between values of the same variable observed at different times.

The dynamics of the difference equation without delays (1.1) can be grasped quickly and intuitively from the shape of the curve $g(x)$. In order to do this, we have to look at $g(x)$ in terms of its landmarks, the equilibrium value x^*, the boundaries of the permanent region M and m, and the pre-image set. It is this change of perspective that makes the mysterious obvious. In the case discussed here of discrete time, nondelay difference equations, we have rigorously demonstrable results. For equations with delay, or for differential equations, there are fewer analytic results, but the qualitative conclusions above can be used as working hypotheses that can guide explorations by numerical methods and analysis.

Far from overthrowing science, the study of chaos opens up new areas for investigation and obliges a more subtle approach to the relations of prediction and understanding. Prediction has often played an important role in science, but science is not prediction. Nor is it necessary for uniform causes to give uniform results. The uniqueness of each ecological site and each individual person overthrows only the most mechanistic, reductionist kind of science and the technocratic ambition to achieve complete control. Rather, we can use our investigative tools to explain patterns of difference under seemingly uniform influences, learn to appreciate the richness of the world, and develop a strategy for coping with uncertainty.[8]

The regular properties of even chaotic equations give us a different take on that ominous butterfly who threatens to flap its wings. Historians and natural scientists have looked at the possibility of small events having big consequences in very different ways long before the arrival of chaos. Claims have been made that if Cleopatra's nose had had a different length (I do not recall now whether the author preferred long or short noses), if King George III had selected a better prime minister, if Rosa Lee Parks had not been so tired that day in Montgomery, then the course of history would have been different. Historians use this to emphasize the unpredictability of history and the absence of lawfulness. Yet, in physical and biological dynamics, instability is a regular part of predictable processes. The phase transitions of materials, the emergence of asymmetry in development, and the course of natural selection when the fitness of a genotype increases with its frequency all allow prediction in the large without prediction in the small. The more insignificant the precipitating factor, the more inevitable the macro-level change. Our task then becomes to examine the structure of the system that makes it unstable or chaotic when its parameters reach some critical values and the processes that bring the parameters to the critical points and to make a determination of the domain of possible outcomes.

In chaotic systems, *anything* cannot happen; only a range of alternatives within a set of constraints can happen. It would take more than the flap of a butterfly's wing to induce monsoon rains in Finland or a drought in the Amazon or equal representation of women on the Harvard faculty. Great quantities of energy and matter are involved in particular configurations for the major events to occur. Only when a system is poised on the brink of a qualitative change can a tiny event set it off. Therefore, the task of promoting change is one of promoting the conditions under which small, local events can precipitate the desired restructuring.

21

Educating the Intuition to Cope with Complexity

The central intellectual problem of our time is that of complexity. The great successes of science so far have still been problems that are conceptually simple, although solving them may have been quite difficult. We answer well the classical question "What is this? What is it made of?" But the great errors both of theory and practice have concerned problems where the complexity was unavoidable. Thus pesticides increased pests. Antibiotics created new diseases. Infectious diseases did not disappear but have reemerged among humans, animals and crops. Economic development has caused poverty.

Faced with ever more urgent problems of complexity, we have four principal modes of investigation:

1. Reduction assumes that the smallest parts of a problem are more fundamental than the whole, and if we know the parts well we can understand the whole. This reductionist focus has been the principal orientation of our science since the seventeenth century. It searches for the smallest particles in isolation and assumes that they will behave in the same way when assembled in the whole. It is an approach that works well in engineering where the parts are built by design and can be tested in the laboratory. In the biological and social sciences it is a useful research tactic but as a philosophy it creates a pattern of knowledge and ignorance that in the long run is harmful and makes us more vulnerable to surprises.

2. The second modality is the statistical democracy of factors. Theory is limited to the selection of the boundaries of a problem and the identification of "factors." Then statistical analysis assigns relative weights to the factors, and it is supposed that the factor, which carries the greatest variance, is also the principal cause. Its greatest power is in organizing information, presenting observations that need explaining, and testing hypotheses. But its omission of theory about the phenomena under study, very useful for avoiding biased preconceptions, tends to impose linear and superficial relations among variables. Often the statistical objects are confused with real world objects.

3. The third modality is simulation. It is based on the great capacity for computing numerical solutions that allow us to avoid the oversimplifications that are required for analytic solutions. But it also demands a lot: precise measurements of many variables and parameters, and exact equations. We are obliged to propose equations even when all we know are the directions of influence among the variables. In the end, we do not know if the results of our calculations pertain to the objects we are trying to model or to the details of the model itself. Simulation gives us numbers and reliable predictions of the state of the system if little changes, but only with difficulty can it explain the reasons for the results we get. We might just be overwhelmed with numbers. It is very expensive to carry out the measurements; we cannot replicate studies of a forest or a lake. Variables, which cannot be measured, are excluded. These are often social variables so that the models promote reductionism.

4. The fourth modality is qualitative and semi-quantitative mathematics, which allows us to include variables that are very different in their physical form even when they belong to different disciplines. It makes fewer assumptions than simulation and indicates the causes of the changes that are observed. Its greatest weakness is that the lack of precision sometimes makes decision making difficult.

Each of these approaches has their uses and limitations. Therefore good research uses a cluster of models of different kinds.

Beyond any problems inherent in the study of complexity, the enterprise of studying it in general suffers from a sharp contradiction in today's world: while it is ever more important to approach problems in a broad transdisciplinary, complex, and theoretical way, the political economy of research pushes us toward ever narrower research and teaching programs. The investors in science demand profitable results in the shortest possible time. The boundaries of university departments are reinforced by economic pressures and demand that students finish their research in the shortest possible time. Legislators demand simple results for making policy. One of the virtues of Cuban science is its very broad focus that permits maximum advances with a minimum of resources. And its greatest weaknesses arise when it allows itself to be too much impressed by Euro-North American science.

We have two major tools for confronting complexity: mathematics and philosophy. Mathematics has various tasks in science. It allows us to make predictions that can be tested. But it is something more: its most important task is *to educate the intuition so that the obscure becomes obvious.* Complexity is overwhelming not because it is intrinsically incomprehensible but because we have posed the problems badly, and with a change of vision it becomes more manageable. We have many historical precedents for this: consider, for example, the geometry problem of proving that if the bisectors of two angles of a triangle are equal then the triangle is isosceles. Within the framework of Euclidean geometry this is a difficult problem, but once we leap to analytic geometry the proof is trivial.

Philosophy has a bad reputation among scientists because it projects an image of irresponsible speculations, enemies of observation and experiment. However, there has been a long tradition of criticism of the dominant trends in science, that warned against the fragmentation of the objects of study, the freezing of dynamic processes into "things," and the imposition of a more or less fundamental ranking according to the sizes of the objects.

Most of the time, philosophers have played a role external to science, criticizing and sometimes proposing programs without carrying them out. Others, such as Kant and Descartes, were able to use their philosophical orientations to illuminate their scientific investigations. An outstanding example, perhaps the first investigation of a complex object as a system, was the masterwork of Karl Marx, *Das Kapital*. When he chose the commodity as the "cell" of capitalism, he didn't present it as the "atom" of the economy, as a fixed and unchanging object that determines the whole, but as a point of convergence of all the economic phenomena, at the same time determined by the whole and determining it. And he was not timid about changing his focus, sometimes to "capital" as such, sometimes to production or labor. These shifts of point of view would have been very confusing if it weren't for his clear sense of dialectical methodology.

My own experience in science comes from evolutionary ecology, which is necessarily complex, from dynamic systems theory, which emphasizes sources, flows, and sinks in a mathematical context, and dialectical materialism. In what follows we use some examples from a few different fields to illustrate general principles. Mathematical representations are introduced only to show how a change of perspective clarifies previously obscure situations.

The dialectical approach begins as a critique of the most common errors, a dissidence within and outside of the scientific project, and then goes on to develop its own approaches as a participant within science. This approach has been explained in many ways and formalized many times. Here we offer a way of presenting and applying part of the dialectical orientation.

The Truth Is the Whole

We begin with Hegel's dictum that the truth is the whole. Clearly, we cannot capture the Whole. We always have to work with relative "wholes." But the proposition has the following practical applications:

- That the problem we are studying is part of something greater than we imagined. It urges us to pose the question in the large, big enough so that a solution fits, and then to justify reducing the problem where necessary to make it manageable instead of posing it in the smallest possible terms in the hope that we can always expand it if necessary. But the experience has been that it is difficult to put an egg together again after we have broken it.

- After we have posed a problem as broadly as we can, we still have to be aware that there is more out there and we might be surprised. Surprises are inevitable in science because we can only study the unknown by treating it as if it were just like the known. This has been successful: the unknowns are often like the knowns, so that science is possible. But they are also different, sometimes very different, so that science is necessary and ordinary common sense is not enough.

- We recognize that the dichotomies we use to divide the world into biological/social, physical/psychological, deterministic/random, qualitative/

quantitative (methods), objective/subjective, and so on fool us in the long run. The most fertile research is along their boundaries, where they interpenetrate. Therefore in teaching we often ask students how phenomena that seem independent affect each other: What might be the effect of nitrogen uptake in wheat on the economic independence of women? How can modern agriculture affect the health of fish? Why does urbanization of the countryside increase the incidence of West Nile virus? How is racism an epidemiological factor? By what pathway might the development of production lead to poverty?

• The Whole that we study includes ourselves, our own scientific activity as an object of study. Once we see our own activity within the scientific process we can ask: How did the pattern of knowledge and ignorance in our own field come about?

Process

Dialectics emphasizes processes more than "things," regarding things as snapshots of process. When we change our focus from objects to processes, we ask two fundamental questions: Why are things the way they are instead of a little bit different? Why are things the way they are instead of very different?

The first is the question of homeostasis, self-regulation. A static, dead system can survive to the extent that it is isolated, evading the perturbations of its surroundings. The Newtonian solution to the problem applies here: things remain the way they are because nothing happens to them, the principle of inertia. But living systems, social as well as biological, are maintained precisely because of the interactions with their surroundings. Therefore we have to look for the forces that maintain things more or less the way we see them in spite of all the perturbations that bombard them from all sides. Here we study the relative equilibrium among opposing processes.

We use abstractions in the form of models as instruments of investigation. Models are intellectual structures that we study instead of studying nature directly. Immediately a contradiction arises: we study the model instead of the original object because it is different, more manageable. But if it is different, how can we claim that what we learn from the model is applicable to reality? Of course, we hope that the model resembles reality in the important aspects and differs only in being more manageable. But we have to confirm that the results correspond to reality and do not come from the details of the model. If we take our model too seriously, if we examine it under a microscope, it only reveals the ink it was printed with.

We construct models with various criteria: they should be realistic, general, precise, and manageable. If we use a small model, it may be more precise since there are fewer variables to measure. But then the dominant processes might appear as external inputs to the system. But when we broaden the model sometimes we lose precision and gain in realism: we notice that the opposing processes are no longer external but arise within the more inclusive system. Since we cannot fully satisfy all the criteria for a good model, we have to choose which criteria to emphasize according to the problem and then change to another model. It is the set of models that brings us closer to understanding reality.

We can make use of concepts from systems theory, though always taking into account that different kinds of systems have different kinds of dynamics. The theory of systems had its origin in engineering, in the design of systems for specific purposes, with well-characterized parts, tested in the laboratory, produced outside the system. The theory proposes that systems have goals to reach and maintain, paths to guide and control, so that its processes are those of optimization. And some systems really are that way. Systems of physical production have distinct elements that capture information, processes that measure "errors," that is, deviations from the goal, and others that affect the changes.

The organism is another kind of system. Natural selection has produced systems that function more or less adequately within normal conditions. But they are different from artificial systems: they are not built from parts made in isolation that develop in mutual interaction. We cannot conceive of a membrane or a liver or even DNA by itself. One of the most common errors is to abstract DNA from its context, assign it an excessive degree of independence and a rank of "fundamental." Every part has multiple functions, and at times these conflict. The system as a whole and its functions evolve on the basis of its past history and random events. We can visualize the processes within an organism as a network of physiological or neurological interactions. Knowledge in these fields points us toward shared processes, branching chains, cycles of synthesis and breakdown, catalysis and inhibition of catalysis. Often the same event induces opposing responses such as inflammatory and anti-inflammatory prostaglandins in response to the accumulation of cellular detritus, or the firing of excitatory and inhibitory neurons. Further, the same molecule behaves differently in different tissues and is associated with different sets of biochemicals.

The ecosystem is also different. Here the dynamics of populations depends on the reproduction, mortality, and migration within a food chain and the flow of matter and energy that provides the structure, organizing the processes of competition, predation, and mutualism. The component populations might have evolved together, but they also might have arisen separately and then come into contact. These systems have their feedbacks that maintain them, and we can study them in terms of equilibrium and non-equilibrium processes, but the system as a whole does not pursue a common goal. Sometimes the adaptive evolution of one species can harm and even eliminate another species or even itself.

Societies represent other kinds of systems, perhaps the most complex. It evolves as a whole but its component parts—different classes and sectors— pursue their own goals. In a class society there is no common national goal or criterion of success. Economies might grow while their populations sink into poverty. And each society influences and is influenced by the other societies in the world system. The analysis has to identify the opposing and cooperating elements, all inserted into the natural world but for that no less social.

In spite of differing in their components and the structure of their processes we can see all of these systems as systems, abstracting "system" from its particulars and recognizing feedbacks and feed-forwards, sources and sinks, stocks and flows, local and global stability, oscillations and chaos. Then it is possible to use some general methodological principles.

In order to study the short-term dynamics of these systems, we abstract away the characteristics of the components as variables and see them only as variables. The system is then a network of variables linked by positive and negative feedbacks. In general these variables are not in equilibrium but in continual movement within limits and around an equilibrium state. Further, each part has its own dynamics, how it responds to outside impacts and erases those impacts each at its own rate. In order to make the discussion more concrete we will use a few simple cases from ecology, physiology, and economics. In each case the models are incomplete, as are all models. We present them only to illustrate a part of our methodology and to capture the essential point while sharpening our approach to other more complete systems.

Figure 21.1 is a negative feedback loop of only two variables that can arise in many kinds of systems. Consider the feedback between predator and prey. In a negative feedback loop, we see one positive and one negative branch. An input that enters the system by way of the predator impacts the predator directly and transmits its impact to the prey with the sign reversed. This generates a negative correlation between predator and prey. But an input that enters by way of the prey changes predator and prey in the same direction, giving a positive correlation. Thus the statistical correlation between predator and prey can be positive or negative depending on where the perturbation enters the system. This helps us identify the source of changes in the system.

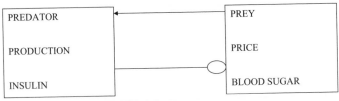

Figure 21.1   Negative Feedback Loop

For example, in a study being led by Caridad Gonzalez and her colleagues at the Institute of Tropical Fruit Research of the Ministry of Agriculture in Cuba on a community of a herbivore (a scale insect on the leaves of Valencia oranges) and its natural enemies (fungi and a wasp) we see that over time the herbivore and its enemies are positively correlated: seasonal changes in the trees change the reproduction of the herbivore so that when the population of the scale insect increases so do its enemies. But if we look at the spatial pattern at a single time, from tree to tree, or branch to branch, there is a negative correlation between the predators and their prey: environmental factors that act first on the fungus or wasp are transmitted to the scale insect in the opposite direction.

We further note that if the input enters at both variables, such as a pesticide, the prey dies because of the direct effect of the poison but benefits because its enemy is also killed. The two pathways within the network oppose each other. But both pathways harm the predator: it is poisoned directly and its food is being killed. It is for this reason—not because predators are more sensitive than herbivores but because of their location in the network—that pesticides often kill off the predators and increase the pest.

Once we understand the negative feedback loop we can apply it to other examples in the system. We interpret the loop as representing capitalist trade.

If variations in trade are driven by natural events such as climate or pests acting on the production of a crop, an increase in the crop reduces prices while a decrease in production increases prices. Therefore there is a negative correlation between production and price. But if the outside world acts on the system by way of the economy, an increase in price increases production, giving us a positive correlation. We found that in world trade during 1961–75 the prices and yields of wheat, rice, barley, and soybeans showed positive correlations: the variation of crop yield depended less on nature than on the economy as a whole in spite of the uncertainty of rainfall and pests.

Finally, the correlation between sugar and insulin levels in the blood also depends on the source of variation. The normal cycle of eating and metabolizing sugar produces a positive correlation between the concentrations of the two while pathologies of the pancreas directly affecting the production of insulin can induce negative correlations.

These examples show how the examination of even a single feedback loop by itself, one small step toward complexity, already indicates new properties of the system. We can call it a "sufficient" system. "Sufficient" here means that if we know the inputs to the variables we can calculate the outputs, and additional information as to the sources of the outputs does not improve the statistical fit. But it is obviously not a complete analysis. The process of abstraction has left out four sets of considerations:

1. The impacts to the variables are treated as external to the system. Thus they are random with respect to the variables we are considering. That is the unrealism of posing a problem too small. In systems that are too reduced the important things come from outside and we cannot do better than statistical associations. In the long run this is misleading. The inputs may arise from human activity, perhaps provoked by the abundance of one of the populations, or from responses of other species in the system, and therefore instead of being random may be correlated with the variables we are observing. A basic error of neoclassical economics is that from the perspective of prices and production, sales come from the domain of "consumer choice," which is independent of society and is simply given for reasons of psychology such as risk aversion. Figure 21.2 shows a model of some of the processes of capitalist economics. We start out with the previous model but insert inventories and demand. When inventories pile up,

sales become urgent. Companies hire public relations firms to increase demand. As in the previous model we have the negative feedback between production and price. But now we also have other feedbacks, both negative and positive. Some of the dynamics of external origin have been internalized, permitting a more complete analysis. For example the cycle:

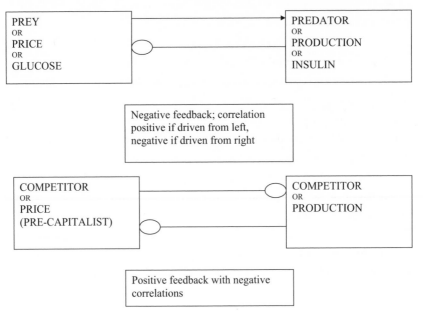

Figure 21.2

Feedbacks in systems

Production→demand —○inventory —○production is positive (the sign is the product of the signs of its links) and can induce the explosive instability we see in the business cycle. The negative cycle inventory —○ price —○ demand —○ inventory is negative and may promote familiar oscillations. In a more extensive model we would focus on those processes that reinforce or ameliorate the conclusions of the reduced model. In the case of blood sugar and insulin, we have so far ignored other factors that influence the concentration of blood sugar.

2. The species in the model of predator and prey are represented by the numbers of individuals, the population size, as an abstraction from the real life of populations. But the population dynamics reach down to act at the indi-

vidual level as well. They affect the development and ages of individuals—the more predators the shorter the life span and the younger and possibly smaller the prey. This affects their fertility, their mobility, their tolerance of dehydration, and their mortality. When we study the epidemiology of dengue fever we count the numbers of mosquitoes and map their distribution as if one *Aedes aegypti* female is like any other. But the mosquito develops in an environment that affects its biology. In addition to counting mosquitoes we have to measure them, because size is an indicator of the temperature and nutrition conditions where they developed and therefore helps us find the kinds of places that contribute most to the population.

3. The model assumes certain system parameters—the rate of reproduction, the predation rate and therefore mortality, the economic organization and level of technology that permits production to respond to prices, the ability of cells to use insulin and take up glucose. Here they are taken simply as given. But they depend on the evolution and past history in each case. The parameters have to be accounted for. We will return to the evolution in the discussion of long-term change.

4. The feedback loops, abstracted from the system, clarify some aspects of the dynamics. But they are embedded in a larger system, and we have to ask how the rest of the system affects what happens in a part. Here it is convenient to use the blood sugar model. In Figure 21.3 we extend the model to include two more variables in the regulation of blood sugar: epinephrine (adrenaline) and anxiety. Here it is also necessary to include the self-inhibition of the variables. This is equivalent to the rate of decomposition or removal from the system. The subsystem (E,A) determines the impact of external inputs on the (G,I) subsystem. If the total feedback of the (E,A) subsystem is strongly negative, both blood sugar and insulin respond strongly as expected. If the feedback is weak (E,A) acts as a sink that absorbs the impacts, leaving (G,I) little changed. But the positive loop between epinephrine and anxiety may be stronger than the auto-inhibitions of E and A separately. Then it is possible for the (E,A) subsystem as a whole to have net positive feedback. This would reverse the effects of changes entering as glucose or insulin: an increase in the consumption of sugar would reduce the levels of both blood sugar and insulin while an input increasing insulin such as pancreatic pathology would reduce the insulin

level and increase glucose. Any factor that reduces the rate of recovery from an occurrence that induces anxiety (its self-inhibition or resilience) could result in this abnormal response of glucose and insulin.

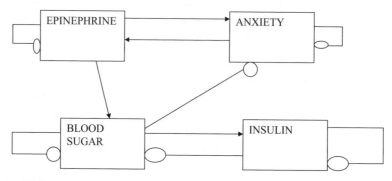

Figure 21.3
Positive feedbacks in large system can reverse expected effects of a pathway.

There are indications that the strength of self-inhibition of the stress hormones is greater in male middle-class teenagers than among working-class youth. After a stressful incident their cortisol peak is lower and is extinguished more rapidly. We do not know yet if this affects the rise and fall of glucose. It is an example of physiology tied to class, showing that our biology is a social biology, and under capitalism we can even take as an object of study the adrenal glands.

Now we insert these physiological processes within their social context. The regulation of blood sugar is not only a biochemical process. The level of glucose depends on the metabolic expenditure of energy. Suppose that a worker in an industry feels exhausted. She/he will need to take a rest or eat something. But not on the assembly line, because if the worker slows down the foreman intervenes to push the worker to speed up again. This increases the metabolic rate and anxiety. But if the shop has a strong union the steward observes the action of the foreman and intercepts it, alleviating the anxiety and allowing the metabolic rate to subside. A possible representation of this process is shown in Figure 21.4.

A positive feedback in the (E,A) subsystem in isolation would be unstable, but within the larger system it can be stabilized. However, if we act to stabilize glucose excessively with an electronic device that measures glucose continuously and intervene to fix the levels of glucose and insulin, then they both dis-

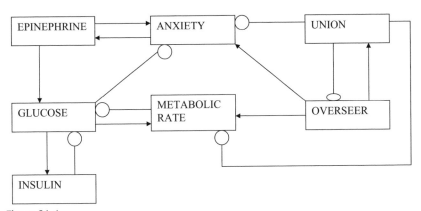

Figure 21.4
A psychological network in the context of social relations

appear as covariables of E and A even though they are still there physically.[1] This serves to isolate (E,A) dynamically and might destabilize it, even provoking a psychological crisis.

A subsystem external to the one we are studying modifies the response of other subsystems to inputs. The system as a whole determines the resistance of all variables to changed conditions. Therefore resistance is a property of the whole network and we can study it in terms of the structure of feedbacks and feed-forwards.

The examples given here illustrate how we can approach complex systems even when we do not know the exact forms of the equations. This approach suggests what we have to observe, what experiments we should carry out, even how to intervene, because the interventions have to be responses to the state of the system by way of pathways of information flow. That makes us part of the system. Consider a model of an epidemic, as in Figure 21.5. The intervention of the Ministry of Health can increase according to the number of cases of a disease, or respond to a serological survey of the population that indicates past exposure. The response can take the form of reducing the rate of contagion, reducing the number of infected people but increasing the number susceptible, or treating the sick and reducing the number infected but increasing the number resistant, or immunization (reducing the number susceptible and increasing the number resistant). Each alternative has its own dynamics.

Taking a few steps back from the analysis of systems of feedback, we can ask the general question What is the cause of phenomenon x? The reduction of glucose may depend on an increase in insulin. But this increase in insulin may depend on an increase in glucose. A preponderance of negative feedback

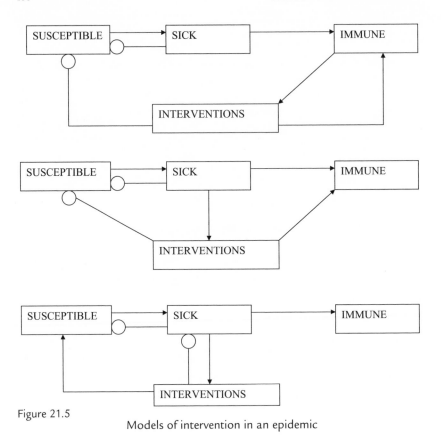

Figure 21.5

Models of intervention in an epidemic

outside the subsystem determines that the effect of insulin on glucose is negative; if most of the input comes into that system through glucose then the correlation between glucose and insulin will be positive. Thus even when the
analysis in isolation indicates a unique cause for the changes in glucose and
insulin that we can verify statistically, the cause of the observed dynamics
resides in the whole system. In very practical terms the truth is the whole. In
teaching, we ask students how a system can give outcomes the opposite of
what we expected: when can it happen that adding nitrogen to a lake reduces
the nitrogen level? When can the stabilization of glucose destabilize the
adrenaline and a patient's mood? When can food aid increase hunger?

The second question is that of development, evolution, or history according to the object of study. It deals with non-equilibrium processes that change
the values of the parameters, the links between variables, the structure of the
network, and in the long run even the variables themselves.

Here we recognize that the "constants" of our models are really variables from the viewpoint of a larger whole, and that when they change they can change the structure of the model and its dynamics. Among those changes that result in qualitative changes in the system are:

- *Changes of parameter.* When parameters change due to external or internal changes, a subsystem may change from having negative to positive net feedback, altering the behavior of glucose, insulin, and other variables. The system may lose its stability and explode out of bounds or oscillate.

- *A feedback loop can change its sign.* For example, under conditions of simple commodity production, if the price the producers receive increases compared to the cost of goods they want to buy, then they tend to reduce production because they can buy it with less product. But under expanded commodity production an increase in relative price leads to more production because it promises greater profit. Thus in the transition from simple to expanded commodity production the feedback loop changes from positive to negative.

- *Links between variables may be added or removed.* A species that exhausts its principal resources may change its behavior to exploit another resource and enter into competition with a new species or be exposed to new diseases. A peasant community may become incorporated into the global market.

- *New elements may enter as variables.* This can happen if a constant begins to vary in response to changes in other variables or if the emissions of one system become great enough to influence the source of its inputs.

- *A variable may differentiate into two variables or distinct variables may merge.* When two populations diverge enough so that they interchange genes at a rate no greater than the mutation rate, they are on the road to speciation. Or when the normal rhythm of mutual aid among peasants is polarized so that some are always the lenders and others always the borrowers, the category "peasant" has to be replaced by the two categories "rich peasants" and "poor peasants," "semi-proletarians" and "semi-bourgeois." Or when urban development forces birds that prefer different habitats have to roost in the same vegetation, two ecological communities may become one.

In some cases we know something of the long-term dynamics. Population genetics offers models of changing gene frequency under natural selection. Then these changes change relations among species. From the perspective of natural selection the variable of interest is gene frequency, even though in models of species interactions the species themselves are often treated as homogeneous. The adaptive value of a gene compared to its allele is a parameter external to the population genetics model. Advances in population genetics have changed the abstraction "adaptive value" from a constant to a variable and from an external variable to a variable in the system influenced by the state of the population itself and its environment.

These are some of the processes of long-term transformation that occur even when in the short run the system seems to be at equilibrium. A mathematical model of short-term dynamics is taken as a given. Therefore by itself it cannot indicate when long-term processes will invalidate it. That knowledge has to come from the specific science it attempts to model.

Finally, we ask how the short-term homeostatic process relates to the processes of long-term change. The short-term processes are usually stronger than the long-term ones but they frequently change direction according to the perturbations of their circumstances. The long-term, disequilibrating processes are usually weaker but are directional and thus in the long run prevail. But there is more: the very same reversible short-term processes also cause irreversible changes. The cycles of regulation of blood sugar can exhaust the pancreas. The "fight or flight" reactions of daily animal life can wear out the adrenals. The predator-prey interactions can impel evolutionary mechanisms to evade predators, and the predator responds to thwart those mechanisms. In the repetitive cycles of microbial reproduction there may be exchanges of DNA or even fusions that lead to evolutionary leaps. The cycles of buying and selling in the capitalist markets result in the concentration of capital. In the long run the short- and long-term processes of self-regulation and disruption of self-regulation are parts of the same system, uniting equilibrium and disequilibrium in the same whole.

The concepts illustrated here are not more difficult than the concepts of biochemistry or thermodynamics, only less well known. In order to confront the complexities of our sciences and our world we have to internalize in our intuition a philosophy of totality, connection within and across levels, dynamics, contradiction, and self-reflectivity—which is dialectics.

22

Preparing for Uncertainty

The world is always surprising us, overthrowing beliefs based alike on tradition, superstition, common sense, or science. It is necessary to understand the ubiquity of surprise in order to prepare for the surprises yet to come.

In recent decades we have learned that the "inert" gases such as argon or neon do in fact form compounds, thus overturning our understanding of the chemical bond; that most of the matter in the universe may not be in stars or planets but exists as interstellar gas, the so-called dark matter; that deep sea thermal vents, where conditions were thought to be unsuitable for life, host a unique and rich biota; that evolution often proceeds by fits and starts rather than by smooth divergence in imperceptible steps; that modern humanity spread over the earth quite recently, perhaps in the last 100,000 years rather than the half million or more years suggested by the classical fossils such as Java Man or Pekin Man; and that dynamic processes do not necessarily approach an equilibrium or a repetitious limit cycle, they may also go to "strange attractors" and show the apparently erratic motion mislabeled "chaos." In such cases improved measurements do not necessarily yield better predictions.

Some of the surprises are merely of intellectual interest. But others have profound human impact when programs and policies based on more or less rational expectations turn out to be wrong. For example, flood control often leads to increased flood damage; high-tech Green Revolution agriculture undermines productive capacity; pesticides increase pests and antibiotics can increase infection; increasing national income by the prevailing pathway of development increases poverty, dependence, and despair; first-draft socialist regimes showed neither the monolithic and rigid inflexibility their adversaries expected, nor the capacity to develop the renovative programs their supporters hoped for, but ended in surrender and collapse; global integration did not give us global harmony but has been accompanied by fragmentation and by nationalist wars mislabeled "ethnic conflict" by an essentialist media; and in public health, the doctrine of the epidemiological transition, that is, the expectation of the secular decline in infectious disease, has been belied repeatedly by the resurgence of malaria, tuberculosis, cholera, diphtheria, rabies, schistosomiasis, and the appearance of new or previously rare or restricted diseases and infections such as Legionnaires' disease, Lyme disease, AIDS, toxic shock syndrome, Lassa fever, Venezuelan hemorrhagic fever, hanta virus, and others.[1] Similar developments in veterinary and plant disease, such as African swine fever, mad cow disease, the distemper-like virus associated with outbreaks of mass die-offs in marine mammals, neurotoxic dinoflagellates that attack fish,[2] tomato gemini virus, bean golden mosaic virus, and variegated chlorosis of citrus suggested a more general phenomenon but were largely ignored by the public health community.

That science is caught by surprise is inevitable. But it is not acceptable to keep making the same mistakes or to ignore the inevitability of surprise and to assume that we have finally arrived at a true understanding.

The Inevitablity of Surprise

The question of why science is caught by surprise can be answered at several levels. At the most general level, science is surprised because the only way we have of studying the unknown is by pretending that it is like the known. The unknown is like the known; this makes science possible. But it is also unlike the known. This makes science necessary. Because of this, all theories eventually prove to be wrong, limited, irrelevant, or inadequate in other ways. As the British Marxist biologist J. B. S. Haldane observed, "The world is not only stranger than we imagine, but stranger than we can imagine."

The knowns that we choose to use for approaching the unknown come from where we are situated in the world. Our human biology sets the size and temporal frame of the familiar and our preferred sensory modalities emphasize visual over auditory or olfactory description. But we also belong to particular cultures, classes, genders, and disciplines within which a shared common sense makes some questions, approaches, and criteria for acceptable answers seem obvious and rules others out of bounds. These constraints are not fixed forever. But we become fully aware of them only after we have overcome them by new instruments or less commonly by conceptual shifts, or when changing social relations make obvious what was previously hidden by a consensus of assumptions.

In addition to these basic epistemological reasons for surprise there are more proximate causes in the fragmentation of knowledge and in philosophical biases such as reductionism, pragmatism, and positivism, which are shared broadly enough to seem to be "just realism."[3]

Knowledge has become a knowledge industry, owned and organized directly by industry and government or guided by them indirectly through the universities. In ways that are not always obvious the organization of the knowledge industry into institutions and fields, its determination of priorities, job definitions and recruitment, and the system of rewards also impinges on the products of that industry.

Lurking behind the constraints of intellectual commitments are unacknowledged conflicts of interest among unequals, expressed in different assumptions, rules about what are legitimate or illegitimate questions, and criteria for acceptable answers. Consider the definition of health. During the height of the sugar boom of Caribbean slavery, an adult slave had a life expectancy of some ten years on a plantation. This was normal good health from the point of view of the planters and of plantation medicine, whereas good nutrition meant mostly sufficient calories for the hard physical labor. The slaves had an alternative view of health expressed in a system of healing partly remembered from Africa, partly borrowed from the indigenous peoples, and partly invented *in situ*. At a later time, the recognition of black lung disease occurred about half a century earlier in Great Britain than in the United States—a consequence of the much stronger British labor movement with its own political party. At the present time, conflicts over the harmful effects of pesticides, electromagnetic fields, or smoking, although expressed as differences of judgment about data, samples, and controls, tend to divide poisoners from poisonees and reveal the partisan nature of even self-consciously objective research.

The Epidemiological Transition

Why did the doctrine of the epidemiological transition seem plausible, what went wrong, and how might the error might be corrected? The expectation that infectious disease would disappear was supported by three lines of evidence:

1. Infectious diseases had been declining over a period of 100 or 150 years. Smallpox, tuberculosis, poliomyelitis, diphtheria, pertussis, leprosy, malaria, and other scourges were diminishing in the more affluent countries and even in some third world areas.

2. Medical and public health tools such as antibiotics, pesticides, immunizations, and water purification provided the means for further advances and even eradication of infection. Technological progress promised to provide more and better tools.

3. Economic development was expected to provide the resources to apply all the new technologies wherever they were needed, and the elimination of poverty would remove the conditions in which epidemics prosper.

Each of these arguments had a certain plausibility, but each was also fundamentally flawed.

1. The extrapolation from only the most recent history is too short a time frame. If we look instead at the longest available historical records we see that diseases come and go. In periods of major change in society the epidemiological pattern also changes. The pandemics of plague occurred in Europe during the decline of Roman society under Justinian and again during the weakening of feudalism in the fourteenth century. The European invasion of the Americas brought with it diseases new to the continent and the decimation of the indigenous population.[4] The decline of the Soviet Union was manifested early in a general decline in life expectancy, and its final collapse saw outbreaks of diphtheria and other infections. Thus historical experience does not justify extrapolation from the most recent and geographically limited experience.

The expectation was narrow in another way as well. Public health and medicine limited itself to human disease. But parasitic infection is a universal phenomenon among living things. Plants and domestic and wild animals are also subject to infectious diseases and epidemics. Plant pathologists monitor closely the appearance and spread of new plant diseases and their extension to new hosts. They observe their waxing and waning with the weather, with changes in technology, and with the fluctuating fortunes of vectors. They follow changes in commercial exchanges of budwood and seed, economic conditions altering the acreage of each crop, and the introduction of new varieties. Studies of natural populations reveal patterns of coexistence of hosts and parasites, and evolutionary genetics traces and observes the adaptations of parasites to new climates or hosts.

Disease must be studied as a general evolutionary ecological phenomenon. But for humans, ecology is a social ecology. In addition to the familiar physical and biological aspects of environment such as temperature and rainfall and the presence of other species we have the social environment, the heterogeneity of human access to resources and subjection to stressors, the division of society into classes, genders, races/ethnicities, occupations, and cultures. Within these, people select, transform, adapt to, and even define their own environments to the extent that their different degrees of freedom permit. The statistical structure of these socially produced environmental elements—their variability in time and space, graininess, predictability, the correlations among them—create the patterns of human ecology. Different societies relate to the rest of nature in different ways and transform their surroundings differently. This is always the case, but we are presently in a period of very rapid and deep changes in our relations with the already transformed nature and with each other.

We need to replace the doctrine of the epidemiological transition with the proposition that we are living in a time of major climatic, vegetational, demographic, technical, social and political change, and that this must also be a time of epidemiological change in which many surprises are likely.[5] The professions of public health, plant pathology, veterinary medicine, and evolutionary ecology are, however, isolated from each other institutionally, physically, intellectually, and economically. Funding programs are tied to each of these areas separately, with little leeway to support transdisciplinary bridges. Their practitioners read different journals and do not always hold one another in the highest regard. Their mutual isolation is reinforced by the sense of urgency that motivates the healing crafts to be impatient with what seems like an irrelevant theoretical detour and by the all too frequent disdain of theoretical researchers toward the "merely applied." All of them share the belief that modern science has so much information to digest and requires so much time acquiring technical skill that narrow specialization is necessary. However, we argue that the major failings of applied science have come about less from ignorance of the parts of a problem than by construing problems too narrowly and failing to look at the whole. This is especially the case when the "whole" spans social and natural science.

Specialization and pragmatism retard the recognition of general problems even when examples of particular cases are well known. Physicians are aware that concurrent infections can complicate diagnosis and treatment. Public health workers know that in poorer countries people often carry several infections simultaneously. But multiple infection has not been faced as a general theoretical problem in epidemiology. Doctors know that some diseases spread from other animals. Veterinarians know that the same disease sometimes affects more than one kind of animal. But there is as yet no general review of the ranges of hosts infected by different groups of parasites, and most public health researchers have never posed questions such as: How many diseases are unique to humans as compared to shared diseases? Are unique diseases more or less virulent than the shared ones? What makes a good vector?

2. The faith in our technical means of cure and prevention has been naively reductionist. Reductionism as a strategy assumes that the smaller the object of study the more "fundamental" it is, and when the smallest parts have been characterized the behavior of the whole is readily understood. Thus a bug in a bottle killed by DDT (a toxicological fact) is interpreted to mean that use of DDT will control the pest (an ecological claim) and therefore that its widespread use would increase food production and alleviate hunger (a sociological and economic expectation).

The linear sequence of steps is plausible. But it is thwarted at each step by the action of variables excluded from consideration that intrude from the larger contexts, such as the price structure or the non-cultivated vegetation, and by excluded processes, such as natural selection, interspecific competition, land concentration, and migration. Thus pesticides initially do kill pests, but also the natural enemies of the pests. The outcome may be that more pests are poisoned but fewer are eaten and pest problems increase. The predators suffer more than their prey because they are harmed both directly by poisoning and indirectly by loss of their food source. Their prey, however, suffer poisoning but are compensated by the poisoning of their predators. The pest does remain unchanged when you apply control measures. Natural selection enters to produce resistance in the pest species. We already know of hundreds of such cases, know some of the conditions that accelerate or retard this adaptation and can estimate the timescale in which this is likely to happen. If the pest population is actually diminished by our intervention, however, it may be replaced by other pests that are now freed from competition.

Heavy pesticide use was part of a larger Green Revolution package that encouraged monocultures. It seems to be a general rule that the larger the area sown to a crop the more pest species attack it.[6] Year-round cultivation often guarantees these pests uninterrupted reproduction. The brown plant hopper, the white fly, army worms, and fruit worms are in a way creations of the Green Revolution.

The emergence of these pests as world problems was not expected because insufficient attention was given to the active responses of networks of interacting species to interventions. By the time that large-scale anti-mosquito and chemotherapy programs were instituted, there were already hundreds of cases known of insects acquiring pesticide resistance. Secondary pests—species that become important pests when pesticide use reduces their natural enemies while monoculture provides them with inexhaustible habitats—were already familiar. Microbial resistance to drugs had been observed. But the broad picture was not taken into account when expectations were advanced about rapid victories.

Such outcomes can no longer be regarded as unfortunate surprises but rather as virtually inevitable outcomes of natural selection.

3. The expectation that modernization would eliminate poverty among and within nations was assumed without question in mainstream discourse. It was part of a development model embedded in Cold War ideology, so that in the diffuse geographic region that called itself the West criticism of the prevailing economic system seemed disloyal. Critics were isolated, and an uncritical consensus was imposed that accepted the World Bank's approach to economic change as if it were the only possible way to develop, almost a law of nature.

This expectation too was not borne out. Poverty has been increasing on a world scale. Burdened by debt, many governments are cutting their outlays for health and sanitation. Environmental practices that create health problems such as deforestation, damming of rivers, or increased irrigation are increasingly encouraged with a sense of economic urgency that thwarts ecological criticism.[7]

The international public health system was caught by surprise by the resurgence of old diseases and the appearance of new ones. The narrow range of experience used for forming expectations and a theoretical framework that was reductionist and pragmatist caused this surprise. Public health officials failed to take into account the rich connectedness of nature/society, the nonlinear complexity and capacity for internally generated dynamics of the objects it sought to manage, and a naive progressivism about technological and economic development. Finally, these biases are rooted both in the long-term history of science and in its contemporary social organization as a knowledge industry. This industry determines the boundaries of the various fields of research, their agendas, and criteria for successful solutions of problems.

Preparations for the Unexpected

All other species as well as humans have to cope with changing conditions. Therefore we can ask how do they and us confront the new? There are basically five ways of preparing for the unexpected: prediction, detection with response, broad tolerance, prevention, and mixed strategies. They are not mutually exclusive: a mixed strategy at the level of enzymes can be part of a broad temperature tolerance at the level of the organism. The short-range end of prediction merges with detection. For instance, when there is measles in Dallas it is no great feat to expect measles in Fort Worth. From the perspective of a population, prediction of the onset of winter is part of the tolerance of a seasonal climate.

Prediction

All predictions are alike in that they assume the future will be like the past. They differ in which past they use and how long ago. Some are so short range they merge with detection while others are projections into a distant future. Some are predictions based on the past performance of the variable of interest, e.g. temperature. Some plants flower when the temperature is high enough to suggest spring. In New England many get caught by a late frost. Thus many insects use the day length rather than the temperature as a predictor of approaching winter and a signal for dormancy. The day length does not vary erratically from day to day but changes in a regular pattern, whereas a few days' cold in summer might fool the insects into premature dormancy. A few warm autumn days might leave them unprepared for winter. Day length is a more reliable signal that it is time to become dormant. These ways of preparing for events before they happen presume that although the variables themselves (day length, temperature) change, the climatic pattern remains constant. Predictions also differ in the precision they offer. Some, like the expected number of cases of a well-understood infection, are intended to be quite precise. The prediction is used to determine how much vaccine to prepare or hospital beds to set aside. Others are more qualitative. For example, though we cannot predict which new diseases will emerge out of the rain forest, we can be quite certain that some unfamiliar infections will appear, that mosquitoes and other flies are likely vectors (eight families of flies bite to obtain mammalian blood), and rodents are likely reservoirs. This knowledge is not useful for preparing a vaccine but can guide a surveillance system. For this kind of prediction you need a broad base of general knowledge of ecology and epidemiology in order to know where to look for emerging problems.

Thus it is reasonable to expect a flu epidemic next winter because we have one every winter and the prevalence of each outbreak remains within historical bounds. This future will be like the past. AIDS is different. The number of cases will change beyond previous levels but the present trends are assumed to continue into the near future so that some projections are possible. To anticipate the health impact of deforestation we cannot assume that either present health conditions or present trends will continue, only that the biological and economic processes that governed the past will continue into the future. And to anticipate the emergence of new diseases we have to apply everything we know about evolutionary ecology as well as development.

The most elementary and short-term prediction is that tomorrow will be like today. There is malaria here now, so there will be malaria here tomorrow. In malarial regions an increase in vector mosquitoes may be a better predictor than the number of present cases of malaria. But it may not be easy to know how many vector mosquitoes there are. Then rainfall, especially abundant rainfall after a dry period, may be a more available predictor than a mosquito census.[8]

At this level of prediction we already have a great deal of information. We are able to use rainfall to predict malaria, monitor rodent populations for plague, census ticks for Lyme disease, and perhaps plankton blooms for cholera.

Another mode of prediction assumes not that tomorrow's variables will be like today's but that today's trends will continue tomorrow. Voles in the sub-deserts of Utah determine how many young to carry at a time not by the present availability of food but by the food's growth rate. This is detected by the vole through substances in the growing tips of the grass that stimulate the endocrines. Many organisms use the change in the environment rather than its present condition. Some mosquitoes lay their eggs above the water surface on vegetation so that the eggs, when soaked after sufficient rain, respond to the rising level of water rather than its mere presence, and birds can use the shortening days rather than the day length itself to prepare for migration. The seeds of some plants in arid habitats are dormant until soaked more than once, indicating a true rainy season rather than an unseasonable sprinkle. We also expected that yesterday's trends would persist when we assumed that the century-long decline of infectious disease would continue.

There are now a number of familiar ecological changes associated with economic development that have predictable epidemiological consequences. Deforestation in the tropics leads to malaria; irrigation allows snails to spread in ditches and increase schistosomiasis; clearing of the land for grain production usually causes explosions of rodent populations and brings unfamiliar viruses such as Venezuelan hemorrhagic fever into contact with people. Here the specific disease is not predicted but the likelihood of some rodent-borne infection is indicated. Similarly, population displacement, refugee camps, periurban sprawl all have their potential epidemiological consequences. Less obvious processes also have their impact. Fertilizer runoff into lakes can lead to plankton blooms and die-offs followed by anoxic conditions in which dragonfly nymphs are killed, removing a major predator of mosquitoes. Especially in narrow estuaries nutrient enrichment from fertilizer and sewage may stimulate plankton blooms that include toxic dinoflagellates or protect the vibrio of cholera.

Prediction becomes less reliable when we attempt to anticipate the consequences of climatic change especially because we are living in an unprecedented time. The environment is changing more rapidly and in more different directions than in any period for which we have records. The atmosphere can change rapidly, but not all processes can keep up with the weather. The slower processes lag behind, so that different parts of the biosphere may no longer fit together. Forests grow slowly, so that during periods of rapid climatic change the trees may be appropriate to previous rather than current climate. Adaptive physiological responses to environmental signals such as day length lose their reliability when the onset of winter is either delayed or advanced. Genetic adaptations may lag behind habitat change and predators may become disconnected from their prey. Therefore the correlated responses among aspects of the biosphere will show a new pattern. The growth of corals can serve as a sink that captures some of the increased carbon dioxide from the atmosphere in the form of calcium carbonate skeletons, but corals grow very slowly. Even if pollution were not poisoning the world's corals, their slow growth would lag behind increases in atmospheric carbon dioxide for a long time. Similarly, increases in carbon would normally result in increased growth of plants and therefore increase in the biomass of forests. But deforestation is overriding any increased growth of trees, and acid rain is inhibiting the growth rate itself at a time when plant physiology would have led us to expect an increased rate. In order to make predictions about these new patterns we have to invoke general biological knowledge and apply principles derived from the whole of living nature rather than trust in the correlations of the past.

The possibility of spread of a vector-borne infectious disease into new regions in response to climatic change and more local human activities cannot be predicted from the tolerance of the vector to physiological conditions alone. It must be studied from a background in biogeography, especially the ecology of invasions and colonization. The survival of a species depends on its relations with other species as well as its own adaptive capacity. Most of the species in a community are not living under conditions that are optimal for them. Some are better adapted to warmer conditions, others to cooler. Therefore when the temperature changes this will make the habitat more suitable for some of the species, less so for others. Some will extend or contract the daily or seasonal time of their activity, others will be found in more or in fewer microsites within the habitat. If conditions get worse for both a predator and its prey, the prey population may increase because greater physiological stress is compensated for by lower predation rates. Indirect effects of temperature change such as turbidity of the water or density of vegetation may alter the effectiveness of predation.

Prediction thus depends on the analysis of community structure as well as physiological tolerances. Legionnaires' disease illustrates this situation. The *Legionella* bacteria has a worldwide distribution, but it is never very abundant because it is not a good competitor in the aquatic community. But when modern technology created the new habitats of water-cooling towers, air conditioning, and large-scale plumbing protected by chlorination and higher temperatures, the competitors of *Legionella* were killed. *Legionella* does not benefit directly from chlorine or heating, but can tolerate these extreme conditions better than its competitors because of its ability to colonize the cells of protozoa and find protection there, and the remains of the killed species of bacteria provide a rich nutritional environment. It is often the case that in extreme, new, or disturbed environments we find otherwise rare species reaching large numbers.

A qualitative description of the relations among the species of a community can be represented by a graph.[9] The constituent species are represented by vertices. These are connected by edges that are either positive or negative according to whether one species increases or decreases the other. We can trace along these edges any external impact as it percolates through the network, increasing along some pathways, damping out on others, and even sometimes reversing direction when positive feedback subsystems cause excessive responses. With a little practice we can form a rapid intuitive sense of what is taking place by examining the qualitative structure of the graph.

A comparative epidemiology could study the pattern of parasites, hosts, vectors, and reservoirs across taxonomic groups. It could ask questions such as, what is the relation between the taxonomic similarity of the parasites and the clinical similarity of their effects? Which groups of parasites have shown evolutionary flexibility with regard to host range or organ affected? Why do rodents harbor so many potential infectors of humans while bears pose no threat to us at all? What happens to a parasite as it passes through different host species? In which vectors does the parasite reduce fitness? Why are there so few mite-transmitted infections compared to those carried by ticks?

Evolutionary epidemiology examines the course of natural selection in diverse species of parasites, vectors, reservoirs, and their natural enemies, and their capacity to adapt to drugs or new environments or spread to new hosts. It asks such questions as: How does selection act when the pathogen itself has a large number of morphologically distinct life forms and exists in a variety of habitats? Which new habitats are suitable for colonization by new pathogens? Long-range prediction of potential directions of evolution for pathosystems depends on a research effort in evolutionary, comparative, and ecological epidemiology.

Detection and Response

Present public health efforts are directed mostly at detection of new outbreaks and rapid response to them. In some cases this has been very successful, and public health efforts have been initiated within days or weeks of an outbreak.

In order to be effective, detection of and response to an outbreak of an unfamiliar infectious disease must be sufficiently rapid compared to its duration. A response that is too slow will lag behind events rather than influence the course of an epidemic. For example, it does not make sense to start an immunization campaign against cholera in a community where the outbreak is already occurring—the time needed to organize an immunization campaign against cholera is too long and the course of the disease is too rapid. But immunization can be effective in protecting neighboring communities before the cholera reaches them.

Detection can be extended toward prediction when we monitor the vector and reservoir populations and even the environmental conditions that favor them. The regular monitoring of rodents, mosquitoes, ticks, algal blooms, and ship's ballast water are, or will have to become, routine parts of detection systems. Surveillance networks with trained technical staffs equipped with inexpensive diagnostic methods are a priority in order for poor countries to participate effectively in worldwide monitoring.

The process of recognizing a new disease is itself complex. Diseases are more readily identified if they affect populations with political influence, if a rare disease becomes common or a local one widespread, if the symptoms are clearly distinct from other known diseases, if the background epidemiology of the community has already been described, if symptoms such as exhaustion which were previously regarded as normal life become socially unacceptable, if there is an organized reporting system for infectious disease, and if diagnostic procedures are available.

Often the affected populations have taken the initiative in forcing health problems unto the agenda of the public health profession. The women's movement in the United States forced attention to toxic shock syndrome. The Black Panther Party in Chicago insisted that the area hospitals work on the clinical and epidemiological aspects of sickle cell anemia. Neighbors near toxic waste dumps or polluting industries have called attention to clusters of leukemia and breast cancer. AIDS research has been demanded and stimulated by the gay community. Public health workers should generalize from this experience and explore ways of collaborating with nonprofesional sections of the public, making use of their numbers, detailed knowledge of their own situations, organizing ability, and creativity rather than treating the public as objects of research, as a passive mass to be reassured or a recalcitrant mass to be cajoled or coerced into particular behaviors.

Tolerance (Reduced Vulnerability)

The course of an infectious disease in an individual depends on exposure to the pathogen, its success in invading the body, the tolerance of the organism for that agent, and professional or self-directed therapy. The rates and probabilities of each of these steps depend on a multiplicity of environmental and social conditions, and when averaged over the population become the parameters of epidemiology.

Many organisms depend for their survival on a broad tolerance of the conditions they might confront. Plant breeders differentiate vertical and horizontal resistance. Vertical resistance confers complete protection, but only against a very specific genotype of pathogen. It is usually conferred by a single gene, and does not last very long. Wheat rusts quickly develop new variants to get around resistant factors, then new wheat varieties are bred, and the cycle begins again. Horizontal resistance is usually more complex. It confers only partial protection, but against a wider range of pathogens, and it is usually polygenic in origin and long lasting. Plants' defenses in nature are usually horizontal. Some individuals germinate early, before the vectors have arrived. Others grow quickly enough to reach flowering even though they are infected. Some deter infection through leaf texture, or inactivate the pathogen chemically, or support bacteria that compete with infectious fungi for nutrients. Dispersal mechanisms allow some plants to escape to as yet uninfected sites. The diversity of means makes it difficult for the pathogen to breach these defenses since it would require doing many things at once. Therefore, though we often find damaged plants in nature, we rarely see decimated populations.

Much medical strategy aims at vertical protection through immunization. This requires prior knowledge of the serotype of the pathogen or the very rapid detection and manufacture of the appropriate antibody. But there are other measures that do not depend on so precise a prediction or so efficient a response. They are all elements of a horizontal strategy that could be in place without waiting for the appearance of particular disease threats. We could institute measures aimed at reducing the immunosuppressive influences of alcohol, stress, drugs, and the burden of multiple infection, such as good nutrition and sanitation, reduction of pollutants, a reasonable pace and variable kind of work, the maintenance of biodiversity, and moderate population densities in housing, work, schools, and public transport, equitably distributed facilities for health care and social support.

Prevention

The strategies of prediction, detection, and tolerance all assume that we have no influence on whether the surprising events happen. At best, we can predict or detect or tolerate them, but they are by definition outside our system, external to the world of public health practice and research in which we act. In contrast, a prevention strategy reaches out into that external world of forests and economies and climates and treats it as part of an enlarged world system of human activity with efforts to influence what happens before it reaches us. Therefore we move from models of response and management to models of positive design.

With a positive design strategy, we examine as many aspects as we can of the relation of our societies with the rest of nature from the perspective of impacts on human and ecosystem health. This requires questioning many of the assumptions that are usually not questioned about agriculture, industry, economics, development, and human settlement. We have to challenge the assumption that things are the way they are because there is no other way to organize social life or that there is only one kind of progress.

A fundamental step in adopting a strategy of positive design is the rejection of the notion that progress occurs along a single axis, from backward to modern, and that the task of the backward is to catch up with the modern in the same way that the present developed countries did. Instead, we have to recognize that the prevailing pathway of development in the world today is at best a successional stage that cannot be maintained. Like any colonizing species, the present world order has very effective dispersal, high rates of growth, and great transforming capacity. It is responsible for the overall pattern of maladjustment between our species and the rest of nature and within our species, and in many ways seems to be destroying the conditions for its own continuation.

Therefore we have to examine what positive design—guided by the criteria of ecosystem and social health—might mean in the areas of agriculture, industrial production, human settlement patterns, land use, demography, and socioeconomic development. In agriculture, it would mean a transition from an industrial to an ecological design, the design of integrated wholes based on gentle technologies.[10] It would include an evolution from the traditional labor-intensive production through the capital-intensive, high-input systems of contemporary agribusiness to low-input, knowledge-intensive production. In places where the high-input agriculture is not yet established we could bypass that stage and move directly to a knowledge-intensive system that makes use of both traditional knowledge and modern ecological science.

The ecological design would transform the random heterogeneity of land use that reflects the history of land tenure to designed heterogeneity in which each patch of the mosaic of land has its own product and also contributes to the other patches, in which forests modulate the flow of water and alter air movement, pastures support pollinators for orchards and provide manure for energy and soil enrichment, flocks of domestic fowl—raised in small areas interspersed with orchards—are used to control insect pests, their droppings feeding worms that then improve the soil of the vegetable beds. Heterogeneity serves as a buffer against fluctuations of weather and prices, provides stable employment, maintains populations of natural enemies of the major pests, conserves fertility, and provides microclimates suitable for different activities.

From the small scale of *minifundia* associated with land hunger through, or bypassing, the large-scale industrial plantings to flexible scales of production that allow both the interaction among different land uses and those kinds of mechanization that are really appropriate, the unit of production, the patch of land use, may be small or large. But the unit of design has to include many patches so as to take advantage of the variability of the landscape and coordinate the use of shared resources among different patches.

The contrast between traditional knowledge as "backward" and scientific knowledge as "modern" has to be rejected in favor of a cooperative effort between farmers and scientists based on mutual respect. The detailed, intimate, very particular knowledge that farmers have of their own circumstances must join with the scientific knowledge that requires some distance from the particular in order to design the gentle technologies adapted to each place.

All positive design requires social equity so that no subpopulations remain especially vulnerable to the "externalities" of dislocation, pollution, loss of community, environmental destruction, or new and/or resurgent infections, or have to absorb most of the impacts of fluctuations in production or prices, or of "natural" disasters. This means that the voices of the vulnerable must be heard from the first stages of planning.

None of this is easy to do. At present an outstanding example is Cuba, which in spite of and because of the current economic crisis leads the world in a commitment to an ecologically rational society.

Mixed strategy

Even the best strategy for facing uncertainty will only be successful part of the time. After we have made our best predictions, alerted our detection systems as best we can, reduced vulnerability, and designed our way of life so as to prevent as much disease as we know how, there will still be surprises. Even our best plans will sometimes turn out to be ineffective or counterproductive. A mixed strategy is part of the adaptive repertoire of many organisms and would be useful here. It includes measures that would be effective for a range of different circumstances. With uncertain weather, a mixed strategy would include crops for drought and for abundant rainfall. In an uncertain economy it would include products for market and for subsistence. In the face of the global uncertainties it would seem advisable to retard the homogenization of world cultures and social systems and to encourage alternative approaches in scientific research and medical care. In science, a mixed strategy would support the most promising approaches but it would also support theoretical standpoints that are not popular and not thought likely to succeed. These must be allowed to develop so that if the mainstream theories prove wrong there are alternatives always available.

Science for a Changing World

There are many obstacles in the way of creating the kind of science we need in order to understand the novel, complex and rapidly changing problems that confront our species. Science has become increasingly a commodity, produced for sale by a knowledge industry. The organization and culture of science is showing many of the problems created in other industries earlier in the Industrial Revolution. This has had several consequences: the choice of research directions is made on the basis of marketability, either to granting agencies or to industries that would use the knowledge to turn into injectable, swallowable, or other consumer goods. Scientists have been losing their autonomy to managers who opt for well-defined, short-term projects that fit into the rigid definitions of their departments or agencies. Production of research is increasingly monopolized. Support is most likely for projects that promise a sure outcome that fits the agendas of its sponsors. The sponsors are often given the neutral label "decision makers," which obscures the partisan nature of policy analysis. More and more scientists are becoming part of an academic proletariat without the job security needed for risky, unconventional, or simply boundary-crossing efforts. An increasing proportion of scientists' efforts go into proposal writing, which has become an art in itself. Budgetary constraints encourage caution rather than innovation. All of this favors highly technical research within well-recognized tracks.

Yet there are also signs of movement in the opposite direction, as it becomes increasingly apparent that our present ways of dealing with the problems of health, agriculture, development, conservation, urban design, and so on are simply insufficient for confronting the rapid and unexpected changes on the horizon. Within academic institutions we see multi-, inter-, and transdiscplinary programs built around issues of science, technology, and society. Outside of academia there are organizations for sustainable development, environmental justice, low-input agriculture, organic farming, public interest environmental research, women's health, preservation of old plant varieties and animal breeds, consumer protection, and conservation in general or of particular habitats. There are community groups watching for clusters of usually rare ailments, a new generation of investigative journalists well informed about environmental and public health problems, labor unions paying more attention than before to the environment on the job and the epidemiology of work, grassroots non-governmental organizations that combine economic and ecological goals, all challenging the prevailing fragmentation and monopoly of knowledge.

There is, then, a growing conflict between the urgent need of our species for the integration and democratization of science, and the economics and sociology of commercialized knowledge that impedes such development. We might attempt merely to predict, detect, or tolerate the outcome of that conflict. Or we could join the struggle to affect what happens.

PART THREE

23

Greypeace

Greypeace
Luke Emaea Drive
Vista Pestosa, FL 09399

Dear Occupant,

I am writing to you because I believe you are one of those special people who care. Although there have been many groups formed to promote the conservation of mountains and forests, nobody seems to care about America's most threatened and neglected habitat, the toxic waste dump.

Toxic waste dumps are a truly American habitat. They were not made by the nibbling of ants or the trampling of elephants' feet; they were totally unknown throughout most of our Earth's history; and they have not been detected on any other planet. This habitat is the Unique creation of Capitalism and Freedom, bequeathed to all other societies as an eternal monument to our initiative. It was one of the first environments to be celebrated by the Romantic poets. To paraphrase Wordsworth's immortal lines:

> I wandered lonely as a cloud
> Of sulphurous hydrocarbon fume
> And not of water vaporized
> As poets carelessly assume,
> My eyes atear, my craft all gone
> And mute the music of my lyres—

'Til my olfaction came upon
A mound, a mount of smould'ring tires
Beside the lake, where grew no trees
In pools of aging PCBs.
Below, an iridescent spring
Did bubble from the sequined sod,
It glowed in every spectral shade
And colours quite unknown to God!
While quite unseen by eyes like mine
A billion molecules peruse
A billion ways to recombine
Unthought of since the Cambrian ooze.
How could I not be touched that day
In heart. And lung. And DNA.

But is this magic place doomed to disappear after so brief a flourishing?

We cannot believe a tragic end awaits so special a place, a place not only valuable in its own right but also a great potential tourist attraction for the millions of people all over the world who yearn for the American Way. And that means DOLLARS.

We believe that millions of Americans appreciate this wondrous place, which is as American as apple alar or superprofits. We believe that if only they knew about the threat to our country's landscape, millions of free-enterprise loving Americans would rally 'round. They would buy our new Greypeace Calendar featuring the Great Toxic Waste Dumps of the United States. They would send for our prefabricated TWD kits, which they can mix with used motor oil, and pour into their backyards to get their own personalized micro-TWDs.

You can become part of the only environmental protection movement that also protects the economy, for, as the great Milton Freidman says, "Environmental protection is fully compatible with the needs of the economy provided you choose the right environment to protect." Remember, no poison is truly wasted if it can find its way home.

Yours truly,

Yuno Yalt, Acting Chief Engineer

24

Genes, Environment, and Organisms

Before the Second World War, and for a short time after it as a consequence of the immense notoriety of the atom bomb project and the promise of nuclear energy, physics and chemistry were the sciences of greatest prestige and the image of what natural science should be. When Americans were polled about the relative prestige of various occupations, they rated chemists and nuclear scientists above all other branches of learning, and even practitioners of such "soft" disciplines as psychology and sociology were rated above mere biologists.[1] The philosophy of science was essentially the philosophy of physics, and in his seminal work on the sociology of science, *Science and the Social Order,* Bernard Barber could write that "biology has not yet achieved a conceptual scheme of very high generality like that of the physical sciences. Therefore it is less adequate as a science."[2]

We have changed all that. It is biology that now fills the science columns of national newspapers, and television's fascination with billions and billions of stars has given way to a concentration on the sex lives of thousands of species of animals. The philosophy of science is now largely a consideration of biological issues; especially those raised by genetics and evolutionary theory. The cleverest science students now choose careers in molecular genetics rather than nuclear physics and it is a fair guess that more people can identify Watson and Crick than they can Bohr and Schrödinger.

In part this new dominance of biology comes from our preoccupation with health, but largely it comes from biology's claim to have become an "adequate science" by fulfilling Barber's demand for a "conceptual scheme of very high generality." At the level of molecules, all life is the same. DNA in its various forms is said to carry the information that determines all aspects of the life of all organisms, from the form of their cells to the form of their desires. The DNA code is "universal" (or nearly so); that is, the same DNA message will be translated into the same protein in every species of living being. At the level of organisms, the apparent profligate variety of shapes and ways of making a living, of nutrition and fornication, are all explained as optimal solutions to problems posed by nature, solutions that maximize the number of genes one will leave to future generations.

Even what appear to be accidental defects are explained by the universal law of optimization of reproduction by natural selection. The naive observer may think that a rotted hole in a tree's trunk is tough luck for a tree, but the informed evolutionary biologist assures us that it is an evolutionarily favored ploy by the tree to attract squirrels who will then spread the tree's seeds far and wide. There is no adversity whose use has not been sweetened by an appeal to natural selection.

The explanation of all of biological phenomena, from the molecular to the social, as special cases of a few overarching laws is the culmination of a program for the mechanization of living phenomena that began in the seventeenth century with the publication in 1628 of William Harvey's *Exercitatio de motu cordis et sanguinis in animalibus* (On the motion of the heart and blood in animals), in which the circulation of the blood is explained in terms of a mechanical pump with a series of pipes and valves. Descartes's elaboration of a general machine metaphor for organisms, in Part V of the *Discours,* made extensive use of the work of Harvey, whom he refers to, with characteristic French hauteur, as "a physician from England, to whom one must give high praise for having broken the ice in this area." But the machine metaphor creates a general program for biological investigation that is circumscribed by just those properties that organisms have in common with machines, objects that have articulated parts whose motions are designed to carry out particular functions. So the program of mechanistic biology has been to describe the bits and pieces of the machine, to show how the pieces fit together and move to make the machine as a whole work, and to discern the tasks for which the machine is designed.

That program has had extraordinary success. We know the structure of living organisms down to the finest details of the internal structure of cells and the folding of molecules, although some important questions remain open, such as an adequate description of the connections in the brains of large organisms like us. We also understand a great deal about the functions of organs, tissues, cells, and a remarkably large number of the molecules that make us up. Nor is there any reason to suppose that what is still unknown will not be revealed by the same techniques and with the same concepts that have characterized biology for the last three hundred years. The program of Harvey and Descartes to reveal the details of the *bête machine* has worked. The problem is that the machine metaphor leaves something out, and naive mechanistic biology, which is nothing but physics carried on by other means, has tried to cram it all in at the expense of a true picture of nature.

The problems of biology are not only the problems of an accurate description of the structure and function of the machines, but also the problem of their *history.* Organisms have history at two levels. Each one of us began life as a single fertilized egg cell, which underwent processes of growth and transformation. The processes of life will continue and we will be continuously transformed, changing the shape of our bodies and minds, until we end "this strange eventful history." In addition to their individual life stories, organisms have a collective history that started three billion years ago with rudimentary agglomerations of molecules, which has now reached its halfway point with tens of millions of diverse species and will end three billion years from now when the Sun consumes Earth in a fiery expansion. Of course, machines, too, have histories, but a knowledge either of the history of technology or of the building of individual machines is not an essential part of the understanding of their workings. The designers of modern cars do not have to consult Daimler's original design for an internal combustion engine nor does my garage mechanic need to know how an automobile assembly plant works. In contrast, a complete understanding of organisms cannot be separated from their histories. So the problem of how the brain functions in perception and memory is precisely the problem of how the neural connections come to be formed in the first place under the influence of sights, sounds, caresses, and blows.

The recognition of the historical nature of biological processes is not new. The problem of bringing the individual and collective histories of organisms into one grand mechanical synthesis already represented an important set of questions for eighteenth-century biology and for the Encyclopedists.[3] The biology of the nineteenth century was consumed with the issue, and the two great monuments of biology of the last century were the Darwinian scheme for evolution and the elaboration of experimental embryology by the German school of *Entwicklungsmechanik*.

The fundamental difficulty of fitting these phenomena into the mechanical synthesis arises from an inconvenient property of historical processes, namely their *contingency*. That is, systems in which history is important are systems in which influences outside the structures themselves play an important role in determining their function. Thus to the extent that those outside forces may vary, the history of the system itself will vary. One does not need to take Tolstoy's extreme anarchic position to agree that the outcome of the battle of Borodino was not determined by the birth either of Napoleon or Kutuzov, nor by the disposition of their troops on September 7, 1812. Any consideration of historical events necessarily demands that we confront the relation between the system that is our object of study and the penumbra of circumstances in which it is embedded, what is inside and what is outside. The relationship between inside and outside is not at issue for the machine, except that what is outside may interfere with its normal functioning. Changes in temperature and violent movements of the base on which it stands are disturbances of the proper motions of a clock, which is why the Admiralty offered a considerable prize for the design of an accurate ship's chronometer. The project to include the life histories of organisms in the machine model then requires that the interaction between the inside and the outside be dealt with and somehow disposed of without compromising the determinist Cartesian program. The embryologists and the evolutionists have taken two quite different approaches to the interaction between the inside and outside, which solve the problem of creating disciplines "of very high generality like that of the physical sciences" but at the expense of seriously distorting our view of living nature and of preventing, in the end, the solution of the very problems these sciences have set for themselves.

The technical word for the process of continual change during the lifetime of an organism is *development*, the etymology of which reveals the theory that underlies its study. Literally, "development" is an unfolding or unrolling, a metaphor that is more transparent in its Spanish equivalent, *desarollo*, and in the German *Entwicklung*, an unwinding. In this view, the history of an organism is the unfolding and revelation of an already immanent structure, just as when we develop a photograph, we reveal the image that is already latent in the exposed film. The process is entirely internal to the organism, the role of the external world being only a provision of a hospitable condition in which the internal process can run its normal course. At most, some special external condition, say the temperature rising above some minimum, may be necessary to trigger the developmental process, which then unfolds by its own internal logic, as the latent photograph becomes manifest when the film is immersed in developing solution.

A characteristic of development theories, whether of the body or the psyche, is that they are *stage* theories. The organism is seen as going through a series of ordered stages, the successful completion of the previous stage being the condition for the initiation of the next. The classical descriptions of animal embryology are in terms of discrete stages, the "two-cell stage," the "four-cell stage," the "blastula [ball of cells] stage," the "neurula [nerve crest] stage." There is then the possibility of arrested development with the system becoming stuck at an intermediate stage, unable to complete its normal life cycle because of an internal fault in the machinery or because the external world has thrown a monkey wrench into the works. Theories of psychic development are classic stage theories. Children must pass successfully through the successive Piagetian stages if they are to understand how to cope with the world of real external phenomena. Freudian theory supposes that abnormality is a consequence of fixation at anal or oral stages on the way to normal genital eroticism. For all these theories, the external world can only trigger or inhibit the normal orderly unfolding of an internally programmed sequence. Developmental biology and psychology finesse the problem of the interplay between inside and outside by denying to the external any creative role.

The claim for the hegemony of internal over external forces in development has been an intellectual commitment since the beginning of developmental biology. Struggles over competing theories of embryogenesis have been carried on entirely within that worldview. The most famous was the debate, at the end of the eighteenth and beginning of the nineteenth century, between preformationism and epigenesis.[4] Preformationists, in a view that strikes us as medieval superstition, held that the adult organism was already present in minuscule, a homunculus within the fertilized egg (indeed, within the sperm), and that the process of development consisted only in the growth and solidification of the tiny transparent miniature.

Epigeneticists, whose view prevailed in modern biology, claimed that only an ideal plan of the adult existed in the egg, a blueprint that was made manifest by the process of organism building. Except that we now identify that plan with physical entities, the genes made of DNA, nothing much has changed in the theory in the last two hundred years. Yet, between a concrete preformation-ism that thought there was a little man in every sperm, and an idealist prefor-mationism that sees the complete specification of the adult already present in the fertilized egg, waiting only to be made manifest, there is not much differ-ence except for the mechanical details. In the claim made at the centenary observance of Darwin's death by one of the world's leading molecular biolo-gists, a co-discoverer of the genetic code, that if he had a large enough comput-er and the complete DNA sequence of an organism he could compute the organism, we hear echoes of the eighteenth century. The trouble with the metaphor of "development" is that it gives an impoverished picture of the actu-al determination of the life history of organisms. Development is not simply the realization of an internal program; it is not an unfolding. The outside matters.

First, even when organisms have a few clearly differentiated "stages" these do not necessarily follow each other in some predetermined order, but the organism, in its lifetime, may pass among the stages repeatedly, depending upon signals from outside. Tropical vines that grow in the deep forest begin life as a germinating seed on the forest floor. In the first stage of growth the vine is *positively* geotropic and *negatively* phototropic. That is, it hugs the ground and grows away from the light toward the dark. This has the effect of bringing the vine to the base of a tree. On encountering a tree trunk, the vine becomes *negatively* geotropic and *positively* phototropic, like most plants, and grows upward along the tree trunk toward the light. In this stage it begins to put out leaves of a characteristic shape. When it gets higher in the tree, where the light intensity is greater, the leaf shape and distance between successive leaves change and flowers appear. Yet higher up, the growing tip of the vine moves out laterally along a branch, again changing its leaf shape, and then changes back to being positively geotropic and negatively phototropic, drops off the branch, and starts to grow straight down toward the forest floor. If it hits another branch, lower down, it starts again an intermediate stage, but if it reaches the ground, it begins its cycle again from the beginning. Depending upon the light intensity and height above the ground, the vine makes different transitions between stages.

Second, the development of most organisms is a consequence of a unique interaction between their internal state and the external milieu. At every moment in the life history of an organism there is contingency of development such that the next step is dependent on the current state of the organism and the environmental signals impinging on it. Simply put, the organism is a unique result of both its genes and the temporal sequence of environments through which it has passed, and there is no way of knowing in advance, from the DNA sequence, what the organism will look like, except in general terms. In any sequence of environments we know of lions give birth to lions and lambs to lambs, but all lions are not alike.

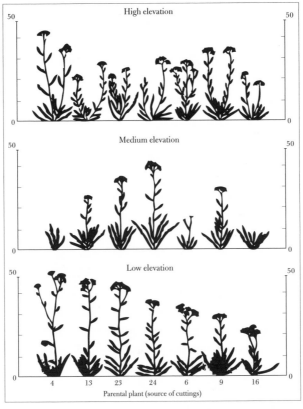

Norms of reaction to elevation for seven different *Achillea* plants (seven different genotypes). A cutting from each plant was grown at low, medium, and high elevations. (Carnegie Institution of Washington)

The consequence of this contingency for the variation among individual organisms is illustrated by a classic experiment in plant genetics.[5] Seven individuals of the plant *Achillea* were collected in California and each plant was cut into three pieces. One piece from each plant was replanted at low elevation (30 meters above sea level), one at intermediate elevation (1,400 meters), and one in the High Sierras (3,050 meters), and the pieces then regrew into new plants. The bottom row of the accompanying picture shows how the pieces of the seven plants grew at low elevation, arranged in decreasing order of their final height. The second row shows the pieces of the same plants grown at intermediate elevation, and the top row is the result of growing pieces at high elevation. The three plants in any vertical column are genetically identical, because they grew from three pieces of the same original plant and therefore carry the same genes.

It is clear that we cannot predict the relative growth of the different plants when the environment is changed. The tallest plant at the low elevation has the *poorest* growth at the intermediate elevation and even fails to flower there. The second largest at the high elevation (plant 9) is intermediate in height at the intermediate elevation, but is the second *smallest* at low elevation. Taken as a whole, there is simply no predictability from one environment to the next. There is no "best" or "largest" genetic type. While we cannot cut people into bits and regrow them in different environments, in every experimental organism where it is possible to duplicate the genetic constitution and test the resultant individuals in different environments, the general result is like that for *Achillea*.

The interaction between genes and environment does not exhaust the sources of variation in development. All "symmetrical" organisms develop asymmetries that fluctuate in direction from individual to individual. The fingerprints of the left and right hands of any individual human being can be distinguished. A fruit fly, no larger than the tip of a lead pencil, having developed while stuck to the inside of a glass culture vessel, has different numbers of sensory bristles on its left and right sides, some flies having more on the left, some more on the right. Moreover, this side-to-side variation is as large as the difference among different flies. But the genes on the left and right sides of a fly are the same, and it seems absurd to think that the temperature, humidity, or concentration of oxygen was different between left and right sides of the tiny developing insect. The variation between sides is a result of random events in

the timing of division and movement of the individual cells that produce the
bristles, so-called developmental noise. Such noise is a universal feature in cell
division and movement and certainly plays a role in the development of our
brains. Indeed, one influential theory of central nervous system development
puts the random growth and hooking up of nerve cells at the base of the entire
process.[6] We simply do not know how much of the difference in cognitive
function between different human beings is a consequence of genetic difference, how much is the result of different life experiences, and how much is the
result of random developmental noise. I cannot play the viola like Pinchas
Zuckerman and I seriously doubt I could have done so had I started at the age
of five. He and I have different nerve connections, and some of these differences were present at birth, but that is not a demonstration that we are genetically different in this respect.

Despite the evidence of environmental and random variation that is lying
about at every hand, developmental biology as a science makes considerable
progress holding on to the metaphor of unfolding by restricting the ambit of
problems that it addresses to just those that can ignore the external and the
indeterminate. Developmental biologists concentrate entirely on how the
front end of an animal is differentiated from the rear end and why pigs don't
have wings, problems that can indeed be approached from the inside of the
organism and which concern some general properties of the machinery. Since
the production of "conceptual schemes of high generality" is the mark of success of a science, what biologist will step off the high road to Stockholm to
wallow in the slough of individual variation? So the limitations of our conceptual schemes dictate not only the form of our answers to questions but which
questions are allowed to be "interesting."

The greatest triumph of nineteenth-century biology was the elaboration of
a mechanistic and materialistic explanation for the history of all of life. The
word *evolution* has the same roots as *development* and signifies, literally, an
unrolling of an already immanent history. Indeed, some pre-Darwinian theories corresponded to the metaphor, the most influential being Karl Ernst von
Baer's fusion of embryology and evolution in his notion of *recapitulation*. In
this scheme, advanced organisms, in their individual development, pass
through a series of stages corresponding to the adults of their less evolved
ancestral forms. That is, their development recapitulates their evolutionary
history. Progressive evolution thus consists in the adding of new stages, but

every species will pass through all the old ones on its way from egg to adult. It is indeed true that at an early embryonic stage we have gill slits like fish, connections between the sides of our heart like amphibians, and tails like puppy dogs, all of which disappear as we mature, so we certainly do carry in our individual histories the traces of our evolution.

Darwinian theory made a radical break with this internalist view. Darwin accepted fully the contingency of evolution and constructed a theory in which both internal and external forces play a role, but in an asymmetrical and alienated way. The first step in the theory is the complete causal separation between the internal and the external. In Lamarckism, with its commitment to the inheritance of acquired characteristics and the incorporation of the external into the organism as a consequence of the organism's own strivings, there is no clear separation of what is inside from what is outside. Darwin's radical difference from Lamarckism was in his clear demarcation of inside and outside, of organism and environment, and his alienation of the forces within organisms from the forces governing their outside world. According to Darwinism, there are mechanisms entirely internal to organisms that cause them to vary one from another in their heritable characteristics. In modern terms, these are mutations of the genes that control development. These variations are not induced by the environment but are produced at random with respect to the exigencies of the outside world. Quite independently, there is an outside world constructed by autonomous forces outside the influence of the organism itself that set the conditions for the species' survival and reproduction. The inside and outside confront each other only through the selective process of differential survival and reproduction of those organic forms that best match *by chance* the autonomous external world. Those that match survive and reproduce, the rest are cast off. Many are called but few are chosen.

This is the process of *adaptation,* by which the collectivity comes to be characterized by just those forms that by chance fit the preexistent demands of an external nature. Nature poses problems for organisms that they must solve or else perish. Nature, love it or leave it. Again, the metaphor corresponds to the theory. By "adaptation" we mean the altering and tuning of an object to fit some preexistent situation, as when traveling Americans use an adapter to make their razors and hairdryers work on European voltages. Evolution by adaptation is when organisms are forced by the demands of an autonomous external world to solve problems that are not of their own making and their only hope is that the internal force of random mutation will, by chance, provide a solution. The organism thus becomes the passive nexus of internal and external forces. It seems almost not to be an actor in its own history.

Darwin's alienation of the environment from the organism was a necessary step in the mechanization of biology, replacing the mystical interpenetration of interior and exterior that was without any material basis. But what is a necessary step in the construction of knowledge at one moment becomes an impediment at another. Whereas Lamarck was wrong to believe that organisms could incorporate the outer world into their heredity, Darwin was wrong in asserting the autonomy of the external world. The environment of an organism is not an independent, preexistent set of problems to which organisms must find solutions, for organisms not only solve problems, they create them in the first place. Just as there is no organism without an environment, there is no environment without an organism. "Adaptation" is the wrong metaphor and needs to be replaced by a more appropriate metaphor like "construction." $\longrightarrow$ juxtaposition

First, though there is indeed an external world that exists independent of any living creature, the totality of that world should not be confused with an organism's environment. Organisms by their life activities determine what is relevant to them. They assemble their environments from the juxtaposition of bits and pieces of the outside world. Just outside my window are patches of dry grass, surrounding a large stone. Phoebes gather the grass to make nests in the rafters of my porch, but the stone is not relevant to them and is not part of their environment. The stone, in contrast, is part of the environment of thrushes, who use it as an anvil to break open snails by rapping them sharply. Not far away is a tree with a large hole in it that is part of the environment of a woodpecker who makes a nest in it, but the hole does not exist in the biological world of the phoebe or thrush. Biologists' descriptions of the "ecological niche" of an organism such as a bird have a revealing rhetoric. "The bird," they say, "eats flying insects in the spring, but switches to small seeds in the fall. It makes a nest of grass, twigs, and mud about two to three feet above the ground in the fork of a tree, in which it raises three to four chicks. It flies south when days get shorter than twelve hours."[7]

Every word is a description of the life activities of the bird, not of an autonomous external nature. It is impossible to judge what the "problems" set by nature are without describing the organism for which these problems are said to exist. In some abstract sense, flying through the air is a potential problem for all organisms, but this problem does not exist for earthworms who, *as a consequence of the genes they carry*, spend their lives underground. Therefore[just as the information needed to specify an organism is not contained entirely in its genes, but also in its environment, so the environmental problems of the organism are a consequence of its genes. Penguins, birds who spend much of their lives underwater, have altered their wings to make them into flippers. At what stage in the evolution of the flying ancestors of penguins did swimming underwater become a "problem" to be solved? We do not know, but presumably their ancestors had already made swimming an important part of their life activities before natural selection favored turning wings into paddles. Fish gotta swim and birds gotta fly. Nor is the origin of flight without its problems. A flightless animal that sprouted rudimentary wings would get no lift from them at all, as one can easily verify by flapping a pair of Ping-Pong paddles. Lift force increases very slowly with the increase of surface area for small wings, and below a certain size there is no lift at all. However, even small thin membranes that can be moved turn out to be excellent devices for dissipating heat, or collecting it from sunshine, and many butterflies use their wings for that purpose. Our present guess is that wings did not originate to solve the problem of flight at all, but were heat-regulatory devices that, when they became large enough, gave the insect some lift and so made flight a new problem to be solved.]

⌊Because organisms create their own environments we cannot characterize the environment except in the presence of the organism that it surrounds.⌉ Using appropriate optical devices it is possible to see that there is a layer of warm moist air surrounding each one of us which moves continually up the surface of our bodies and off the tops of our heads. This layer, present in all organisms that live in air, is a result of the production of heat and water by our metabolism. As a consequence we carry around with us our own atmosphere. If the wind should blow and strip away that boundary layer we would be exposed to the outside world and know how cold it really is out there. That is the meaning of the wind-chill factor.

⌊ Second, every organism, not just the human species, is in the constant process of changing its environment, both creating and destroying its own means of subsistence.⌋ It is part of the ideology of the environmental movement that alone among species, human beings are in the process of destroying the world they inhabit and that undisturbed nature is in unchanging harmony and balance. There is nothing here but Rousseauian romance. Every species consumes its own resources of space and nutriment, and in the process produces waste products that are toxic to itself and its offspring. Every act of consumption is an act of production, and every act of production an act of consumption. Every animal, when it breathes in precious oxygen, exhales poisonous carbon dioxide, poisonous to itself, but not to plants, who thrive on it. As Mort Sahl once observed, no matter how cruel and unfeeling we may be, every time we breathe we make a flower happy. Every organism deprives its fellows of space and, when it feeds and digests, excretes toxic waste products into its own neighborhood.

In some cases as a matter of their normal function, organisms make it impossible for their own offspring to succeed them. When the stony farms of New England were abandoned in the westward rush after 1840, the untilled fields were at first occupied by herby weeds and then were taken over by pure stands of white pine. In the early 1900s it was thought that the pines would be a steady source of income from wood and pulp, but they failed to replace themselves and gave way to hardwoods, immediately when cut and slowly when left alone. The problem is that pine seedlings are intolerant of shade and cannot grow in a forest, even a forest of pines. The adult pines create a condition that is inimical to their own offspring, so that they can survive as a species only if some of the seeds can, like the eighteenth- and nineteenth-century children of European farmers, colonize newly opened areas where they are not oppressed by their parents. But all organisms also produce the conditions necessary for their existence. Birds make nests, bees hives, and moles burrows. When plants put down roots they change the texture of the soil and excrete chemicals that encourage the growth of symbiotic fungi that help the plant's nutrition. Fungus-gardening ants gather and chew up leaves, which they seed with the spores of mushrooms that they eat. At every moment every species is in the process of creating and re-creating, both beneficially and detrimentally, its own conditions of existence, its own environment.

Some may object that some important elements of the outer world are thrust on organisms by the very laws of nature. After all, gravitation would be a fact of nature even if Newton had never existed. But the relevance to an organism of external forces, even of gravitation, is coded in its genes. We are oppressed by gravity, acquiring flat feet and bad backs by virtue of our large size and upright posture, both consequences of the genes we have inherited. Bacteria, living in a liquid medium, do not experience gravity, but they are subject to another "universal" physical force, Brownian motion. Because they are so small, bacteria are buffeted about by the random thermal motions of molecules in the liquid medium; a force which, fortunately, does not send us reeling from one side of the room to the other. All natural forces operate effectively in particular ranges of size and distance so that organisms, as they grow and evolve, may move from the domain of one set of forces to another. All the organisms that now exist have evolved and must survive in an atmosphere that is 18 percent oxygen, an extremely reactive and chemically powerful element. But the earliest organisms did not have to cope with free oxygen, which was absent from the aboriginal atmosphere, an atmosphere with high concentrations of carbon dioxide. It is organisms themselves that have produced the oxygen, through photosynthesis, and have depleted the carbon dioxide to the fraction of a percent that it now represents by trapping it in vast deposits of limestone, coal, and petroleum. The proper view of evolution then is of coevolution of organisms and their environments, each change in an organism being both the cause and the effect of changes in the environment. The inside and the outside do indeed interpenetrate, and the organism is both the product and location of that interaction.

The constructionist view of organism and environment is of some consequence to human action. A rational environmental movement cannot be built on the demand to save the environment, which, in any case, does not exist. Clearly, no one wants to live in a world that smells and looks worse than at present, or in which life is even more solitary, poor, nasty, brutish, and short than it is now. But that wish cannot be realized by the impossible demand that human beings stop changing the world. Remaking the world is the universal property of living organisms and is inextricably bound up in their nature. Rather, we must decide what kind of a world we want to live in and then try to manage the processes of change as best we can to approximate it.

25
The Dream of the Human Genome

1.

FETISH . . . An inanimate object worshipped by savages on account of its supposed inherent magical powers, or as being animated by a spirit.
—Oxford English Dictionary

Scientists are public figures, and like other public figures with a sense of their own importance, they self-consciously compare themselves and their work to past monuments of culture and history. Modern biology, especially molecular biology, has undergone two such episodes of preening before the glass of history. The first, characteristic of a newly developing field that promises to solve important problems that have long resisted the methods of an older tradition, used the metaphor of revolution. Tocqueville observed that when the bourgeois monarchy was overthrown on February 24, 1848, the Deputies compared themselves consciously to the Girondins and the Montagnards of the National Convention of 1793:

> The men of the first Revolution were living in every mind, their deeds and words present to every memory. All that I saw that day bore the visible impress of those recollections; it seemed to me throughout as though they were engaged in acting the French Revolution rather than continuing it.

The romance of being a revolutionary had infected scientists long before Thomas Kuhn made Scientific Revolution the shibboleth of progressive knowledge. Many of the founders of molecular biology began as physicists,

steeped in the lore of the quantum mechanical revolution of the 1920s. The
Rousseau of molecular biology was Erwin Schrödinger, the inventor of the
quantum wave equation, whose *What Is Life?* was the ideological manifesto
of the new biology. Molecular biology's Robespierre was Max Delbruck, a
student of Schrödinger, who created a political apparatus called the Phage
Group, which carried out the experimental program. A history of the Phage
Group written by its early participants and rich in the consciousness of a revolutionary tradition was produced twenty-five years ago.[1]

   The molecular biological revolution has not had its Thermidor, but on the
contrary it has ascended to the state of an unchallenged orthodoxy. The self-
image of its practitioners and the source of their metaphors have changed
accordingly to reflect their perception of transcendent truth and unassailable
power. Molecular biology is now a religion, and molecular biologists are its
prophets. Scientists now speak of the "Central Dogma" of molecular biology,
and Walter Gilbert's contribution to the collection *The Code of Codes* is titled
"A Vision of the Grail." In their preface, Daniel Kevles and Leroy Hood
address the metaphor with straight faces and no quotation marks:

> The search for the biological grail has been going on since the turn of the
> century, but it has now entered its culminating phase with the recent creation
> of the human genome project, the ultimate goal of which is the acquisition of
> all the details of our genome. . . . It will transform our capacities to predict
> what we may become. . . .
>
>   Unquestionably, the connotations of power and fear associated with the
> holy grail accompany the genome project, its biological counterpart. . . .
> Undoubtedly, it will affect the way much of biology is pursued in the twenty-
> first century. Whatever the shape of that effect, the quest for the biological
> grail will, sooner or later, achieve its end, and we believe that it is not too early
> to begin thinking about how to control the power so as to diminish—better
> yet, abolish—the legitimate social and scientific fears.[2]

   It is a sure sign of their alienation from revealed religion that a scientific community with a high concentration of Eastern European Jews and atheists has chosen for its central metaphor the most mystery-laden object of medieval Christianity.

   As there were legends of the Saint Graal of Perceval, Gawain, and
Galahad, so there is a legend of the Grail of Gilbert. It seems that each cell of

my body (and yours) contains in its nucleus two copies of a very long molecule called deoxyribonucleic acid (DNA). One of these copies came to me
from my father and one from my mother, brought together in the union of
sperm and egg. This very long molecule is differentiated along its length into
segments of separate function called genes, and the set of all these genes is
called, collectively, my genome.

What I am, the differences between me and other human beings, and the
similarities among human beings that distinguish them from, say, chimpanzees,
are determined by the exact chemical composition of the DNA making up my
genes. In the words of a popular bard of the legend, genes "have created us
body and mind."[3] So when we know exactly what the genes look like we will
know what it is to be human, and we will also know why some of us read *The
New York Review of Books* while others cannot get beyond the *New York Post*.
"Genetic variations in the genome, various combinations of different possible
genes . . . create the infinite variety that we see among individual members of a
species," according to Joel Davis in *Mapping the Code*.[4] Success or failure,
health or disease, madness or sanity, our ability to take it or leave it alone—all
are determined, or at the very least are strongly influenced, by our genes.

The substance of which genes are made must have two properties. First, if
the millions of cells of my body all contain copies of molecules that were originally present only once in the sperm and once in the egg with which my life
began, and if, in turn, I have been able to pass copies to the millions of sperm
cells that I have produced, then the DNA molecule must have the power of
self-reproduction. Second, if the DNA of the genes is the efficient cause of my
properties as a living being, of which I am the result, then DNA must have the
power of *self-action*. That is, it must be an active molecule that imposes specific form on a previously undifferentiated fertilized egg, according to a
scheme that is dictated by the internal structure of DNA itself.

Because this self-producing, self-acting molecule is the ground of our
being, "precious DNA" must be guarded by a "magic shield" against the "hur-
ricane of forces" that threaten it from the outside, according to Christopher
Wills, by which he means the bombardment by the other chemically active
molecules of the cell that may destroy the DNA. It is not idly that DNA is
called the Grail. Like that mystic bowl, DNA is said to be regularly self-renew-
ing, providing its possessors with sustenance *"sans serjant et sans seneschal,"*
and shielded by its own knights from hostile forces.

How is it that a mere molecule can have the power of both self-reproduction and self-action, being the cause of itself and the cause of all the other things? DNA is composed of basic units, the *nucleotides,* of which there are four kinds, adenine, cystosine, guanine, and thymine (A, C, G, and T) and these are strung one after another in a long linear sequence that makes a DNA molecule. So one bit of DNA might have the sequence of units . . . CAAATTGC . . . and another the sequence . . . TATCGCTA . . . and so on. A typical gene might consist of 10,000 basic units, and since there are four different possibilities for each position in the string, the number of different possible kinds of genes is a great deal larger than what is usually called "astronomically large." (It would be represented as a 1 followed by 6,020 zeros.) The DNA string is like a code with four different letters whose arrangements in messages thousands of letters long are of infinite variety. Only a small fraction of the possible messages can specify the form and content of a functioning organism, but that is still an astronomically large number.

The DNA messages specify the organism by specifying the makeup of the proteins of which organisms are made. A particular DNA sequence makes a particular protein according to a set of decoding rules and manufacturing processes that are well understood. Part of the DNA code determines exactly which protein will be made. A protein is a long string of basic units called amino acids, of which there are twenty different kinds. The DNA code is read in groups of three consecutive nucleotides, and to each of the triplets AAA, AAC, GCT, TAT, etc., there corresponds one of the amino acids. Since there are sixty-four possible triplets and only twenty amino acids, more than one triplet matches the same amino acid (the code is "redundant"). Another part of the DNA determines when in development and where in the organism the manufacture of a given protein will be "turned on" or "turned off." By turning genes on and off in the different parts of the developing organism at different times, the DNA "creates" the living being, "body and mind."

But how does the DNA re-create itself? By its own dual and self-complementary structure (as the blood of Christ is said to be renewed in the Grail by the dove of the Holy Ghost). The string of nucleic acids in DNA that carries the message of protein production is accompanied by another string helically entwined with it and bound to it in a chemical embrace. This DNA doppelganger is matched nucleotide by nucleotide with the message strand in a complementary fashion. Each A in the message is matched by a T on the complementary strand, each C by a G, each G by a C, and each T by an A.

Reproduction of DNA is, ironically, an uncoupling of the mated strands, followed by a building up of a new complementary strand on each of the parental strings. So the self-reproduction of DNA is explained by its dual, complementary structure, and its creative power by its linear differentiation.

The problem with this story is that although it is correct in its detailed molecular description, it is wrong in what it claims to explain. First, DNA is not self-reproducing; second, it makes nothing; and third, organisms are not determined by it.

DNA is a dead molecule, among the most nonreactive, chemically inert molecules in the living world. That is why it can be recovered in good enough shape to determine its sequence from mummies, from mastodons frozen tens of thousands of years ago, and even, under the right circumstances, from twenty-million-year-old fossil plants. The forensic use of DNA for linking alleged criminals with victims depends on recovering undegraded molecules from scrapings of long-dried blood and skin. DNA has no power to reproduce itself. Rather, it is produced of elementary materials by a complex cellular machinery of proteins. Although it is often said that DNA produces proteins, proteins (enzymes) produce DNA. The newly manufactured DNA is certainly a *copy* of the old, and the dual structure of the DNA molecule provides a complementary template on which the copying process works. The process of copying a photograph includes the production of a complementary negative which is then printed, but we do not describe the Eastman Kodak factory as a place of self-reproduction.

No living molecule is self-reproducing. Only whole cells may contain all the necessary machinery for "self"-reproduction and even they, in the process of development, lose that capacity. Nor are entire organisms self-reproducing, as the skeptical reader will soon realize if he or she tries it. Yet even the sophisticated molecular biologist when describing the process of copying DNA lapses into the rhetoric of "self-reproduction." Christopher Wills, in the process of a mechanical description of DNA synthesis, tells us that "DNA cannot make copies of itself *unassisted*" (emphasis added) and further that "for DNA to duplicate [itself], the double helix must be unwound into two separate chains."

Not only is DNA incapable of making copies of itself, aided or unaided, but it is incapable of "making" anything else. The linear sequence of nucleotides in DNA is used by the machinery of the cell to determine what sequence of amino acids is to be built into a protein, and to determine when and where the protein is to be made. But the proteins of the cell are made by other proteins, and without that protein-forming machinery *nothing* can be made. There is an appearance here of infinite regress (What makes the proteins that are necessary to make the protein?), but this appearance is an artifact of another error of vulgar biology, that it is only the genes that are passed from parent to offspring. In fact, an egg, before fertilization, contains a complete apparatus of production deposited there in the course of its cellular development. We inherit not only genes made of DNA but an intricate structure of cellular machinery made up of proteins.

It is the evangelical enthusiasm of the modern Grail Knights and the innocence of the journalistic acolytes whom they have catechized that have so fetishized DNA. There are, too, ideological predispositions that make themselves felt. The more accurate description of the role of DNA is that it bears information that is read by the cell machinery in the productive process. Subtly, DNA as information bearer is transmogrified successively into DNA as blueprint, as plan, as master plan, as master molecule. It is the transfer onto biology of the belief in the superiority of mental labor over the merely physical, of the planner and designer over the unskilled operative on the assembly line.

The practical outcome of the belief that everything we want to know about human beings is contained in the sequence of DNA is the Human Genome Project in the United States. Its international analogue is the Human Genome Organization (HUGO), called by one molecular biologist "the UN for the human genome."

These projects are administrative and financial organizations rather than research projects in the usual sense. They have been created over the last five years in response to an active lobbying effort, by scientists such as Walter Gilbert, James Watson, Charles Cantor, and Leroy Hood, aimed at capturing very large amounts of public funds and directing the flow of those funds into an immense cooperative research program.

The ultimate purpose of this program is to write down the complete ordered sequence of nucleotides A, T, C, and G that make up all the genes in the human genome, a string of letters that will be three billion elements long. The first laborious technique for cutting up DNA nucleotide by nucleotide and identifying each nucleotide in order as it is broken off was invented fifteen years ago by Allan Maxam and Walter Gilbert, but since then the process has become mechanized. DNA can now be squirted into one end of a mechanical process and out the other end will emerge a four-color computer printout announcing "AGGACTT. . . ." In the course of the genome project yet more efficient mechanical schemes will be invented and complex computer programs will be developed to catalog, store, compare, order, retrieve, and otherwise organize and reorganize the immensely long string of letters that will emerge from the machine. The work will be a collective enterprise of very large laboratories, "Genome Centers," which are to be specially funded for the purpose.

The project is to proceed in two stages. The first is so-called physical mapping. The entire DNA of an organism is not one long unbroken string but is divided up into a small number of units, each of which is contained in one of a set of microscopic bodies in the cell, the chromosomes. Human DNA is broken up into twenty-three different pairs of chromosomes, while fruit flies' DNA is contained in only four chromosomes. The mapping phase of the genome project will determine short stretches of DNA sequence spread out along each chromosome as positional landmarks, much as mile markers are placed along superhighways. These positional makers will be of great use in finding where in each chromosome particular genes may lie. In the second phase of the project, each laboratory will take a chromosome or a section of a chromosome and determine the complete ordered sequence of nucleotides in its DNA. It is after the second phase, when the genome project, *sensu strictu,* has ended, that the fun begins, for biological sense will have to be made, if possible, of the mind-numbing sequence of three billion instances of A, T, C, and G. What will it tell us about health and disease, happiness and misery, the meaning of human existence?

The American project is run jointly by the National Institutes of Health (NIH) and the Department of Energy in a political compromise over which should have control over the hundreds of millions of dollars of public money that will be required. The project produces a glossy-paper newsletter distributed free, headed by a coat of arms showing a human body wrapped Laocoön-like in the serpent coils of DNA and surrounded by the motto "Engineering, Chemistry, Biology, Physics, Mathematics." The Genome Project is the nexus of all sciences. My latest copy of the newsletter advertises the free loan of a 23-minute video on the project "intended for high school age and older," featuring, among others, several of the contributors to *The Code of Codes,* and a calendar of fifty "Genome Events."

None of the authors of the books under review seems to be in any doubt about the importance of the project to determine the complete DNA sequence of a human being. "The Most Astonishing Adventure of Our Time," say Jerry E. Bishop and Michael Waldholz; "The Future of Medicine," according to Lois Wingerson; "Today's most important scientific undertaking," dictating "The Choices of Modern Science," Joel Davis declares in *Mapping the Code.*

Nor are these simply the enthusiasms of journalists. The molecular biologist Christopher Wills says that "the outstanding problems in human biology . . . will all be illuminated in a strong and steady light by the results of this undertaking"; the great panjandrum of DNA himself, James Dewey Watson, explains, in his essay in the collection edited by Kevles and Hood, that he doesn't "want to miss out on learning how life works"; and Walter Gilbert predicts that there will be "a change in our philosophical understanding of ourselves." Surely, "learning how life works" and "a change in our philosophical understanding of ourselves" must be worth a lot of time and money. Indeed, there are said to be those who have exchanged something a good deal more precious for that knowledge.

2.

Unfortunately, it takes more than DNA to make a living organism. Even the organism does not compute itself from its DNA. A living organism at any moment in its life is the unique consequence of a developmental history that results from the interaction of and determination by internal and external forces. The external forces, what we usually think of as "environment," are themselves partly a consequence of the activities of the organism itself as it produces and consumes the conditions of its own existence. Organisms do not find the world in which they develop. They make it. Reciprocally, the internal forces are not autonomous, but act in response to the external. Part of the internal chemical machinery of a cell is only manufactured when external conditions demand it. For example, the enzyme that breaks down sugar lactose to provide energy for bacterial growth is only manufactured by bacterial cells when they detect the presence of lactose in their environment.

Nor is "internal" identical with "genetic." Fruit flies have long hairs that serve as sensory organs, rather like a cat's whiskers. The number and placement of those hairs differ between the two sides of a fly (as they do between the left and right sides of a cat's muzzle), but not in any systematic way. Some flies have more hairs on the left, some more on the right. Moreover, the variation between sides of a fly is as great as the average variation from fly to fly. But the two sides of a fly have the same genes and have had the same environment during development. The variation between sides is a consequence of random cellular movements and chance molecular events within cells during development, so-called developmental noise. It is this same developmental noise that accounts for the fact that identical twins have different fingerprints and that the fingerprints on our left and right hands are different. A desktop computer that was as sensitive to room temperature and as noisy in its internal circuitry as a developing organism could hardly be said to compute at all.

The scientists writing about the Genome Project explicitly reject an absolute genetic determinism, but they seem to be writing more to acknowledge theoretical possibilities than out of conviction. If we take seriously the proposition that the internal and external codetermine the organism, we cannot really believe that the sequence of the human genome is the Grail that will reveal to us what it is to be human, that it will change our philosophical view of ourselves, that it will show how life works. It is only the social scientists and social critics, such as Daniel J. Kevles, who comes to the Genome Project from his important study of the continuity of eugenics with modern medical genetics; Dorothy Nelkin, both in her book with Laurence Tancredi and in her chapter in Kevles and Hood; and, most strikingly, Evelyn Fox Keller in her contribution to *The Code of Codes,* for whom the problem of the development of the organism is central.

Nelkin, Tancredi, and Keller suggest that the importance of the Human Genome Project lies less in what it may reveal about biology, and whether the project may in the end lead to a successful therapeutic program for one or another illnesses, than in its validation and reinforcement of biological determinism as an explanation of all social and individual variation. The medical model that begins, for example, with a genetic explanation of the extensive and irreversible degeneration of the central nervous system characteristic of Huntington's chorea may end with an explanation of human intelligence, of how much people drink, how intolerable they find the social condition of their lives, whom they choose as sexual partners, and whether they get sick on the job. A medical model of all human variation makes a medical model of normality, including social normality, and dictates a therapeutic or preemptive attack on deviance.

There are many human conditions that are clearly pathological and that can be said to have a unitary genetic cause. As far as is known, cystic fibrosis and Huntington's chorea occur in people carrying the relevant mutant gene irrespective of diet, occupation, social class, or education. Such disorders are rare: 1 in 2,300 births for cystic fibrosis, 1 in 3,000 for Duchenne's muscular dystrophy, 1 in 10,000 for Huntington's disease. A few other conditions occur in much higher frequency in some populations but are generally less severe in their effects and more sensitive to environmental conditions, as for example sickle cell anemia in West Africans and their descendants, who suffer severe effects only in conditions of physical stress. These disorders provide the model on which the program of medical genetics is built, and they provide the human interest drama on which books like *Mapping Our Genes* and *Genome* are built. In reading them, I saw again those heroes of my youth, Edward G. Robinson curing syphilis in *Dr. Ehrlich's Magic Bullet,* and Paul Muni saving children from rabies in *The Story of Louis Pasteur.*

It is said that a wonder-rabbi of Chelm once saw, in a vision, the destruction by fire of the study house in Lublin, fifty miles away. This remarkable event greatly enhanced his fame as a wonderworker. Several days later a traveler from Lublin, arriving in Chelm, was greeted with expressions of sorrow and concern, not unmixed with a certain pride, by the disciples of the wonder-rabbi. "What are you talking about?" asked the traveler. "I left Lublin three days ago and the study house was standing as it always has. What kind of a wonder-rabbi is that?" "Well, well," one of the rabbi's disciples answered, "burned or not burned, it's only a detail. The wonder is he could see so far." We live still in an age of wonder-rabbis, whose sacred trigram is not the ineffable YWH but the ever-repeated DNA. Like the rabbi of Chelm, however, the prophets of DNA and their disciples are short on details.

According to the vision, we will locate on the human chromosomes all the defective genes that plague us, and then from the sequence of the DNA we will deduce the causal story of the disease and generate a therapy. Indeed, a great many defective genes have already been roughly mapped onto chromosomes and, with the use of molecular techniques, a few have been very closely located and, for even fewer, some DNA sequence information has been obtained. But causal stories are lacking and therapies do not yet exist; nor is it clear, when actual cases are considered, how therapies will flow from a knowledge of DNA sequences.

The gene whose mutant form leads to cystic fibrosis has been located, isolated, and sequenced. The protein encoded by the gene has been deduced. Unfortunately, it looks like a lot of other proteins that are a part of cell structure, so it is hard to know what to do next. The mutation leading to Tay-Sachs disease is even better understood because the enzyme specified by the gene has a quite specific and simple function, but no one has suggested a therapy. In contrast, the gene mutation causing Huntington's disease has eluded exact location, and no biochemical or specific metabolic defect has been found for a disease that results in catastrophic degeneration of the central nervous system in every carrier of the defective gene.

A deep reason for the difficulty in devising causal information from DNA messages is that the same "words" have different meanings in different contexts and multiple functions in a given context, as in any complex language. No word in English has more powerful implications of action than "do." "Do it now!" Yet in most of its contexts "do" as in "I do not know" is periphrastic and has no meaning at all. Though the periphrastic "do" has no *meaning*, it undoubtedly has a linguistic *function* as a placeholder and spacing element in the arrangement of a sentence. Otherwise, it would not have swept into general English usage in the sixteenth century from its Midlands dialect origin, replacing everywhere the older "I know not."

So elements in the genetic messages may have meaning, or they may be periphrastic. The code sequence GTAAGT is sometimes read by the cell as an instruction to insert the amino acids valine and serine in a protein, but sometimes it signals a place where the cell machinery is to cut up and edit the message; and sometimes it may be only a spacer, like the periphrastic "do," that keeps other parts of the message an appropriate distance from each other. Unfortunately, we do not know how the cell decides among the possible interpretations. In working out the interpretive rules, it would certainly help to have very large numbers of different gene sequences, and it is possible to sometimes suspect that the claimed significance of the genome sequencing project for human health is an elaborate cover story for an interest in the hermeneutics of biological scripture.

Of course, it can be said, as Gilbert and Watson do in their essays, that an understanding of how the DNA code works is the path by which human health will be reached. If one had to depend on understanding, however, we would all be much sicker than we are. Once, when the eminent Kant scholar Lewis Beck was traveling in Italy with his wife, she contracted a maddening rash. The specialist they consulted said it would take him three weeks to find out what was wrong with her. After repeated insistence by the Becks that they had to leave Italy within two days, the physician threw up his hands and said, "Oh, very well, Madam. I will give up my scientific principles. I will cure you today."

Certainly an understanding of human anatomy and physiology has led to a medical practice vastly more effective than it was in the eighteenth century. These advances, however, consist in greatly improved methods for examining the state of our insides, of remarkable advances in microplumbing, and of pragmatically determined ways of correcting chemical imbalances and of killing bacterial invaders. None of these depends on a deep knowledge of cellular processes or on any discoveries of molecular biology. Cancer is still treated by gross physical and chemical assaults on the offending tissue. Cardiovascular disease is treated by surgery whose anatomical bases go back to the nineteenth century, by diet and by pragmatic drug treatment. Antibiotics were originally developed without the slightest notion of how they do their work. Diabetics continue to take insulin, as they have for sixty years, despite all the research on the cellular basis of pancreatic malfunction. Of course, intimate knowledge of the living cell and of basic molecular processes may be useful eventually, and we are promised over and over that results are just around the corner. But as Vivian Blaine so poignantly complained in *Guys and Dolls*:

> *You promised me this*
> *You promised me that.*
> *You promised me everything*
> *under the sun.*
>
> *. . .*
>
> *I think of the time gone by*
> *And could honestly die.*

Not the least of the problems of turning sequence information into causal knowledge is the existence of large amounts of polymorphism.

Whereas the talk in most of the books under review is of sequencing the human genome, every human genome differs from every other. The DNA I got from my mother differs by about 0.1 percent, or about 3 million nucleotides, from the DNA I got from my father, and I differ by about that much from any other human being. The final catalog of "the" human DNA sequence will be a mosaic of some hypothetical average person corresponding to no one. This polymorphism has several serious consequences. First, all of us carry one copy, inherited from one parent, of mutations that would result in genetic diseases if we had inherited two copies. No one is free of these, so the catalog of the standard human genome after it is compiled will contain, unknown to its producers, some fatally misspelled sequences that code for defective proteins or no protein at all. The only way to know if the standard sequence is, by bad luck, the code of a defective gene is to sequence the same part of the genome from many different individuals. Such polymorphism studies are not part of the Human Genome Project and attempts to obtain money from the project for such studies have been rebuffed.

Second, even genetically "simple" diseases can be heterogeneous in their origin. Sequencing studies of the gene that codes for a critical protein in blood clotting has shown that hemophiliacs differ from people whose blood clots normally by any one of 208 different DNA variations, all in the same gene. These differences occur in every part of the gene, including bits that are not supposed to affect the structure of the protein.

The problem of telling a coherent causal story, and of then designing a therapy based on knowledge of the DNA sequence in such a case, is that we do not know even in principle all the functions of the different nucleotides in a gene, or how the specific context in which a nucleotide appears may affect the way in which the cell machinery interprets the DNA; nor do we have any but the most rudimentary understanding of how a whole functioning organism is put together from its protein bits and pieces. Third, because there is no single, standard, "normal" DNA sequence we all share, observed sequence differences between sick and well people cannot, in themselves, reveal the genetic cause of a disorder. At the least, we would need the sequences of many sick and many well people to look for common differences between sick and well. But if many diseases are like hemophilia, common differences will not be found and we will remain mystified.

3.

The failure to turn knowledge into therapeutic power does not discourage the advocates of the Human Genome Project because their vision of therapy includes *gene* therapy. By techniques that are already available and need only technological development, it is possible to implant specific genes containing the correct gene sequence into individuals who carry a mutated sequence, and to induce the cell machinery of the recipient to use the implanted genes as its source of information. Indeed, the first case of human gene therapy for an immune disease—the treatment of a child who suffered from a rare disorder of the immune system—has already been announced and seems to have been a success. The supporters of the Genome Project agree that knowing the sequence of all human genes will make it possible to identify and isolate the DNA sequences for large numbers of human defects which could then be corrected by gene therapy. In this view, what is now an ad hoc attack on individual disorders can be turned into a routine therapeutic technique, treating every physical and psychic dislocation, since everything significant about human beings is specified by their genes.

Gene implantation, however, may affect not only the cells of our temporary bodies, our *somatic* cells, but the bodies of future generations through accidental changes in the *germ* cells of our reproductive organs. Even if it were our intention only to provide properly functioning genes to the immediate body of the sufferer, some of the implanted DNA might get into and transform future sperm and egg cells. Then future generations shall also have undergone the therapy in absentia and any miscalculations of the effects of the implanted DNA would be wreaked on our descendants. David Suzuki and Peter Knudtson make it one of their principles of "genethics" (they have self-consciously created ten of them) that

> while genetic manipulation of human somatic cells may lie in the realm of
> personal choice, tinkering with human germ cells does not. Germ-cell therapy,
> without the consent of all members of society, ought to be explicitly forbidden.

Their argument against gene therapy is a purely prudential one, resting on the imprecision of the technique and the possibility that a "bad" gene today might turn out to be useful some day. This seems a slim base for one of the Ten Commandments of biology for the techniques may get a lot better and mistakes can always be corrected by another round of gene therapy. The vision of power offered to us by gene therapists makes gene transfer seem rather less permanent than a silicone implant or a tummy tuck. The bit of ethics in *Genethics* is, like a Unitarian sermon, nothing that any decent person could quarrel with. Most of the genethic principles turn out to be advice about why we should not screw around with our genes or those of other species. While most of their arguments are sketchy, Suzuki and Knudtson are the only authors among those under review who take seriously the problems presented by genetic diversity among individuals, and who attempt to give the reader enough understanding of the principles of population genetics to think about these problems.

Most death, disease, and suffering in rich countries do not arise from muscular dystrophy and Huntington's chorea, and, of course, the majority of the world's population is suffering from one consequence or another of malnutrition and overwork. For Americans, it is heart disease, cancer, and stroke that are the major killers, accounting for 70 percent of deaths, and about sixty million people suffer from chronic cardiovascular disease. Psychiatric suffering is harder to estimate, but before the psychiatric hospitals were emptied in the 1960s, there were 750,000 psychiatric inpatients. It is now generally accepted that some fraction of cancers arise on a background of genetic predisposition. That is, there are a number of genes known, the so-called oncogenes, that have information about normal cell division. Mutations in these genes result (in an unknown way) in making cell division less stable and more likely to occur at a pathologically high rate. Although a number of such genes have been located, their total number and the proportion of all cancers influenced by them is unknown.

In no sense of simple causation are mutations in these genes *the* cause of cancer, although they may be one of many predisposing conditions. Although a mutation leading to extremely elevated cholesterol levels is known, the great mass of cardiovascular disease has utterly defied genetic analysis. Even diabetes, which has long been known to run in families, has never been tied to genes and there is no better evidence for a genetic predisposition to it in 1992 than there was in 1952 when serious genetic studies began. No week passes without the announcement in the press of a "possible" genetic cause of some human ill that upon investigation "may eventually lead to a cure." No literate public is unassailed by the claims. The *Morgunbladid* of Reykjavik asks its readers rhetorically, "*Med allt í genunum?*" (Is it all in the genes?) in a Sunday supplement.

The rage for genes reminds us of tulipomania and the South Sea Bubble in Charles Mackay's *Extraordinary Popular Delusions and the Madness of Crowds*. Claims for the definitive location of a gene for schizophrenia and manic-depressive syndrome using DNA markers have been followed repeatedly by retraction of the claims and contrary claims as a few more members of a family tree have been observed, or a different set of families examined. In one notorious case, a claimed gene for manic depression, for which there was strong statistical evidence, was nowhere to be found when two members of the same family group developed symptoms. The original claim and its retraction both were published in the international journal *Nature,* causing David Baltimore to cry out at a scientific meeting, "Setting myself up as an average reader of *Nature,* what am I to believe?" Nothing.

Some of the wonder-rabbis and their disciples see even beyond the major causes of death and disease. They have an image of social peace and order emerging from the DNA data bank at the National Institutes of Health. The editor of the most prestigious general American scientific journal, *Science,* an energetic publicist for large DNA-sequencing projects, in special issues filled with full-page multicolored advertisements from biotechnology equipment manufacturers, has visions of genes for alcoholism, unemployment, domestic and social violence, and drug addiction. What we had previously imagined to be messy moral, political, and economic issues turn out, after all, to be simply a matter of an occasional nucleotide substitution. Though the notion that the war on drugs will be won by genetic engineering belongs to Cloud Cuckoo Land, it is a manifestation of a serious ideology continuous with the eugenics of an earlier time.

Daniel Kevles has quite persuasively argued in his earlier book on eugenics that classical eugenics became transformed from a social program of general population improvement into a family program of providing genetic knowledge to individuals facing reproductive decisions.[5] But the ideology of biological determinism on which eugenics was based has persisted, and, as is made clear in Kevles's excellent short history of the Genome Project in *The Code of Codes,* eugenics in the social sense has been revivified. This has been in part a consequence of the mere existence of the Genome Project, with its accompanying public relations and the heavy public expenditure it will require. These alone validate its determinist *Weltanschauung*. The publishers declare the glory of DNA and the media showeth forth its handiwork.

4.

The nine books cited in the notes for this chapter are only a sample of what has been and what is to come. The cost of sequencing the human genome is estimated optimistically at $300 million (ten cents a nucleotide for the three billion nucleotides of the entire genome), but if development costs are included it surely cannot be less than a half-billion in current dollars. Moreover the genome project *sensu strictu* is only the beginning of wisdom. Yet more hundreds of millions must be spent on chasing down the elusive differences in DNA for each specific genetic disease, of which some three thousand are now known, and some considerable fraction of that money will stick to entrepreneurial molecular geneticists. None of the authors has the bad taste to mention that many molecular geneticists of repute, including several of the essayists in *The Code of Codes,* are founders, directors, officers, and stockholders in commercial biotechnology firms, including the manufacturers of the supplies and equipment used in sequencing research. Not all authors have Norman Mailer's openness when they write advertisements for themselves.

It has been clear since the first discoveries in molecular biology that "genetic engineering," the creation to order of genetically altered organisms, has an immense possibility for producing private profit. If the genes that allow clover plants to manufacture their own fertilizer out of nitrogen in the air could be transferred to maize or wheat, farmers would save great sums and the producers of the engineered seed would make a great deal of money. Genetically engineered bacteria grown in large fermenting vats can be made into living factories to produce rare and costly molecules for the treatment of viral diseases and cancer. A bacterium has already been produced that will eat raw petroleum, making oil spills biodegradable. As a consequence of these possibilities, molecular biologists have become entrepreneurs. Many have founded biotechnology firms funded by venture capitalists. Some have become very rich when a successful public offering of their stock has made them suddenly the holders of a lot of valuable paper. Others find themselves with large blocks of stock in international pharmaceutical companies that have bought out the biologist's mom-and-pop enterprise and acquired their expertise in the bargain.

No prominent molecular biologist of my acquaintance is without a financial stake in the biotechnology business. As a result, serious conflicts of interest have emerged in universities and in government service. In some cases graduate students working under entrepreneurial professors are restricted in their scientific interchanges, in case they may give away potential trade secrets. Research biologists have attempted, sometimes with success, to get special dispensations of space and other resources from their universities in exchange for a piece of the action. Biotechnology joins basketball as an important source of educational cash.

Public policy, too, reflects private interest. James Dewey Watson resigned as head of the NIH Human Genome Office as a result of pressure put on him by Bernardine Healey, director of the NIH. The immediate form of this pressure was an investigation by Healey of the financial holdings of Watson or his immediate family in various biotechnology firms. But nobody in the molecular biological community believes in the seriousness of such an investigation, because everyone including Dr. Healey knows that there are no financially disinterested candidates for Watson's job. What is really at issue is a disagreement about patenting the human genome. Patent law prohibits the patenting of anything that is "natural," so, for example, if a rare plant were discovered in the Amazon whose leaves could cure cancer, no one could patent it. But, it is argued, isolated genes are not natural, even though the organism from which they are taken may be. If human DNA sequences are to be the basis of future therapy, then the exclusive ownership of such DNA sequences would be money in the bank.

Dr. Healey wants the NIH to patent the human genome to prevent private entrepreneurs, and especially foreign capital, from controlling what has been created with American public funding. Watson, whose family is reported to have a financial stake in the British pharmaceutical firm Glaxo, has characterized Healey's plan as "sheer lunacy," on the grounds that it will slow down the acquisition of sequence information.[6] (Watson has denied any conflict of interest.) Sir Walter Bodmer, the director of the Imperial Cancer Research Fund and a major figure in the European genome organization, spoke the truth that we all know lies behind the hype of the Human Genome Project when he told the *Wall Street Journal* that "the issue [of ownership] is at the heart of everything we do."

The study of DNA is an industry with high visibility, a claim on the public purse, the legitimacy of a science, and the appeal that it will alleviate individual and social suffering. So its basic ontological claim, of the dominance of the Master Molecule over the body physical and the body politic, becomes part of general consciousness. Evelyn Fox Keller's chapter in *The Code of Codes* brilliantly traces the percolation of this consciousness through the strata of the state, the universities, and the media, producing an unquestioned consensus that the model of cystic fibrosis is a model of the world. Daniel Koshland, the editor of *Science*, when asked why the Human Genome Project funds should not be given instead to the homeless, answered, "What these people don't realize is that the homeless are impaired. . . . Indeed, no group will benefit more from the application of human genetics."[7]

Beyond the building of a determinist ideology, the concentration of knowledge about DNA has direct practical social and political consequences, what Dorothy Nelkin and Laurence Tancredi call "The Social Power of Biological Information." Intellectuals in their self-flattering wish-fulfillment say that knowledge is power, but the truth is that knowledge empowers only those who have or can acquire the power to use it. My possession of a Ph.D. in nuclear engineering and the complete plans of a nuclear power station would not reduce my electric bill by a penny. Thus with the information contained in DNA, there is no instance where knowledge of one's genes does not further concentrate the existing relations of power between individuals and between the individual and institutions.

When a woman is told that the fetus she is carrying has a 50 percent chance of contracting cystic fibrosis, or for that matter that it will be a girl although her husband desperately wants a boy, she does not gain additional power just by having that knowledge, but is only forced to make decisions and act within the confines of her relation to the state and her family. Will her husband agree to or demand an abortion, will the state pay for it, will her doctor perform it? The slogan "a woman's right to choose" is a slogan about conflicting relations of power, as Ruth Schwartz Cowan makes clear in her essay "Genetic Technology and Reproductive Choice: An Ethics for Autonomy" in *The Code of Codes*.

Increasingly, knowledge about the genome is becoming an element in the relation between individuals and institutions, generally adding to the power of institutions over individuals. The relations of individuals to the providers of health care, to the schools, to the courts, to employers are all affected by knowledge, or the demand for knowledge, about the state of one's DNA. In the essays by Henry Greeley and Dorothy Nelkin in *The Code of Codes,* and in much greater detail and extension in *Dangerous Diagnostics,* the struggle over biological information is revealed. The demand by employers for diagnostic information about the DNA of prospective employees serves the firm in two ways. First, as providers of health insurance, either directly or through their payment of premiums to insurance companies, employers reduce their wage bill by hiring only workers with the best health prognoses.

Second, if there are workplace hazards to which employees may be in different degrees sensitive, the employer may refuse to employ those it judges to be sensitive. Not only does such employment exclusion reduce the potential costs of health insurance, it shifts the responsibility of providing a safe and healthy workplace from the employer to the worker. It becomes the worker's responsibility to look for work that is not threatening. After all, the employer is helping the workers by providing a free test of susceptibilities and allowing them to make more informed choices of the work they would like to do. Whether other work is available at all, or worse paid, or more dangerous in other ways, or only in a distant place, or extremely unpleasant and debilitating is simply part of the conditions of the labor market. So Koshland is right after all. Unemployment and homelessness do indeed reside in the genes.

Biological information has also become critical in the relation between individuals and the state, for DNA has the power to put a tongue in every wound. Criminal prosecutors have long hoped for a way to link accused persons to the scene of a crime when there are no fingerprints. By using DNA from a murder victim and comparing it with DNA from dried blood found on the person or property of the accused, or by comparing the accused's DNA with DNA from skin scrapings under the fingernails of a rape victim, prosecutors attempt to link criminal and crime. Because of the polymorphism of DNA from individual to individual, a definitive identification is, in principle, possible. But, in practice, only a bit of DNA can be used for identification, so there is some chance that the accused will match the DNA from the crime scene even though someone else is in fact guilty.

Moreover, the methods used are prone to error, and false matches (as well as false exclusions) can occur. For example, the FBI characterized the DNA of a sample of 225 FBI agents and then, on a retest of the same agents, found a large number of mismatches. Matching is almost always done at the request of the prosecutor, because tests are expensive and most defendants in assault cases are represented by a public defender or court-appointed lawyer. The companies who do the testing have a vested commercial interest in providing matches, and the FBI, which also does some testing, is an interested party.

Because different ethnic groups differ in the frequency of the various DNA patterns, there is also the problem of the appropriate reference group to which the defendant is compared. The identity of that reference group depends in complex ways on the circumstances of the case. If a woman who is assaulted lives in Harlem near the borderline between black, Hispanic, and white neighborhoods at 110th Street, which of these populations or combination of them is appropriate for calculating the chance that a "random" person would match the DNA found at the scene of the crime? A paradigm case was tried last year in Franklin County, Vermont. DNA from bloodstains found at the scene of a lethal assault matched the DNA of an accused man. The prosecution compared the pattern with population samples of various racial groups, and claimed that the chance that a random person other than the accused would have such a pattern was astronomically low.

Franklin County, however, has the highest concentration of Abenaki Indians and Indian-European admixture of any county in the state. The Abenaki and Abenaki-French Canadian population are a chronically poor and underemployed sector in rural Franklin County and across the border in the St. Jacques River region of Canada, where they have been since the Western Abenaki were resettled in the eighteenth century. The victim, like the accused, was half Abenaki, half French-Canadian and was assaulted where she lived, in a trailer park where about one-third of the residents are of Abenaki ancestry. It is a fair presumption that a large fraction of the victim's circle of acquaintance came from the Indian population. No information exists on the frequency of DNA patterns among Abenaki and Iroquois, and on this basis the judge excluded the DNA evidence. But the state could easily argue that a trailer park is open to access from any passerby and that the general population of Vermont is the appropriate base of comparison. Rather than objective science we are left with intuitive arguments about the patterns of people's everyday lives.

The dream of the prosecutor, to be able to say, "Ladies and gentlemen of the jury, the chance that someone other than the defendant could be the criminal is one in 3,426,327" has very shaky support. When biologists have called attention to the weaknesses of the method in court or in scientific publications they have been the objects of considerable pressure. One author was called twice by an agent of the Justice Department, in what the scientist describes as intimidating attempts to have him withdraw a paper in press.[8] Another was asked questions about his visa by an FBI agent attorney when he testified, a third was asked by a prosecuting attorney how he would like to spend the night in jail, and a fourth received a fax demand from a federal prosecutor requiring him to produce peer reviews of a journal article he had submitted to the *American Journal of Human Genetics*, fifteen minutes before a fax from the editor of the journal informed the author of the existence of the reviews and their contents. Only one author of the books cited, Christopher Wills, discusses the forensic use of DNA, and he has been a prosecution witness himself. He is dismissive of the problems and seems to share with prosecutors the view that the nature of the evidence is less important than the conviction of the guilty.

Both prosecutors and defense forces have produced expert witnesses of considerable prestige to support or question the use of DNA profiles as a forensic tool. If professors from Harvard disagree with professors from Yale (as in this case), what is a judge to do? Under one legal precedent, the so-called "*Frye* rule," such a disagreement is cause for barring the evidence which "must be sufficiently established to have gained general acceptance in the particular field in which it belongs."[9] But all jurisdictions do not follow *Frye*, and what is "general acceptance" anyway? In response to mounting pressure from the courts and the Department of Justice, the National Research Council (NRC) was asked to form a Committee on DNA Technology in Forensic Science, to produce a definitive report and recommendations. They have now done so, adding greatly to the general confusion.[10]

Two days before the public release of the report, the *New York Times* carried a front-page article by one of its most experienced and sophisticated science reporters, announcing that the NRC Committee had recommended that DNA evidence be barred from the courts. This was greeted by a roar of protest from the committee, whose chairman, Victor McKusick of Johns Hopkins University, held a press conference the next morning to announce that the report approved of the forensic use of DNA substantially as it was now practiced. The *Times*, acknowledging an "error," backed off a bit, but not much, quoting various experts who agreed with the original interpretation. A member of the committee was quoted as saying he had read the report "fifty times" but had not intended to make the criticisms as strong as they actually appeared in the text.

One seems to have hardly any other choice but to read the report for oneself. As might be expected the report says, in effect, "none of the above," but in substance it gives prosecutors a pretty tough row to hoe. Nowhere does the report give wholehearted support to DNA evidence as currently used. The closest it comes is to state:

> The current laboratory procedure for detecting DNA variation. . . is *fundamentally* sound [emphasis added]. . . .
> It is now clear that DNA typing methods are a most powerful adjunct to forensic science for personal identification and have immense benefit to the public.

and further that

> DNA typing is capable, *in principle,* of an extremely low inherent rate of false results [emphasis added].

Unfortunately for the courts looking for assurances, these statements are immediately preceded by the following:

> The committee recognizes that standardization of practices in forensic laboratories in general is more problematic than in other laboratory settings; stated succinctly, forensic scientists have little or no control over the nature, condition, form, or amount of sample with which they must work.

Not exactly, on the one hand, the ringing endorsement suggested by Professor McKusick's press conference. On the other hand, no statements call for the outright barring of DNA evidence. There are, however, numerous recommendations which, taken seriously, will lead any moderately businesslike defense attorney to file an immediate appeal of any case lost on DNA evidence. On the issue of laboratory reliability the report says:

> Each forensic-science laboratory engaged in DNA typing must have a formal, detailed quality assurance and quality-control program to monitor work.

and

Quality-assurance programs in individual laboratories alone are insufficient to
ensure high standards. External mechanisms are needed. . . .
      Courts should require that laboratories providing DNA typing evidence
have proper accreditation for each DNA typing method used.

The committee then discusses mechanisms of quality control and accreditation in greater detail. Since no laboratory currently meets those requirements and no accreditation agency now exists, it is hard to see how the committee's report can be read as an endorsement of the current practice of presenting evidence. On the critical issue of population comparisons the committee actually uses legal language sufficient to bar any of the one-in-a-million
claims that prosecutors have relied on to dazzle juries:

Because it is impossible or impractical to draw a large enough population to
test directly calculated frequencies of any particular profile much below 1 in
1,000, there is not a sufficient body of empirical data on which to base a claim
that such frequency calculations are reliable or valid.

"Reliable" and "valid" are terms of art here, and Judge Jack Weinstein,
who was a member of the committee, certainly knew that. This sentence
should be copied in large letters and hung framed on the wall of every public defender in the United States. On balance, the *New York Times* had it
right the first time. Whether by ineptitude or design the NRC Committee
has produced a document rather more resistant to spin than some may
have hoped.
      In order to understand the committee's report, one must understand the
committee and its sponsoring body. The National Academy of Sciences is a
self-perpetuating honorary society of prestigious American scientists, founded during the Civil War by Lincoln to give expert advice on technical matters.
During the Great War, Woodrow Wilson added the National Research
Council as the operating arm of the Academy, which could not produce from
its own ranks of eminent ancients enough technical competence to deal with
the growing complexities of the government's scientific problems. Any arm of
the state can commission an NRC study and the present one was paid for by
the FBI, the NIH Human Genome Center, the National Institute of Justice,
the National Science Foundation, and two nonfederal sources, the Sloan
Foundation and the State Justice Institute.

Membership in study committees almost inevitably includes divergent prejudices and conflicts of interest. The Forensic DNA Committee included people who had testified on both sides of the issue in trials and at least two members had clear financial conflicts of interest. One was forced to resign near the end of the committee's deliberations when the full extent of his conflicts was revealed. A preliminary version of the report, much less tolerant of DNA profile methods, was leaked to the FBI by two members of the committee, and the Bureau made strenuous representations to the committee to get them to soften the offending sections. Because science is supposed to find objective truths that are clear to those with expertise, NRC findings do not usually contain majority and minority reports, and in the present case a lack of unanimity would be the equivalent of a negative verdict. So we may expect reports to contain contradictory compromises among contending interests, and public pronouncements about a report may be in contradiction to its effective content. *DNA Technology in Forensic Science* in its formation and content is a gold mine for the serious student of political science and scientific politics.

There is no aspect of our lives, it seems, that is not within the territory claimed by the power of DNA. In 1924, William Bailey published a *Washington Post* article about "Radithor," a radioactive water of his own preparation, under the headline, "Science to Cure All the Living Dead. What a Famous Savant Has to Say About the New Plan to Close Up the Insane Asylums, Wipe Out Illiteracy, and Make Over the Morons by His Method of Gland Control."[11] Nothing was more up-to-date in the 1920s than a combination of radioactivity and glands. Famous savants, it seems, still have access to the press in their efforts to sell us, at a considerable profit, the latest concoction.

Epilogue

The promise of great advances in medicine, not to speak of our knowledge of what it is to be human, is yet to be realized from sequencing the human genome. Although the DNA carrying the normal form of a gene has been put into the bodies of people suffering from a variety of genetic disorders, there is not a single case of successful gene therapy in which a normal form of a gene has become stably incorporated into the DNA of a patient and has taken over the function that was defective. There was, for example, an early report that normal DNA sprayed into the lungs of a cystic fibrosis patient was taken up by cells and resulted in partial recovery, but the optimism was premature. An alternative method has been to graft genetically normal cells or tissue into a patient in the hope that the cells will proliferate and take over normal function. A case has been reported of a considerable lowering of cholesterol level in a patient suffering from an extreme form of inherited hypercholesterolemia after liver cells with the normal form of the gene were implanted. Unfortunately, the lowered level was still pathologically high and we await news of further progress. There seems no fundamental reason why such methods should not work sometimes, but the trick has not yet been discovered. Over and over, reports of first isolated successes of some form of DNA therapy appear in popular media, but the prudent reader should await the *second* report before beginning to invest either psychic or material capital in the proposed treatment.

One of the issues raised around the original Human Genome Project was that it seemed to pay no attention to the known genetic variation from individual to individual and from group to group. Whose genome was going to be represented in the human genome? As a result of agitation around this issue a small fraction of the budget of the project was diverted to studying genetic variation. One outcome was the formation of the Human Genome Diversity Project, a cooperative project of a number of human geneticists led by L. L. Cavalli-Sforza of Stanford University, to characterize genetic variation across the species. Originally the intent was to obtain a picture of the genetic patterns in a great diversity of small or disappearing populations, but it was protested that such a study was all very well for anthropologists but not for a random sample of humanity who are mostly living in the densely populated regions. The other issue was that sampling of a variety of indigenous populations from around the world was a form of exploitation of these people with no advantage to them at all, despite the value that technologically advanced countries might reap from the knowledge. A combination of these two objections resulted finally in the abandonment of the project.

But even then the main problems posed for the Genome Project by genetic polymorphism are not solved. We will still not know whether the bit of genome sequenced from a particular donor carries one copy of a defective sequence. We will still not know, from comparing sequences from a large number of sick and well people, which of the many nucleotide differences between them is responsible for the abnormality. That is not to say that a diversity project would be useless. It would greatly increase the observed repertoire of DNA sequences carried by well and by sick people and so help us to avoid being led astray from too narrow a base of comparison. For example, there are over two hundred different nucleotide changes, any of which can cause hemophilia. Most of these have been discovered by sequencing the relevant gene in people from different regions of the world. The genetic array of hemophilias in Calcutta is not the same as in Germany. Thus, the study of diversity could provide us with the raw material we need to understand what makes a hemophiliac, but in the end the molecular biology of the gene and protein must be explored. That is, we need to understand how the different nucleotide changes cause a deficiency or absence of the needed clotting protein, or, if the protein is present but abnormal in its structure, how such structural variation interferes with the clotting reaction. Knowing that a gene variation is at the root of disorder is useless unless it is possible to provide a story of physical mediation that can be translated into therapeutic action.

The main developments in genome research have revolved around the generation of the sequence information itself, and the application of that information to the production of pharmaceutical treatments. Just as for cloning, the course of human genome research cannot be understood outside the context of commercial interest.[12]

The Human Genome Project, funded by the NIH and the Department of Energy, had commercial competition. Early in the project Craig Venter, one of the cleverer participants, fell out with the directors over a strategic issue. Of the three billion nucleotides in the human genome, it is estimated that only about 5 percent are really in genes that code for proteins used by the organism. The remaining 95 percent are said to be in "junk" DNA without function, which is to say that nobody happens to know if it has a function. If it really is junk, as Venter pointed out not unreasonably, then sequencing it should be a secondary objective for a project whose claim to legitimacy is to cure human disease and understand human nature. He proposed that the Human Genome project could save a lot of time and money using a method of his invention that would pick out only the genic DNA. When the directors of the project disagreed, he quit and set up business for himself.

Venter changed his mind about what was worth doing. His Institute for Genomic Research joined with a scientific-instrument producer, the Perkin-

Elmer Corporation, to sequence the entire genome, junk and all, using hundreds of newly designed automated sequencers. It was estimated that they would become available on the open market for a mere $300,000 apiece. The total projected cost was only about $250 million and the total time needed was originally estimated to be three years, if the robots really worked. In March 1999 the competition between the public and private sequencing projects was intensified by the announcement that the public project intended to finish 90 percent of the sequence by the spring of 2000, while Venter's timetable was still aiming at completion in the middle of 2001. In the end both the public and private consortia announced their sequences simultaneously by prior agreement in February 2001.

There is a good deal more at stake than the profit from some machines or a contract to determine the sequence. Since the early 1990s the courts have held that a gene sequence is patentable, even though it is a bit of a natural organism. (At the end of 1998, the CEO of one genome company, Human Genome Sciences, a former professor at the Harvard Medical School, wrote that his corporation had filed over 500 gene patent applications.)[13] One value of a patent on a gene sequence lies in its importance in the production of targeted drugs, either to make up for the deficient production from a defective gene or to counteract the erroneous or excessive production of an unwanted protein. In the first instance, the protein coded by the gene may itself be the drug, in which case it can be produced by transferring the gene to a bacterium or other cell, and growing the protein in mass quantities in fermenters. A classic example is the production of human insulin to supplement the lack of its normal production in diabetics. Alternatively, the cell's production of a protein coded by a particular gene, or the physiological effect of the genetically encoded protein, could be affected by some molecule synthesized in an industrial process and sold as a drug. The original design of this drug and its ultimate patent protection will depend upon having rights to the DNA sequence that specified the protein on which the drug acts. Were the patent rights to the sequence in the hands of a public agency like the NIH, a drug designer and manufacturer would have to be licensed by that agency to use the sequence in its drug research, and even if no payment were required the commercial user would not have a monopoly but would face possible competition from other producers.

A promising case of a drug developed from a knowledge of the genetic control of protein synthesis is Herceptin, registered, produced, and marketed by

Genentech for the treatment of breast and ovarian cancers. One form of these cancers is the consequence of the duplication of the HER-2 gene, which results in the overproduction of a protein that greatly stimulates cell division. Herceptin is an antibody molecule that specifically blocks this stimulation of cell division. It remains to be seen how profitable Herceptin will be, but the present value of possessing such a drug is estimated at about five billion dollars.[14]

Currently a number of genomics companies are involved in possible drug production, in collaboration with major pharmaceutical companies. None has yet made any money by selling a drug based on genome sequencing, but their prospectuses all predict a profit soon. Before a pharmaceutical company can make any money on the production and sale of a drug, clinical trials must satisfy both medical practitioners and the FDA that a drug is both effective and safe, and even then the costs of production and marketing may exceed what can be taken in. There is also the possibility of commercial success in diagnostic testing, but it remains in the future. For example, using the DNA sequence, a test has been developed for the BRCA1 mutation that is involved in a small fraction of breast cancers.

It may turn out, in the end, that the providers of capital have been as deluded by the hype of the human genome as has anyone else. Judging by the results so far the prudent investor may be better off spending a week at Saratoga. Only a foolhardy person would predict that no gene therapies will ever be commercially successful. Even at Saratoga long shots pay off once in a while.

It was impossible to say in 1992 how far the Human Genome Project or drug therapies based on it would have developed in seven years. What became clear very quickly, however, was the future of the forensic applications of DNA technology. The report of the National Academy of Sciences was doomed to the dustbin. At first the Department of Justice and other law enforcement agencies were quite happy with the report because it gave a generalized approval, in principle, to the use of DNA profiles in identification. But more and more courts began to rule DNA evidence inadmissible when the detailed analysis of the report was brought out at trials. The problem of genetic differences between ethnic groups was particularly damaging to prosecution calculations of how likely the crime scene DNA was to match a random innocent person. It soon became obvious that prosecutorial agencies were going to press for some action that would validate DNA evidence. And so they did. The National Academy of Sciences, through its subsidiary, the

National Research Council, is obliged to carry out any inquiry for which it has competence when that inquiry is requested and paid for by an entity of the federal government. The result is that it is sometimes asked to visit the same question again if the clients are not satisfied with the first outcome. The most notorious case was a report showing that high-protein dog food was bad for pets' kidneys, a result that was unsatisfactory to a leading producer then engaged in a high-pressure advertising campaign for its high-protein dog diet. The political clout of the dog food cannery was sufficient that three successive reports were called for, all unsatisfactory, before the company and their government representative finally gave up.

With the dog food case as a procedural precedent, the director of the FBI asked for a new report on forensic DNA in 1993 and money was also provided by other interested agencies. There was no great problem in predicting the outcome of the committee's deliberations once the membership was known, since by 1993 everyone in the field had expressed a clear view on the matter. I wrote to the president of the National Academy offering to save everyone a lot of time and money by writing the report if he would just send me the list of the committee members, but he did not take my suggestion. Before the committee had even met, the chairman, an eminent geneticist, gave a speech at a meeting of the Forensic Science Association, in which he assured the representative of the FBI that everything would turn out well. The two main issues of contention, quality control of crime laboratories and the difficulties that laypersons have in understanding probability statements, were neatly finessed in the report. All laboratories that sequence DNA have problems of cross-contamination between samples. This becomes particularly acute when a minute sample of DNA, say from a scraping of a bit of dried blood, is compared against a large sample of blood taken from a suspect. If these are not handled with great care and attention, DNA from the large sample may wind up contaminating the small one. Moreover, a lot of DNA comparison is not done in the relatively sophisticated central crime laboratory of the FBI but in local state and county forensic facilities. The FBI laboratory itself had repeatedly refused to allow independent assessors to observe their procedures or submit to blind proficiency tests. Yet the best the committee could recommend was that "laboratories should adhere to high quality standards . . . and make every effort to be accredited for DNA work."[15] Well, perhaps not *every* effort.

As to the problem of juror comprehension of probability statements, the recommendation was that "behavioral research should be carried out to identify any conditions that might cause a trier of fact to misinterpret evidence on DNA profiling and to assess how well various ways of presenting expert testimony on DNA can reduce any such misunderstandings." This recommendation neatly ignores the already extensive literature showing that laypersons often misunderstand probability statements even when they are presented in a one-to-one interview. So, for example, studies funded by the NIH of the results of genetic counseling found that couples who were told that they had one chance in four of having an affected child would sometimes respond that they were not worried because they were only having two children.

As might be expected, with the new report in hand prosecutorial agencies no longer had problems in court with the admissibility of DNA evidence. However, one of the unforeseen results of the present general admissibility of DNA evidence, unforeseen by eager prosecutors, is the important role that DNA typing has had in the exculpation of accused people, including large numbers who have served prison time for crimes they did not commit. The Innocence Project has now used DNA recovered from crime scenes and samples taken from accused people to free more than three hundred falsely convicted victims of a system that has depended strongly on picking out people in police lineups.

26

Does Culture Evolve?

In his well-known essay "Two Cultures," C. P. Snow reported a gap between the literary and natural-scientific cultures. Acknowledging that "a good deal of the scientific feeling" is shared by some of his "American sociological friends," Snow was well aware that there was a degree of artificiality in limiting the number of cultures to the "very dangerous" one of two. Yet, he based his binarist decision largely on the cohesion of the natural-scientific and literary communities that made of them cultures "not only in an intellectual but also in an anthropological sense."[1] The intellectual division of labor and the development of disciplinary languages certainly seem to substantiate his reference to two incommensurate cultures. Anyone who has sat on a university committee reviewing grant proposals from, and consisting of citizens of, each of the cultures must have observed the pattern of who accuses whom of using jargon and be convinced that at least the academic version of Snow's gap, that between the humanities and the natural sciences, has widened into a seemingly unbridgeable abyss. It has become commonplace that the two cultures have nothing in common.

Perhaps, however, too much has been made of this abyss. Members of the literary culture, and of the humanities in general, may be appalled at the thought of scientists mucking around on cultural terrain and subjecting it to "scientific analysis." But natural scientists seem more irritated than intimidated by the apparent independence of human culture from scientific study. And social scientists expressing their discontent about being dangled over the abyss helped prompt Snow to take "A Second Look" and to acknowledge the "coming" of a "third" social-scientific culture with the potential to "soften" the communication difficulties between the other two.[2] Cultural anthropologists, moreover, at least those with a "scientific" rather than a "relativist" bent, could point to a long tradition in their discipline of attempting to bridge the abyss by subjecting culture and its "evolution" to scientific study.

The idea that culture evolves antedated the Darwinian theory of organic evolution and, indeed, Herbert Spencer argued in support of Darwin that, after all, everything else evolves.[3] Of course, the validation of the theory of organic evolution has in no way depended on such argument by generalization. It is Darwinism that became the theory of evolution, and, standing Spencer on his head, an inspiration for theories of cultural evolution since 1859. There has been a long and bloody Hundred Years' War among cultural anthropologists over whether human culture can be said to evolve, a war in which the contending parties alternate in their periods of hegemony over the contested territory. That struggle has, in part, been a philosophical consequence of a diversity in the understanding of what distinguishes an evolutionary from a "merely" historical process. In greater part, however, it can only be understood as a confrontation between the drive to scientize the study of culture and the political consequences that seem to flow from an evolutionary understanding of cultural history.

Until the last decade of the nineteenth century, partly under the influence of Darwinism, but also as an extension of pre-Darwinian progressivist views that characterized a triumphant industrial capitalism, anthropological theory was built on an ideology of evolutionary progress. Lewis Henry Morgan's construal of the history of culture as the progress from savagery through barbarism to civilization was the model. In the 1890s Boas successfully challenged the racism and imperialism that seemed the inevitable consequences of Morgan's progressivist views and set an anti-evolutionist tone which characterized cultural anthropology until after the Second World War. Beginning with the celebration in 1959 of the hundredth anniversary of the publication of *Origin of Species,* there was a demand from within anthropology to reintroduce an evolutionary perspective into the cultural history from which it had been purged by the Boasites, a demand later given collateral support by the development within biology of sociobiological theories of human nature. But again the implication that there were "higher" and "lower" stages of human culture, an implication that seemed built into any evolutionary theory, could not survive its political consequences, and so by 1980 cultural anthropology once again returned to its Boasian model of cultural change, cultural differentiation, and cultural history, but without cultural evolution.

In his Preface to the manifesto of cultural evolution *redivivus, Evolution and Culture,* Leslie White bitterly attacked the Boas tradition, conflating it with general creationist anti-evolutionism:

> The repudiation of evolutionism in the United States is not easily explained. Many nonanthropological scientists find it incredible that a man who has been hailed as "the world's greatest anthropologist" . . . , namely Franz Boas, a man who was a member of the National Academy of Sciences and President of the American Association for the Advancement of Science, should have devoted himself assiduously and with vigor for decades to this antiscientific and reactionary pursuit.[4]

But why does White insist, illogically and counterfactually, that a denial of cultural evolution is anti-evolutionism *tout court?* There is a hint in the word "antiscientific," but all is explicitly revealed two pages later: "The return to evolutionism was, of course, inevitable *if . . . science was to embrace cultural anthropology.* The concept of evolution has proved itself to be too fundamental and fruitful to be ignored indefinitely by *anything calling itself a science"* (emphasis added).[5] Thus the demand for a theory of cultural evolution is really a demand that cultural anthropology be included in the grand twentieth-century movement to scientize all aspects of the study of society, to become validated as a part of "social science." The issue was particularly pressing for cultural anthropologists because they were engaged in an institutional struggle for support of their research and academic prestige with members of their own academic departments who practiced the undoubtedly scientific activity of physical anthropology.

But the demand for a theory of cultural evolution also arose from among the natural sciences, particularly among evolutionary biologists for whom the ability to explain all properties of all living organisms, using a common evolutionary mechanism, is the ultimate test of the validity of their science. Ever scornful of what they acronymiously dubbed the SSSM (the "standard social

science model" based on Durkheim's axiom), evolutionary biologists doubt-
ed not that the scientific analysis and understanding of the place and evolution of culture in the life history of *Homo sapiens* was properly the province of
students of human evolution. The advent of culture was, after all, a biological
adaptation and it must therefore be explicable by biological science. Yet a
combination of two inhibiting factors kept the forays of evolutionary biologists into the cultural realm to a minimum at least from the end of the Second
World War into the mid-1970s. These were the close link between biologically based pseudoscientific social and cultural theories and genocide, and the
lack of a properly comprehensive theory. This latter problem, as most recent
cultural evolutionists agree, was finally solved with the concluding chapter of
E. O. Wilson's *Sociobiology*, which provided the impetus for the latest round
of attempts to subject human history to evolutionary explanation. There,
Wilson sketched the certainty that, as he put it a few years later in *On Human
Nature,* the appropriate instrument for closing the "famous gap between the
two cultures" is "general sociobiology, which is simply the extension of population biology and evolutionary theory to social organization."[6]

Though rather adamant about their scientific right to explain not just the
evolution of human cultural capacities but also cultural evolution, biologists
are also rather uneasy about their self-imposed obligation to do so. For they
wager the *raison d'être* of science on establishing the validity of the principle
of reductionism: in order for science to remain tenable, it must have universal
explanatory power; and this means "nesting" the human sciences in the great
hierarchy of sciences. If evolutionary biology cannot explain human culture,
then perhaps its explanations of other phenomena ought to be reexamined.
Intrigued by the challenge, Wilson noted that reduction is "feared and resent-
ed" by too many in the human sciences,[7] and, in a bold Napoleonic metaphor,
he sniffed "a not unpleasant whiff of grapeshot" in the thought that the applicability of sociobiology to human beings is a battle on which hangs the fate of
"conventional evolutionary theory."[8] Thrilled by the challenge and inspired
by the apparent potential of the sociobiological synthesis, an increasing number of scientists attempted to build on Wilson's blueprint in order to bridge
the abyss and lay claim to the territory on the other side.

Some members of the social sciences, those who preferred to be recognized as *bona fide* scientists and not just as members of a "third" culture, were
meanwhile growing uneasy over the proliferation of opposing theories and

models that had apparently brought the production of social-scientific knowledge to a standstill. Such social scientists began to question their own SSSM and turned increasingly to the new and seemingly infallible sociobiological synthesis for the models and explanatory mechanisms that would put their own disciplines on proper scientific footing. Alexander Rosenberg, for example, bemoans the inability of the social sciences to live up to John Stuart Mill's hope for them, namely, to be based on explanatory laws. In a telling formulation he claims that

> the social sciences would be of only passing interest, only entertaining diversions. like an interesting novel or an exciting film, unless they too stood the chance of leading to the kind of technological achievements characteristic of natural science. For a social science conceived as anything less practical in ultimate application would simply not count as knowledge, on my view. And if it does not count as *knowledge*, disputes about its methods and concepts are no more important than learned literary criticism or film reviews are to our uninformed enjoyment of the books and movies we like.[9]

Rosenberg expects this to be rectified as soon as the social sciences are treated as life sciences, and he optimistically predicts that the study of human behavior, once set on a biological footing, "will admit of as much formally quantified and mathematical description as the most mathematical economist could hope for." Against all claims for their uniqueness he insists that the traditional social sciences have been "superseded" by, and will only become truly scientific when subsumed under, sociobiology.[10]

More recently, anthropologist John Tooby and psychologist Leda Cosmides have also chastised the social sciences for their "self-conscious stance of intellectual autarky"; their "disconnection from the rest of science has left a hole in the fabric of our organized knowledge where the human sciences should be." The lack of progress in the social sciences has been caused by their "failure to explore or accept their logical connections to the rest of the body of science—that is, to causally locate their objects of study inside the larger network of scientific knowledge."[11]

This desideratum is the cornerstone of the journal *Politics and the Life Sciences*, whose editors and contributors insist that the social sciences must be nested within the life sciences. The hopes for a synthesis implicit in the journal's name were expressed by Richard Shelly Hartigan in a flattering review of Richard D. Alexander's *The Biology of Moral Systems*. Predicting marital bliss, Hartigan confidently asserts that "the lengthy divorce of the natural from the human sciences is about to end with reunion. Though the nuptials may be delayed awhile, the parties are at least getting to know each other again more intimately."[12] The reunion consists of articles devoted to the "Darwinian" explanation of such topics as social alienation, the nuclear arms race, the legal process, social stratification, oral argument in the Supreme Court, the relation between human intelligence and national power, and even feminism.[13]

These examples could be multiplied, but as this brief overview indicates, the biggest engineering project attempting to bridge the gap between the cultures of the natural and the human sciences over the last few decades has been initiated by natural scientists, anxious perhaps about having wagered their *raison d'être* on the success of their imperialist venture; and it has quickly drawn the participation of those social scientists optimistic about overcoming their inferiority complex and gaining respectability by grounding their own disciplines in the natural sciences. The bridge itself is the concept of "cultural evolution" whose scientific girders are the categories and explanatory laws either directly borrowed or derived from a narrowly selectionist approach to the study of biological evolution.

At the outset we must make clear what the issue of cultural evolution is *not* about. First, there is no question that culture as a phenomenon has evolved from the absence of culture as a consequence of biological change. Whether other primates have culture on some definition, the insectivores, from which the primates evolved, do not, so at some stage in biological evolution culture appeared as a novelty. Second, no one challenges the evident fact that human cultures have changed since the first appearance of *Homo sapiens,* but not even the most biologistic theory proposes that major changes within the phenomenon of culture—say, the invention of an alphabet or of settled agriculture—was a consequence of genetic evolution of the human central nervous system. Human culture has had a history, but to say that culture is a consequence of a historical process is not the same as saying that it evolves. What constitutes an evolutionary process as opposed to a "merely" historical one? What explanatory work is done by claiming that culture has evolved?

Leslie White's *cri de coeur* accusing the Boasians of aligning themselves with anti-evolutionist creationism confounds two quite different issues. The mid-nineteenth-century struggle against evolution, mirrored in modern Christian creationism, was not over whether the succession of life-forms from earlier times to the present has some law-like properties that give some shape to that history. Rather it was, and remains, a denial that organismic forms have had a history at all, that there has been significant change in species and that present-day life-forms arose from others quite unlike them. But no one denies that culture has had a history, that industrial production arose from societies that were at a previous time pastoralist and agricultural. Not even the most literal of fundamentalists thinks that God created the motorcar on the sixth day. Ironically, it is a form of traditional Christianity that simultaneously denies an intelligible history to organic life as a whole while asserting a directionality to human history, the ascent toward final redemption from the depths of the Fall.

White's identification of the struggle over cultural evolution with the struggle over organic evolution, if it is more than a deliberate piece of propaganda in a battle for academic legitimacy, is really a struggle over the nature of historical processes. At base, it is meant to be a rejection of the proposition that human cultural history is just one damn thing after another, claiming that, on the contrary, there is an underlying nomothetic process. But in asserting the claim that culture evolves White claimed more than what was necessary. History may indeed be law-like in some sense, but does that make a historical process evolutionary? There may be law-like constraints on historical change like Ibn Khaldun's rule that "Bedouins can gain control only over flat territory," but we do not therefore characterize the *Muqaddimah* as providing an "evolutionary" theory of history, any more than Hegel's third kind of history, the philosophical, is claimed to be a theory of evolution.[14]

It might be asserted that for theories to qualify as evolutionary they must consist of more than mere constraints and prohibitions; rather, they must be characterized by generative laws or mechanisms whose operations produce the actual histories. But the *Muqaddimah* offers laws of the origin, transformation, differentiation, and eventual extinction of political formations: "Dynasties of wide power and large royal authority have their origin in religion based either on prophethood or truthful propaganda"; "The authority of the dynasty at first expands to its limit and then is narrowed down in successive stages, until the dynasty dissolves and disappears"; "With regard to the amount of prosperity and business activity in them, cities and towns differ in accordance with the different size of their population."[15] These are not simply empirical generalizations. Each is derived as the necessary consequence of basic properties of human motivation, just as the war of all against all is derived by Hobbes from the basic assumptions that human beings are, by nature, self-expanding in their demands and that the resources for their expansion are limited. The ease with which the concept of the "evolution of culture" has been employed in anthropology and human evolutionary biology finds no parallel in the discourse of contemporary historians. When François Furet and Mona Ozouf write, in their Preface to *A Critical Dictionary of the French Revolution*, that "ignoring the evolution of historiography means overlooking an important aspect of the event itself," they mean only that historiography has changed, that is, that it has had a history.[16]

It might be that "evolution" and "history" are meant to be separated by questions of scale and grain. Modes of production, familial and other group relationships, forms of political organization, levels of technology are seen as general properties of human social existence. They are also "culture" and they are said to "evolve" whereas spatio-temporally individualized sequences, like the events in France from the Estates General to Thermidor, are only instantiations of classes of cultural phenomena, schemata that are repeated in different places and at different times. So Leslie White makes the distinction between the particularity of micro (historical) events and the generality of macro (evolutionary events): "I should like to call the temporal particularizing process, in which events are considered significant in terms of their uniqueness and particularity, 'history' and call the temporal generalizing process which deals with the phenomena as classes rather than particular events, 'evolution.' "[17] But if this is what is meant to discriminate evolution from mere history, then the cultural evolutionist departs radically from theories of evolution of the physical world. For Darwinism, not only organic life as a whole, but each species and each population in each species evolves. The standard model of organic evolution begins with the evolutionary forces that cause local populations to change over relatively short times, and derives the evolution of individual species in time from changes in populations that comprise them. Moreover, in its usual reductionist form, evolutionary theory explains the evolution of life as a whole as a mechanical consequence of the rise and fall of individual species. So why, if human culture evolves, has not Bedouin culture evolved, or the Middle East, or the state called Saudi Arabia?

The attempt to differentiate "cultural evolution" from "history" brings us to the edge of a different kind of abyss—one that is broader and older, though obscured by the more visible one between the human and natural sciences. This abyss cuts across established disciplinary boundaries and separates nomological and historical modes of explanation. Civil wars always inflict the deepest wounds. And the battles *within* the human sciences (between historians emphasizing contingency and particularity and social scientists insisting on general laws and models) and *within* the natural sciences (between biologists who insist on the contingency, the historicity of evolution and those who view evolution as a lawful process of selection and adaptation) are by virtue of the proximity of the antagonists frequent, intense, and have perhaps the longest lasting effects.

Snow's depiction of the abyss along disciplinary lines makes those battles appear as perhaps bitter, but nevertheless only intradisciplinary squabbles, as merely different perspectives on common problems. Yet the cross-disciplinary affinities of "historians" versus "scientists" are nowhere more evident than in the issue that both claim as their own: that which appears to one group as "cultural evolution," to the other as "human histories." The ease, for example, with which confirmed selectionists among evolutionary biologists and those social scientists similarly concerned with explanatory laws have found common cause in the concept of cultural evolution indicates that on fundamental ontological and epistemological issues there is no abyss between them. That ease finds its counterpart in the ease with which the two authors of this essay, a historian and a geneticist, agree on a historical approach to cultural change. The differences between these two perspectives are incommensurable, not because of disciplinary boundaries, but because they involve different conceptions about the nature of "scientific" inquiry, different ontological and epistemological assumptions, and accordingly different modes of explanation.

Darwinian theorists of cultural evolution universally agree that selection is *the* explanatory law, the key to explaining all "evolutionary" or "historical" developments at any sociocultural and historical coordinates. In this way human history is reduced to a unitary process, its complex dynamics to a rather singular logic, and the particularity of historical time is reduced to "empty abstract time" (Walter Benjamin).[18]

We begin with different assumptions about historical objects and, accordingly, about historical time. We view historical phenomena as particulars embedded in particular sociocultural forms, each with its own systemic properties and discrete logic of production and reproduction, its own dynamics of stasis and change. Each sociocultural form therefore has its own time and history, to borrow an appropriate phrase from Louis Althusser. Because every historical phenomenon has its own particular locus in a particular sociocultural constellation with its own concrete and particular time and history, there is no one transhistorical law or generality that can explain the dynamics of all historical change. Our contention is that cultural evolutionary theories have not been (nor will be) able to meet even their own claims to explain the past and predict the future. And this is because of the problematic assumptions about the nature of culture and the problematic conflation of historical and evolutionary processes.

The Forms of Evolutionary Theory

Models of the evolution of phenomena are traditionally models of the temporal change in the nature of ensembles of elements. The individual elements in the ensemble can be physical objects like organisms or stars or properties like size or chemical composition or syntactic structure. So when we speak of the "evolution of human beings" we mean a change in the composition of the ensemble of physical individuals that we identify individually as human, but we can as well consider the "evolution of European painting" as a change in the ensemble of materials, techniques, subjects, and design principles that characterize the production of that art. Whether it is physical objects or attributes or artifacts, it is not any individual element, but the composition of the ensemble that is at the center of interest.

Evolutionary theories as they have been constructed for the physical world and as they have been taken over into human social phenomena can be classified according to two properties. First, they may be either transformational or variational. In a transformational theory, the ensemble of elements changes in time because each of the elements in the ensemble undergoes roughly the same secular change during its individual history. That is, the evolution of the ensemble is a result of the developmental pattern of each individual. The transformational model characterized all evolutionary theories until Darwin, and has remained the model for the evolution of the physical universe since Kant and Laplace produced the Nebular Hypothesis for the origin of the Solar System. The collection of stars in the cosmos has been evolving because every star is individually undergoing an aging process from its birth at the Big

Bang, through a sequence of nuclear reactions until it exhausts its nuclear fuel and then collapses into a dead mass. It is this model that is embodied in the very word *evolution*, an unfolding or unrolling of a history that is already immanent in the object.

The alternative, invented by Darwin to explain organic evolution, is a variational evolutionary scheme. In variational evolution, the history of the ensemble is not a consequence of the uniform unfolding of individual life histories. Rather, variational evolution through time is a consequence of variation among members of the ensemble at any instant of time. Different individuals have different properties and the ensemble is characterized by the collection of these properties and their statistical distribution. The evolution of the ensemble occurs because the different individual elements are eliminated from the ensemble or increase their numbers in the population at different rates. Thus the statistical distribution of properties changes as some types become more common and others die out. Individual elements may indeed change during their lifetime, but if they do, these changes are in directions unrelated to the dynamic of the collection as a whole and on a time scale much shorter than the evolutionary history of the group. So the developmental changes that characterize the aging of every living organism are not mirrored in the evolution of the species. Every human being may become grayer and more wrinkled with age, but the species as a whole has not become so in five million years of evolution from its common ancestor with other primates. Organic evolution is then a consequence of a twofold process: the production of some variation in properties among individual elements followed by the differential survival and propagation of elements of different types. Moreover, the production of the variation is causally independent of its eventual fate in the population. That is what is meant by the claim that organic evolution is based on "random" variation. It is not that the changes in individual properties are uncaused, or the consequence of some force outside of normal physical events. Rather, it is that the forces of change internal to organisms, leading to the production of variant individuals, are causally random with respect to the external forces that influence the maintenance and spread of those variants in the population.

The invention of the variational scheme for organic evolution, with its rigorous separation of internal developmental forces from external culling forces, is the major epistemological break achieved by Darwin. All other evolutionary schemes that had been postulated until the appearance of the *Origin of the*

Species in 1859, whether of the evolution of the cosmos, of organisms, of language, or of ideas, were transformational. The Darwinian variational scheme, with its denial of the causal role of individual developmental histories was a negation of evolution as it had previously been understood. The retention of the term *evolution* by Darwinists, while stripping it utterly of its former structural implication, has led to a considerable confusion and ambiguity in subsequent arguments about cultural evolution, for there has been no agreement among cultural evolutionists about just what sort of evolution they mean.

The choice of a transformational, developmental theory of evolution implies properties of the process that are not integral to, although they may be present in a variational theory: directionality and staging. In an unfolding process the possibility of each successive transformation is dependent on the completion of a previous step of transformation to provide the initial state for the next change. It is not necessary that the complete unfolding be predictable from the very origin of the system because successive steps may be contingent. There may be more than one local unfolding possible from a given state, and these alternatives may be chosen, contingent on various external circumstances. Transformational theories, nevertheless, usually assume a very restricted contingency, putting very strong constraints on which states may succeed each other, and in what order. Indeed, the standard theory of embryonic development which provides a metaphorical basis for developmental theories of evolution assumes that there is one and only one possible succession of states. Thus, there is one direction, or at most a few alternative possible directions of change immanent in the nature of the objects. Directionality does not in itself imply that change is monotone or that there is a repeated cycling among states along some simple axis, yet again and again transformational theories take the form of a "Law of Increase of. . . ," complexity, efficiency, control over resources or energy, of Progress itself.

A variational theory, in contrast, does not have directionality built into it because the variation on which the sorting process operates is not intrinsically directional, and changes in the statistical distribution of types in the ensemble are assumed to be the consequence of external circumstances that are causally independent of the variation. Nevertheless, one-way directionality has penetrated Darwinism by means of a claim about natural selection. If the differential numerical representation of different types in a species occurs not by chance events of life and death, but because the properties of some organisms confer on them greater ability to survive and reproduce in the environment in which they find themselves, might there not be some properties that would confer a general advantage over most or all environments? Such properties, then, ought to increase across the broad sweep of organisms and over the long duration of evolutionary history, putting aside any particularities of history. So, for example, it has been claimed that complexity has increased during organic evolution, since complex organisms are supposed somehow to be able to survive better the vagaries of an uncertain world. Unfortunately no agreement can be reached on how to measure complexity independent of the explanatory work it is supposed to do. It is, in fact, characteristic of directionality theories that organisms are first arrayed along an axis from lower to higher and then a search is instituted for some property that can be argued to show a similar ordering.

From directionality it is only a short step to a theory of stages. Transformational developmental theories are usually described as a movement from one stage to the next in the sequence, from savagery to barbarism to civilization, from artisanal production to competitive industrial capitalism to monopoly capital. Development begins by some triggering, starting the process from its germ, but there are thought to be a succession of ordered stages through which each entity must pass, the successful passage through one stage being the condition for moving on to the next. Variation among individual entities then arises because there is some variation in the speed of these transitions, but primarily because of attested development, the failure to pass on to the next stage. Freudian and Piagetian theories are of this nature. It should be no surprise to anthropologists that transformational evolutionary theories of culture identify present-day hunters and gatherers as being in an attested stage of cultural evolution.

The second property that distinguishes among evolutionary schemes is the mortality of the individual objects in the ensemble. Members of the ensemble may be either immortal, or at least have potential lifetimes that are of the same order as the ensemble as a whole, or they may be mortal or at least have lifetimes significantly shorter than the duration of the entire collection whose evolution is to be explained. The lifetime of the material universe is the same as the lifetime of the longest lived of individual stars. Individual organisms, in contrast, invariably have their entrances and their exits, but the species may persist. The classification of an evolutionary system as either mortal or immortal is independent of whether it is transformational or varia-

tional and the construction of an evolutionary theory for a domain of phenomena—culture, for example—will require model assumptions about both of
these properties. Two of the schemata are illustrated by phenomena to which
the concept of evolution is commonly applied. Stellar evolution is a transformational evolution of a system composed of immortal objects; organic evolution is variational and its objects, individual organisms, are mortal. Although
we do not ordinarily think of it in such terms, an example of an evolutionary
process that is variational but whose objects are immortal is any separation of
a mixture of physical materials by sieving, as for example in panning for gold.
The lighter particles are washed away, leaving the flakes of gold behind so that
the concentration of gold becomes greater and greater as the process continues, yet the same bits of gold are present at the end of the process as at the
beginning. Pre-Darwinian theories of organic evolution were transformational, the entire species evolving as a consequence of slow directional changes in
individuals who were, nevertheless, mortal.

The mortality of the individual objects in an evolutionary process raises a
fundamental problem, namely, how the changes in the composition of the
ensemble that occur within the lifetime of short-lived elements are to be accumulated over the long-term evolution of the group. Whether the evolution is
variational or transformational, there must be some mechanism by which a
new generation of successors retains some vestige of the changes that
occurred in a previous time. In the classical vulgar example of Lamarckian
transformational evolution, if the ancestors of giraffes slightly elongated their
necks to reach up into trees, all the effort would have been wasted, for after
their deaths their offspring would need to repeat the process *ab initio*. Nor
does the variational scheme of Darwin solve the problem. Were slightly
longer-necked variant giraffes to survive better or to leave more offspring than
their short-necked companions, and so enrich the proportion of the longer
variant in the species, no cumulative change would occur over generations
unless the bias introduced by the sieving process in one generation were
somehow felt in the composition of the next. That is, it demands some mechanism of inheritance of properties, in the broadest sense. Beyond the observation that offspring had some general resemblance to their parents, neither
Darwin nor Lamarck had the benefit of a coherent theory of inheritance, so
they had to content themselves with a variety of ad hoc notions about the passage of characteristics, all of which had in common that the properties of indi-

vidual organisms were somehow directly influenced by the properties of their biological parents at the time of conception. Theorists of cultural evolution, conscious of the need for a theory of inheritance, yet deprived of any compelling evidence for particular law-like mechanisms for the transgenerational passage of cultural change, are in a much more difficult position, although they do not seem to have realized it, because they do not even know whether an actor-to-actor, not to speak of a parent-to-offspring, model of the passage of culture has any general applicability.

The Paradigms of Cultural Evolutionary Theory: Transformational Theories of Cultural Evolution

A remarkable feature of the history of attempts to create a theory of cultural evolution is the disjuncture between the powerful impetus given to those attempts by the triumph of Darwinism and the form those essays took until recently. Darwin's substitution of the variational scheme of evolution for a transformational one eliminated the need for the postulation of intrinsic directional forces driving the process of change and consequently avoided the need for a theory of progress. If directionality and its special variant, progress, are claimed to be features of a variational evolutionary scheme, they must be imported by means of a force not inherent in the variational process itself. If there is directionality, it must come from outside organisms, as a claim, for example, about the nature of environments and their histories. Differential reproduction and survival of randomly generated variants contains no intrinsic direction. Developmentalist, transformational theories of evolution, in contrast, are directional by necessity because the motive mechanism is some form of unfolding of an already immanent program.

Beginning with Edward Burnett Tylor's *Primitive Culture* (1871) and Lewis Henry Morgan's *Ancient Society* (1877), cultural evolutionary theory, called forth by the historical phenomenon of Darwinism, ignored the structure of Darwinian explanation, and remained transformational for nearly a hundred years. Nearly all of the theories of cultural evolution have had more in common with Herbert Spencer's *Progress: Its Law and Cause* (1857) than with Darwin's *Origin*. First, they have been dominated by notions of progress and direction. This accent on direction and progress has even been used to characterize organic evolution itself. In the most important manifesto of cultural evolutionism since its revival after the Second World War, *Evolution and Culture,* Marshall Sahlins provides a diagram (26.1) of the evolution, reproduced here, not of culture but of all animal life. Superimposed on the upward trend along the axis of "Levels of General Progress," identified by Sahlins as "general evolution," are minor diversifications within a level of progress, symptomatic of "specific evolution" (mere history, perhaps).[19] Whereas diagrams like this were icons of nineteenth-century evolutionism, notions of general progress in biology have been expunged from current descriptions of organic evolution. In the modern practice of reconstructing phylogenetic relationships, the antonym of "primitive" is not "advanced," but "derived."

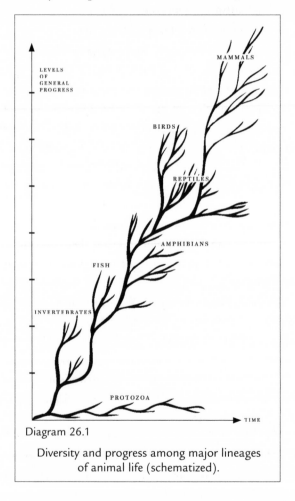

Diagram 26.1

Diversity and progress among major lineages
of animal life (schematized).

Second, given a commitment to directionality and progress, it then becomes necessary to decide what criteria should be used to determine progress aside from later as against earlier. In theories of organic evolution, recurrent attempts to use the notion of progress have foundered on this issue. It is clear from the fossil record that there has been no increase in the duration of species since the earliest record of multicellular organisms. Nor would anyone be so foolish as to predict that vertebrates will outlast the bacteria, should a major catastrophe overtake all life on earth. Increasing complexity has been a favorite of progressivist theorists both for organic evolution and for cultural and political structures, but there is no agreement among physical scientists on how complexity is to be measured and there is the recurrent danger that it will be conveniently defined, post hoc, to put *Homo sapiens* at the top. Sahlins dismisses that shibboleth of bourgeois economic theory, efficiency, as a measure on the grounds that "an organism can be more efficient than another and yet remain less highly developed."[20] By "highly developed" he means having more parts and subparts, more specialization of parts and more effective integration and, subserving these, the transformation of more total energy. Exactly how that cashes out in the great progress from fishes to reptiles in the diagram is not made clear. It is clear, however, what work is done in the domain of culture. Industrial capitalism certainly turns over more calories per capita than does the economy of the Yanamamo of the Orinocan rain forest, and almost any description of a European polity of 1999 will show it to have more parts and subparts with greater specialization than a fief in thirteenth-century Europe, although the question of the relative integration of feudal and bourgeois society as a whole can be debated. Nor can this characterization of an increasing level of cultural progress be attacked on the grounds that some earlier cultures, say Athenian democracy, as most would agree, were more progressive than Carolingian feudalism. The combination of general and specific evolution allows for local exceptions, especially if cultures in different parts of the world are undergoing independent evolutionary trajectories because accidents of geography prevent any effective contact between them or because catastrophic historical events have left a culture without a sufficient population to sustain it. It is only the long sweep of human cultural history that is meant to be progressive. The problem with such a theory is that it is hard to imagine any observation that could not be rationalized. The mere numerosity of the human species makes it impossible to return to feudal agricultural production, although a global nuclear war with a 95 percent mortality rate might do the trick. Would that be an example of specific or general cultural evolution?

Third, transformational evolution demands a mechanism or, at the very least, a set of empirical law-like regularities that are characteristic of all times and places, even if these cannot be generated from lower level mechanical principles. Transformational theories of cultural evolution, to the extent that they attempt to generate putative trends from some lower-level principles at all, usually do so from middle-level laws of the same ontological status as Ibn Khaldun's generative rules, rather than deriving them explicitly from properties of human beings and their consequent interactions in assemblages, as Hobbes did. *Evolution and Culture* provides a "Law of Cultural Dominance" that assures that more advanced cultures will spread and replace the less advanced when they come in contact, and a "Law of Evolutionary Potential" that asserts that the more specialized and adapted to local circumstances a culture is, the less likely it is to progress to a higher stage. Beyond appealing to the reasonable notion that cultures that control more energy are likely to take over those that control less, provided they do not destroy themselves in the meantime, and the rather more ideological prejudice that progress comes from struggle, no lower-level mechanisms are adduced that generate these laws.

Although transformational theories do not have carefully articulated lower-level mechanisms providing the mediation for the law-like higher-level properties that are claimed, there is general agreement on elements that would go into such a theory of mediation. Human beings have certain properties:

1. They have great physical power to alter their surrounding circumstances.

2. They have self-reflexive consciousness so they can assess and react to their own psychic states.

3. They can imagine and plan what does not yet exist, so they can invent novelties.

4. They have a recursive linguistic function that allows them to communicate complex hypothetical structures and causal assertions.

5. They are always born and develop psychically in group contexts.

These properties are sufficient to allow groups of human beings to generate a variety of artifacts, activities, and group relations to decide how well these satisfy their physical and psychic desires, to consciously plan and alter their activities and beliefs, and to pass information about these activities and beliefs between individuals and across generational boundaries, and they generate the possibility of coercing or convincing other groups to adopt particular patterns of activity.

The problem with this list of properties of human beings and the powers that derive from them is that they contain no assertions about the nature of the transformation of individual properties into group properties and structures, or the way in which individuals are transformed by the group, or the manner in which group properties have their own dynamic relationships. That is, there is no social theory or psychosocial theory. Of course, a completely atomistic and reductionist evolutionary theory would not require such a social theory, but no transformational theory of cultural evolution denies the relevance of social and psychosocial causes. There is simply no agreement on what these are or how they would generate the "laws" of directionality and progress. It has remained for variational theories of cultural evolution to play the reductionist game.

The Paradigms of Cultural Evolutionary Theory: Variational Theories of Cultural Evolution

Variational models for cultural evolution have appeared in the last twenty years as a concomitant of the invention of sociobiology and its transformation into evolutionary psychology. It was the intention of sociobiology to give an orthodox Darwinian explanation of the origin of major features of human culture like religion, warfare, family structure, and so on as manifestations of the higher reproductive rate of individuals with certain behavioral properties, but not to explain changes that have occurred in the forms of those phenomena during the process of human history. Indeed, the chief evidence offered for the origin of these features through biological, genetic evolution was precisely that they were universal. All human cultures have religion, all engage in warfare, and E. O. Wilson claimed that male domination in society would persist indefinitely.[21] The ambition to extend classical Darwinism to the explanation of all aspects of species life, including species social behavior, resulted in an immense popularity of adaptive evolutionary thinking in fields like economics, political science, and psychology that were in search of more "scientific" explanatory schemes. One result of this intellectual fashion was, ironically, the creation of formal Darwinian models of differentiation and temporal change of social institutions, but without the biological genetic content of organic evolution. It is important to stress that Darwinian theories of the evolution of human cultural diversity in time and space are emphatically not theories that this diversity is based in genetic differences and that genetic evolution is at the base of the change from agricultural to industrial societies, or the development of the centralized state. Instead, a variety of theories of cultural evolution have been created that are *isomorphic* with the skeletal structure of Darwinian evolutionary theory, substituting for its various concrete biological elements analogical features from culture.

The skeletal structure of the Darwinian variational scheme for organic evolution consists of three assertions:

1. Individual organisms within populations vary from one another in their characteristics. This variation arises from causes within organisms that are orthogonal to their effects on the life of the organism (Principle of Random Variation).

2. Offspring resemble their parents (and other relatives) on the average more than they resemble unrelated organisms (Principle of Heredity).

3. Some organisms leave more offspring than others (Principle of Differential Reproduction). The differential reproduction may be a direct causal consequence of the characteristics of the organism (natural selection), or it may be a statistical variation that arises from purely random differential survival. This latter possibility is often ignored in vulgar expositions of Darwinian evolution, and all changes are ascribed to natural selection, but it is now certain that a great deal of evolution, especially molecular evolution, is a consequence of stochastic variations in reproduction.

If there is no variation among organisms, then even if different individuals leave different numbers of offspring, nothing will change. If there were no heredity of characteristics, then even if different organisms left different numbers of offspring, there would be no effect on the characteristics of the next generation. Finally, if different organisms all left exactly the same number of offspring no change would be expected in the composition of the population. In order to produce a scheme of cultural evolution that is isomorphic with the Darwinian variational structure there must be analogs of its elements.

The production of those analogs has occupied a great many people in a variety of disciplines over the last few decades. With so many competing models produced, it is hardly surprising that there is a great deal of spirited debate among the authors of the large and expanding literature on cultural evolution.[22] But however full of sound and fury, this debate is essentially an intramural affair. For beneath all the differences in details, there is a paradigmatic unity among Darwinian theories of cultural evolution based on the assumption that cultural evolution can and must be explained in terms isomorphic with the three principles of Darwin's variational scheme. Before they can proceed with that explanation, however, cultural evolutionists undertake a cleanup project, accomplished through sleights of conceptual hand, that clears away anything between the "biological" and the "cultural" that might have a constitutive effect in the production and "evolution" of cultural forms. This entails the disappearance of the social or, at least, depriving the social of causal efficacy and then the neutralizing of culture.

The easiest way to make society disappear is simply to dissolve it by definitional fiat into a mere population. E. O. Wilson writes: "When societies are viewed strictly as populations, the relationship between culture and heredity can be defined more precisely."[23] Robert Boyd and Peter Richerson state rather categorically that "cultural evolution, like genetic evolution in a sexual species, is always a group or population phenomenon"; and in a later work: "Because cultural change is a population process, it can be studied using Darwinian methods."[24] A more nuanced way of dissolving society into a collection of atomistic individuals is to create a choice between two extreme alternatives. Melvin Konner correctly rejects the society-as-organism metaphor by contrasting the cell that is devoted "entirely to the survival and reproduction of the organism" with "the purposes of the individual human [that] are wedded to the survival and reproduction of the society only transiently and skeptically." But he overdraws the consequences of this obvious insight and concludes that evolution "has designed the individual with a full complement of independence and a canny ability to subvert, or at least try to subvert, the purposes of society to its own. Every time a human being gets fed up with his or her society or church or club or even family, and voluntarily changes affiliation, we have another factual disproof of the central metaphor of social and political science."[25] Here he assumes that the repudiation of the obviously false metaphor of society as organism is a justification for an equally obviously false atomistic individualism that renders society a mere population.

However accomplished, the dissolution of societies into populations or, as in more nuanced approaches, the reduction of differential social power to the status of a subordinate variable, precludes the possibility that social systems might have properties unique to them as organized systems, that is, that social relations might be characterized by structures of unequal power that affect individual social behavior and the fitness of cultural traits.[26] This dissolution means, in turn, that social hierarchy and inequality are explained as just the consequence of the differential cultural fitness of individuals or of the cultural traits they bear, rather than, say, as a consequence of antagonistic and exploitative social relations.[27]

Having taken the crucial preliminary step of dissolving society, the next step is, perhaps surprisingly, to neutralize culture as well. In order to qualify as an instance of a variational theory of evolution, culture must be proven to consist of isolatable, individual entities, and to be only the sum of its parts. It is thus necessary to refute any and all claims that cultures have unique and discrete properties and a system-specific logic that require them to be analyzed each on its own terms. This is sometimes done by definitional fiat aimed at another superorganismic strawman. E. O. Wilson, for example, insists that "cultures are not superorganisms that evolve by their own dynamics." Culture, concurs Jerome Barkow, "is not a 'thing,' not a concrete, tangible object. It isn't a cause of anything. To describe behavior as 'cultural' tells us only that the action and its meaning are shared and not a matter of individual idiosyncrasy."[28]

The definitional fiats that posited population-like models of culture received at least two slight challenges. Discontented with an excessively atomistic view of culture, Bernardo Bernardi constructs a constellation of "anthropemes" consisting of "ethnemes," themselves subdivided into "idioethnemes" and "socioethnemes"; and Martin Stuart-Fox divides memes into mentemes.[29] Though these attempts appear to reject the notion of isolated, individual memes and to aim at systematic complexity, they fall short. Tellingly, in suggesting the division of the meme into mentemes, Stuart-Fox consciously attempted to construct a categorial analogy with modern linguistic terminology. But he did not follow up this overture and consider Saussure's fundamental insight on which modern linguistics is based, namely that meaning is system-specific, that each term (sign) acquires its historically specific meaning by virtue of its place within a discrete set of differential relations. By neglecting this insight, attempts such as Stuart-Fox's and Bernardi's focus only on the aggregate rather than the systemic. Only additive in method, they treat memes as aggregates of smaller entities, as cultural molecules composed of cultural atoms—which effects only a slight displacement of their ontological individualism, reproducing it at the level of compounds.

Coevolutionists have also made overtures to the systemic character of culture by removing it from a tight genetic leash and insisting that culture evolves relatively autonomously on its own cultural track. But regardless of the number of evolutionary tracks advocated, all theories of cultural evolution pay only lip service to the complexity of culture: because they persist in treating culture as merely the sum total of individual cultural units at a given stage in the selection process, as a kind of "state of the 'memes' " at a given point in time, they deny culture any system-specific characteristics; and this, in turn, allows all cultures to be explained according to the same (transhistorical and therefore ahistorical) selectionist logic.

With society and culture reduced to mere aggregates and deprived of any systemic and system-specific characteristics, the ground is prepared for the construction of a scheme of cultural evolution that is isomorphic with the Darwinian variational structure. This, as mentioned above, requires the construction of cultural analogs to the three fundamental principles of the Darwinian variational scheme.

First, a decision has to be made about the Principle of Random Variation, about the identity of the objects that have variation, heredity, and differential reproduction. Are these objects individual human beings who are the bearers of different cultural characteristics and who pass on those characteristics to other human beings by various means of social and psychological communication, and who have differential numbers of cultural "offspring"? This is the approach generally favored by those focusing on behavior and defining cultural in behavioralist terms. Or are they the characteristics themselves with properties of heredity and differential reproduction? This is the more common approach in recent years, especially among the "coevolutionists" who have taken an "ideational" view of culture using so-called trait-based models of the evolutionary process.

An example of the former is Cavalli-Sforza's and Feldman's theory of cultural transmission, while Dawkins's "memes" are an example of the latter.[30]

Either way, a fundamental problem results from the assumption that these cultural units, say, the idea of monotheism or the periphrastic "do," somehow spread or disappear in human populations, namely, no theory of cultural evolution has provided the elementary properties of these abstract units. Presumably they are mortal and so need rules of heredity. But, for a variational theory, it must be possible to count up the number of times each variant is represented. What is the equivalent for memes of the number of gene copies in a population? Perhaps it is the number of individual human beings who embody them, but then the death of a human carrier means the loss of a meme copy and so memes do, after all, have the problem of heredity. A major problem of creating a variational theory of cultural evolution is that the task of building a detailed isomorphism has not been taken seriously enough.

Once the individual units are settled upon, little time is spent determining the sources of variation in those units, the "cultural analogs of the forces of natural selection, mutation, and drift that drive genetic evolution."[31] Following a quick definitional determination of the sources of variation—randomness and drift, selection, and perhaps the addition of a uniquely cultural source such as intentionality—the next step is to find the cultural analogs to the Principle of Heredity.

Most cultural evolutionists simply accept as given that culture is a system of heredity or at least of unidirectional transmission. Boyd and Richerson state axiomatically that "Darwinian methods are applicable to culture because culture, *like genes*, is information that is transmitted from one individual to another" (emphasis added). In a later essay they turn inheritance into the defining characteristic of cultural evolutionary theory: "The idea that unifies the Darwinian approach is that culture constitutes a system of inheritance"; and after a brief discussion that moves from inheritance through the "population-level properties" of culture that makes it "similar. . . to gene pools," they conclude that "because cultural change is a population process, it can be studied using Darwinian methods."[32]

To be sure, however, Boyd and Richerson spoke a bit too inclusively. Whereas some cultural evolutionists use "inheritance" and "transmission" interchangeably, others are uneasy about the genetic and parental overtones of "inheritance" and prefer "transmission." But both terms refer to a process of descent that occurs in the same unidirectional manner between an active donor and a passive recipient. The semantic advantage of "transmission" is that it drops the genetic connotational baggage of "inheritance" while preserving the portrayal of cultural change as a unidirectional process of descent with modification and selection.

Whether conceptualized as "heredity" or "transmission," the problematic issue is that both terms require the establishment of some laws of the heredity of units or their characteristics if human individuals are the units. We then require the details of the passage of culture to new individuals, by analogy with the Mendelian mechanism of the passage of genetic information from parent to offspring by way of DNA. In making this analogy, however, the biological model implies constraints that have not been apparent to cultural evolutionists. We say that parents "transmit" their genes (or at least copies of their genes) to their offspring, so models of cultural evolution begin with models of the "transmission" of cultural traits from one set of actors to others by analogy with the transmission of genes. Parents may transmit traits to their children, or teachers to their pupils, or siblings and other peers to one another by a variety of simple rules. The outcomes of evolutionary models of this kind turn out to be extremely sensitive to the postulated rules of transmission, and since there is no firm basis on which to choose the rules, almost anything is possible. But there is a deeper problem. Is culture "transmitted" at all? An alternative model, one that accords better with the actual experience of acculturation, is that culture is not "transmitted" but "acquired." Acculturation occurs through a process of constant immersion of each person in a sea of cultural phenomena, smells, tastes, postures, the appearance of buildings, the rise and fall of spoken utterances. But if the passage of culture cannot be contained in a simple model of transmission, but requires a complex mode of acquisition from family, social class, institutions, communications media, the workplace, the streets, then all hope of a coherent theory of cultural evolution seems to disappear. Of course, it was simpler in the Neolithic, but there was still the family, the band, the legends, the artifacts, the natural environment.

Some dissenters present serious challenges to the inheritance/transmission model even though they remain faithful to its explanatory principle. Martin Daly questions the value of the inheritance model because he finds no cultural analog to the gene, because cultural traits "are not immutable" like genetic traits, because cultural "transmission need not be replicative," because the recipients are not "simply vessels to be filled," and because "social influence" makes the processes of cultural change less regular than is implied by the term "transmission."[33] Though Daly and others raise perfectly legitimate and important questions about inheritance and transmission analogies, they deprive their insights of real force by still maintaining that cultural change is a process that can and must be explained in terms isomorphic with "the evolutionary model of man."[34]

This assumption brings us to the third analogical element in theories of cultural evolution, the Principle of Differential Reproduction. Whether they define the units as cultural atoms or cultural molecules, whether they speak of cultural change as inheritance or of transmission to passive recipients or to active acquisitors, they all insist that cultural change is a process of *descent* with modification—and as such it has all the attributes of a variational evolutionary process eligible for Darwinian, that is, selectionist explanation. To all cultural evolutionists may be extended that which Martin Stuart-Fox said of himself, namely that they *"take for granted* (a) the scientific status of the synthetic theory of evolution and (b) that this theory provides *the most likely model* on which to base a theory of cultural evolution" (emphasis added).[35]

However, the forces that cause the differential passage of culture across generations and between groups seem not to be encompassed by the reductionist model in which individual actors have more cultural offspring by virtue of their persuasiveness or power or the appeal of their ideas, or in which memes somehow outcompete others through their superior utility or psychic resonance. Atomistic models based on the characteristics of individual humans or individual memes can be made, but they appear as formal structures with no possibility of testing their claim to reality. How are we to explain the disappearance of German and French as the languages of international scientific discourse, and their universal replacement by English without terms like "Nazi persecution of Jews," "industrial output," "military power in the Cold War," or "gross national product"? That is, no variational theory of cultural change can be adequate if it attempts to create a formal isomorphism with Darwinist individualism.

Historical, political, social, and economic phenomena must be dismantled in order to be molded into the raw material for selectionist theories of cultural evolution. This is effected through the dissolution of social systems with structural asymmetries of power into individuals and through the reduction of cultural systems to eclectic aggregates of differentially reproduced memes. This dual process strips historical phenomena of their sociocultural particularity. Once transformed in this way, they may be subjected to nomological explanation as individual instances of the exogenous, because transhistorical, law of selection. Even the recognition given by William Durham and others to the systemic character of culture and to the possibility that social asymmetries of power might affect cultural transmission and fitness are drained of content by the fundamental assumptions of the cultural evolutionist paradigm: the definition of culture as an aggregate of individual, heritable units and the selectionist explanation of its evolution. And in these assumptions lies the self-validating circularity of cultural evolutionary theories: selectionist explanation requires individual, heritable units of culture; and reduction of culture to an aggregate of such units renders it susceptible to selectionist explanation—whose scientific status had been taken for granted from the very beginning.

As its etymology suggests, any "theory" is a way of looking at the world, and what one sees is that which is visible through one's particular set of theoretical lenses. Cultural evolutionary theories, however, base (and wager) their claim to break through all theoretical biases and to attain scientific status on their verifiability, their ability to postdict past and predict future cultural evolution. If, with the emergence of the hegemony of the physical sciences, the cornerstone of a scientific theory has been the elimination of the historical, its touchstone has been its predictive capacity—a matter that cultural evolutionists address with increasing confidence.

We have already encountered Alexander Rosenberg's optimism about the use of mathematical models in the new sociobiologically based social sciences and his confidence in their predictive capacities.[36] The same optimism is prevalent among the contributors to *Politics and the Life Sciences* who are convinced that the predictive powers of the new evolutionary political science will render it capable of informing policy decisions. Certain that Darwinian models of cultural evolution can produce "a useful retrodiction of ethnography," Lumsden and Wilson were somewhat circumspect, anticipating only predictions of "short-term changes in the forms of ethnographic distributions." Nevertheless, they remained—and Wilson has become ever more— optimistic that "the history of our own era can be explained more deeply and more rigorously with the aid of biological theory," and that this approach might enable us to look "down the world-tube of possible future histories."[37]

Similarly, Boyd and Richerson quickly overcame their initial caution to assert that "Darwinian models can make useful predictions."[38]

Though they wager the validity of their theories on their predictive capacities, theorists of cultural evolution rig the explanatory game in a variety of ways. One is by covering all bets. This can be done by playing with probabilistic explanations. In the gambling hall, probabilities only provide the odds, but probabilistic predictions of cultural evolution are guaranteed winners, since they encompass all possibilities. Because of our evolved capacity to reason we could be soberly advancing down the road toward wisdom, courage, and compassion; or because of our innate capacity for aggression we could be headin' for nuclear Armageddon—or anything in between. Or it can be done by constructing a historical analog to randomly drift in theories of biological evolution—the catchall explanation of that which cannot be subsumed under selection.

A second way to rig the game is with postdictive readjustment. The cultural evolutionist, like the economist, is "an expert who will know tomorrow why the things he predicted yesterday didn't happen today."[39] The gambler's losses might be recouped in a later game, but cannot be undone. In cultural evolutionary explanation and prediction, however, the game may be replayed indefinitely until the model is successfully readjusted. Combined with probabilistic explanations, postdictive readjustment renders the model invulnerable by disarming its weaknesses.

The irony here is that the constant recourse to postdictive readjustments brings the science of cultural evolution into the neighborhood of "just plain history"—almost. The difference is that the faith in the scientific status of the law of selection erects a third safeguard for theories of cultural evolution. This belief precludes as "not scientific" any non-evolutionary—that is, historical— explanation of cultural change. But because cultural evolutionary theories are based on a unitary, transhistorical principle, they produce explanations that are too broad to be either falsifiable or explanatory.

Historians, cultural evolutionists argue, are too close to the fray, and their timescales too short—which leads them into all kinds of unimportant detours and false starts that appear to the historical eye as enterprises of great pith and moment. To gain proper perspective, cultural evolutionists draw back, occasionally indulging in imaginary space travel, in order to attain a sufficiently distant viewpoint from which to view the human species as one among many and to avoid the "anthropocentrism" that would exempt culture (a biological adaptation) from biological explanation. But distance can also be deceiving.

From their distant viewpoint cultural evolutionists willingly see only the broad patterns of cultural evolution and ignore the inconvenient and contingent details of history that do not fit into those patterns. This conscious oversight produces theories of cultural evolution that are explicitly or implicitly progressivist: since culture is a successful and cumulative adaptation that breaks free of natural selection, the more culture the better for human welfare and survival. This linear logic points to the contemporary West with the most advanced level of science and technology (the ultimate cultural adaptations insuring human welfare and survival) as the current pinnacle of cultural evolution. But the road to modern Western civilization has taken a series of abrupt and thoroughly unpredictable turns. What general theory of cultural evolution could postdict the collapse of the Roman Empire and the Dark Ages? Or the emergence on a distant frontier of the Eurasian landmass of a new geo-cultural entity, a "continent" called Europe? Or that in a very brief historical time span this new culture would overtake much more advanced Asian cultures and establish itself as the most powerful and dominant in the world, with one of its tiny "populations," the English, having acquired an empire on which the sun never set? But the result of all those unpredictable turns, the late modern West, which should be the pinnacle of cultural evolution, has been the epitome of barbarism (which only a small group of *fin de siècle* artists and intellectuals, members of the "literary culture," suspected).

From their distant viewpoint, cultural evolutionists may ignore acts of barbarism in Western history like the genocide of Native Americans or the Nazi Holocaust as just specks of dust on the plain of history, momentary aberrations irrelevant to the question of cultural evolution. Alternatively, they may subject both to the same explanatory principle as just two examples of human aggression explained through some selectionist variation or combination of inclusive fitness, innate aggression, the stress of overpopulation, and/or the need for *Lebensraum*. But to explain the character, causes, and consequences of these two forms of genocide according to the same transhistorical principle would lead to a gross misunderstanding of each and would tell us little about their historically and politically significant differences. Such an approach, for example, is far too broad either to postdict the success of Nazism or to predict the ongoing consequences of the Nazi period, of the historical memory that continues to affect significantly the history not only of Germany and Europe, but also of the Middle East. Whether they forcibly subsume disparate historical phenomena under a transhistorical explanatory principle or write off as mere contingencies historically significant events that cannot be so subsumed, cultural evolutionary theories cannot answer the many crucial questions pertaining to the particularity, the uniqueness, of all historical phenomena. In failing to live up to their own claims to be able to explain history, including that of our own era, "more deeply and more rigorously," cultural evolutionary theories also fail to live up to their further claim to explain history more "usefully"—to explain Nazism, for example, with sufficient precision to prevent its reoccurrence and to develop appropriate policies to deal with its consequences.

It is therefore no use to fall back on yet another safeguard, the claim that the field is still young, the models are still being built, and one day. . . . The problem is more serious than "not yet enough time." Cultural evolutionary theories are carefully constructed, logically consistent, and very neat. Their neatness, however, is achieved either by dismissing as inessential to cultural evolution the contingencies that are so essential to historical change or by subsuming them to a single transhistorical principle of explanation. But this formulaic treatment is fully inappropriate to the labyrinthine pathways, the contingent complexity, the many nuances, and general messiness of history. And it results in linear explanations that approach closely enough to history to allow the distant observer to mistake proximity for causality. These analytical lines are actually false tangents—briefly nearing, though never touching, the contours of history.

We conclude, finally, by returning to the question of whether any useful work is done by considering cultural evolution as distinct from the history of human societies. Transformational theories of cultural evolution have the virtue that they at least provide a framework of generality with which to give human long-term history the semblance of intelligibility. But the search for intelligibility should not be confused with the search for actual process. There is no end of ways to make history seem orderly. Variational isomorphisms with Darwinian evolution suffer from the inverse problem. Rather than being so flexible as to accommodate any historical sequence, they are too rigid in structure to be even plausible. They attempt to mimic, for no reason beyond the desire to appear scientific, a theory from another domain, a theory whose structure is anchored in the concrete particularities of the phenomena that gave rise to it.

27

Is Capitalism a Disease?:
The Crisis in U.S. Public Health

The scientific tradition of the "West," of Europe and North America, has had its greatest success when it has dealt with what we have come to think of as the central questions of scientific inquiry: "What is this made of?" and "How does this work?" Over the centuries, we have developed more and more sophisticated ways of answering these questions. We can cut things open, slice them thin, stain them, and answer what they are made of. We have made great achievements in these relatively simple areas, but have had dramatic failures in attempts to deal with more complex systems. We see this especially when we ask questions about health. When we look at the changing patterns of health over the last century or so, we have both cause for celebration and for dismay. Human life expectancy has increased by perhaps thirty years since the beginning of the twentieth century and the incidence of some of the classical deadly diseases has declined and almost disappeared. Smallpox presumably has been eradicated; leprosy is rare; and polio has nearly vanished from most regions of the world. Scientific technologies have advanced to the point where we can give very sophisticated diagnoses, distinguishing between kinds of germs that are very similar to one another.

But the growing gap between rich and poor make many technical advances irrelevant to most of the world's people. Public health authorities were caught by surprise by the emergence of new diseases and the reappearance of diseases believed to be eradicated. In the 1970s, it was common to hear that infectious disease as an area of research was dying. In principle, infection had been licked; the health problems of the future would be degenerative diseases, problems of aging, and chronic diseases. We now know this was a monumental error. The public health establishment was caught short by the return of malaria, cholera, tuberculosis, dengue, and other classical diseases. But it was also surprised by the appearance of apparently new infectious diseases: the most threatening of which is AIDS, but also Legionnaires' disease, Ebola virus, toxic shock syndrome, multiple drug resistant tuberculosi, and many others. Not only was infectious disease not on the way out, but old diseases have come back with increased virulence and totally new ones have emerged.

How did this happen; why was public health caught by surprise? Why did the health professions assume that infectious disease would disappear and why were they so wrong? Infectious disease had been declining dramatically in Europe and North America for the last 150 years. One of the simplest kinds of predictions is that things will continue the way they have been going. Health professionals argued that infectious disease would disappear because we were inventing all kinds of new technologies for coping with them. We now can carry out diagnoses so rapidly that some diseases that might kill a person in two days can be identified in the laboratory soon enough to permit treatment. Instead of spending weeks culturing bacteria, we can use DNA to distinguish between pathogens that may have very similar symptoms. More important, we had developed a new arsenal of antimicrobial weapons, drugs, and vaccines, as well as pesticides to get rid of mosquitoes and ticks that are disease carriers. We came to understand that, through mutation and natural selection, microorganisms can present a recurring threat. We assumed that whatever the microbial changes, the disease-causing mechanism would remain the same while we developed ever newer weapons against it. It was, we believed, a war between us and the microbes, in which we would have the upper hand because our weapons were growing stronger and ever more effective. Another cause for optimism—at least this was the argument put forth by the World Bank and the International Monetary Fund—was that economic development would eliminate poverty and produce affluence, making all the new technologies universally available. Finally, the demographers noted that though most infectious diseases are most deadly to children, we have an aging population, so the proportion of people likely to catch such illnesses will be smaller. One thing missing from this hypothesis was that one reason children are so vulnerable is that they have not developed the immunities that go along with exposure; older people have reduced susceptibility precisely because they have been exposed. But if there are fewer children, older people will have a lower level of immunity and will contract diseases at an older age. Indeed, some diseases, like mumps, are more serious in adults than they are in children.

So what was wrong with our epidemiological assumptions? We need to recognize that the historical mindset in medicine and related sciences was dangerously—and ideologically—limited. Nearly all who engaged in public health prediction took too narrow a view, both geograpically and temporally. Typically, they looked only at a century or two instead of the whole sweep of human history. Had they looked at a wider time frame, they would have recognized that diseases come and go when there are major changes in social relations, population, the kinds of food we eat, and land use. When we change our relations with nature, we also change epidemiology and the opportunities for infection.

The Plague in Europe

Plague erupted in Europe for the first time in the sixth century during the decline of the Roman Empire under Justinian. Europe suffered from social disruption and declining production. The sanitary facilities of the great ancient cities were crumbling; under those circumstances, when plague was introduced it swept through the population with devastating effects. Plague reappeared in the fourteenth century during a developing crisis of feudalism, causing a population decline even before plague became widespread. The standard history of this plague occurrence is that sailors landing in ports along the Black Sea brought plague with them from Asia in 1338; it then spread westward and, in a short time, reached Rome, Paris, and London. In other words, plague spread because it had been introduced from elsewhere. But it seems more likely that plague had entered Europe many times before but really didn't take off. It only became successful when the population became more vulnerable, when the human ecosystem could not confront a disease spread by rats at a time when the social infrastructure that would have controlled rats had crumbled.

An Ecological Proposal

When we at look at other diseases, we see that they rose and fell with historical change and circumstance. So, instead of a doctrine of the epidemiological transition, which held that infectious disease would simply disappear as countries developed, we need to substitute an ecological proposal: With any major change in the way of life of a population (such as population density, patterns of residence, means of production), there will also be a change in our relations with pathogens, their reservoirs, and with the vectors of diseases. The new hemorrhagic fevers appearing in South America, Africa, and elsewhere almost all seem to be related to increased contact with rodents that humans don't normally meet, caused by the clearing of land for the production of grain in particular. Grain is also rodent food; rodents survive by eating seeds and grasses. When a forest is cleared and grain is planted, we also eliminate the coyotes, jaguars, snakes, and owls that eat rodents. The net result is an increase in rodent food and a reduction in rodent mortality. The rodent population grows. Now, these disease-deliverers are social animals. They nest and build communities; when a new generation emerges, the young adults go out looking for homes elsewhere—often wandering into warehouses and people's homes, facilitating the transmission of diseases.

Another human activity, irrigation, is especially related to the breeding of snails, who transmit liver fluke disease, and mosquitoes, who spread malaria, dengue, and yellow fever. When irrigation proliferates, as it did, for example, after the construction of Egypt's Aswan Dam, habitats for mosquitoes were created. Rift Valley fever, which had occasionally erupted in Egypt, can now be found fulltime. The development of giant cities in the third world has created new environments for the spread of dengue, transmitted by the same mosquito that transmits yellow fever (*Aedes aegypti*). It has adapted to life around the edges of cities. A poor competitor against other varieties of mosquito in the forest, these mosquitoes are able to breed in abandoned lots, in puddles, water barrels, and old tires—in the special environment that we create in the giant cities in the tropics. Dengue and yellow fever are particularly threatening because of the growth of urbanization in the tropics with megacities like Bangkok, Rio de Janeiro, Mexico City, and others with populations of ten to twenty million. As human population grows, there are new opportunities for diseases. For instance, you need a few hundred thousand in a population before it can sustain measles. If there are fewer, measles can infect the entire population; those who survive will be resistant. But if there aren't enough new babies to maintain the disease, it will disappear and have to be reintroduced. But in a population of a quarter-million people, there will be enough new babies who are not resistant that the disease can sustain itself in the population. Consider this: if we know there are diseases that require a quarter-million people to be self-sustaining, what diseases will emerge in crowded populations of ten or twenty million? Clearly, as life conditions change so do opportunities for disease.

Yet another kind of myopic thinking in the public health community arose from the fact that doctors are concerned with human diseases but have not paid much attention to diseases of wildlife or of domestic animals or plants. Had they done this, they would have had to confront the reality that all organisms carry diseases. Diseases come from the invasion of an organism by a parasite. When an infection takes place, it may or may not produce symptoms. But all organisms deal with parasites and, from the point of view of the parasite, invading an organism is a way of escaping from competition in the water or in the soil. For instance, the bacteria that causes Legionnaires' disease lives in water. It is found all over the world but is never very common because it is a poor competitor. It has very finicky dietary requirements, so normally humans don't encounter it. However, it has two things going for it. It can tolerate high temperatures and is resistant to chlorine. It withstands chlorine by hiding out inside an amoeba. In a convention center, hotel, or truck stop, water is both heated and chlorinated. And if it's a good hotel, we may find a showerhead that gives a fine spray of tiny droplets, perfect for carrying the bacteria into the furthest corners of your lungs. What we've done is to create the ideal environment for Legionnaires' disease. The chlorine and the high temperature kill their competitors, the remains of which form a coating on the inside of the pipes that is marvelously rich in the food that the Legionnaires' bacterium loves.

If we look at other organisms, we see a constant jockeying for position between parasites and hosts. The more common a species is, the more attractive it is to new invasions by parasites. Humans are very common, and thus offer wonderful opportunities for invasion. When we observe disease patterns, we see that cholera, for example, spread from the Eastern Hemisphere into the Americas, entering Peru and then traveling up to Central America.

But a similar route was followed by a disease of orange trees, by viruses of beans and tomatoes, as well as by wildlife diseases. What we see, then, is a constant coevolution between pathogens and hosts across all animal and plant life rather than a situation unique to humans. Surely, we would have a much better understanding of potential dangers if we understood human illness from this perspective.

Transmission of Diseases

What kinds of insects spread viruses to people? Nearly all of them are mosquitoes or flies, or belong to a second group that includes ticks, fleas, and lice. These are the two main groups that overwhelmingly spread human virus diseases, even though there are hundreds of thousands of other kinds of insects. There are very few diseases spread by beetles, none that I know of by butterflies or dragonflies. Why? Are there circumstances under which they might become transmitters of diseases? Among plants, the major distributors of plant viruses belong to a totally different group of insects—to aphids. However, both groups have similar mouths and subsist by sucking liquid from their hosts: the mosquito sucking blood, the aphid sucking sap. If you have ever sucked something through a straw, you know that after a while a vacuum builds up and in order to be able to continue slurping the liquid, you have to be able to return liquid. Similarly, the salivary glands of mosquitoes and of aphids return liquid to their hosts when they take up the blood or the sap, and in that liquid you find the viruses. That's why when we study viruses, we look at the salivary glands of mosquitoes, or of ticks, or anything else. We can begin to encounter these generalizations when we step back from looking at the particular details of a given disease and try to get a broad picture. But this wasn't done.

The Failure to Study Evolution and Society

Another kind of scientific narrowness—a self-imposed intellectual constriction—is the failure to study evolution. Evolution tells us immediately that organisms respond to the challenges of their environment. If the challenge is, for example, an antibiotic, organisms will respond by adapting to those antibiotics. In agriculture, we know of hundreds of cases of insects that have become pesticide-resistant; in medicine, increasing numbers of microorganisms have become resistant to antibiotics meant to fight them. Some microbes have become resistant to antibiotics even before they are used! This happens when an antibiotic is released on the market with a new trade name but in fact is hardly different from its predecessor. It may look different, but if it acts on the bacteria in the same way, it will be met by the same defenses. It is not enough to look at the agent of disease; we have to look at what makes populations vulnerable. Conventional public health failed to look at world history, to look at other species, to look at evolution and ecology, and, finally, to look at social science. There is a growing body of literature that says that the poor and oppressed are more vulnerable to nearly all health hazards. But we still don't recognize class differences in the United States. Researchers discuss differences in income or a mother's education level or even socioeconomic status. But U.S. epidemiology does not deal with class, even when class is the best predictor of life expectancy, of old-age disability, or the frequency of heart attacks. As a predictor of coronary disease, it is better to measure class position than to measure cholesterol.

Other Explanations

Why do we wear these intellectual blinders that have so hobbled the study and practice of public health in this country? First, there are a multiplicity of long-term intellectual biases. Take, for example, American pragmatism. Americans pride themselves on their practicality. "Theory" is almost a dirty word. When we are overwhelmed with the urgency of a population that is sick, of kids that are dying, it becomes a luxury to ask about evolution. This overwhelming sense of urgency is one of the reasons why doctors don't look at diseases of tomatoes, don't ask about competition between different kinds of mosquitoes, and certainly don't look at historical factors. There is an inevitable tunnel vision built into the urgency of carrying out applied clinical or epidemiological work.

A second reason is the Western scientific tradition of reductionism, which says that the way to understand a problem is to reduce it to its smallest elements and change things one at a time. This is very successful when the question is, "What is this made of?" Then we can isolate it, cut it out of an organism, put it in the blender or under the microscope. In fact, we have been marvelously successful at identifying what things are made of. That is why we have had a growing, if irrational, sophistication about small phenomena and events throughout the whole of scientific enterprise. Why is it we are so successful at giving individual emergency treatment and so ineffectual in stopping or preventing malaria, in anticipating its return, or dealing broadly with the health of whole populations? We are marvelously successful at breeding a wheat plant that can better use nitrogen to produce more grain but much less successful at alleviating hunger in the countryside.

Four Hypotheses

So the typical failure has been a refusal to look at complexity. The successes have been successes of the small, where we could focus on isolated elements. In the United States, even though we spend more than any other country on health care, we have among the worst results among the industrial countries; certainly we are behind the Europeans and, in many ways, also behind Japan when the usual indicators of health are considered. This is something that worries public health people: Why, they ask, do we spend so much and have so little to show for it compared to other countries?

Here are four hypotheses:

One, we don't actually get more health care; we just spend more for it. We know that something like 20 percent of our health care bill is in administration, that is, the cost of billing and the like. The rate of profit of the pharmaceutical industry is greater than that of capitalism as a whole, and much of that is in the United States. Doctors' salaries are huge, as are charges for hospital rooms. The consequence is that "investment" per patient is enormous.

Two, even when we do get more health care, it is not always good health care. Now, this seems paradoxical because we have more MRIs and more CT scans and more dialysis machines than most other countries. So why is our health not better? Medical decisions are not always made for medical reasons. There are a lot of incentives for making decisions as to which kind of techniques to use, what kinds of interventions—when to carry out heart surgery, for example—which give rise to differences in medical procedure among countries. We do a lot more implanting of pacemakers than Europe and perform more cesarean sections and hysterectomies. A hospital buys an expensive machine to attract both doctors and patients. But once on hand, it has to be used. You can't allow an MRI machine to sit idle in the hospital, so doctors are encouraged to use it if only to amortize the institution's investment. Another is that in order to keep the "batting average" of a surgeon high, he or she has to perform enough operations (several hundred a year) to keep skill levels up. An isolated hospital with only one heart transplant every three or four months is not a safe place to go. The wise patient will seek out a hospital with a highly regarded cardiac service equipped with the latest technology. But to win that prestige, skills must be maintained, so there's an incentive to keep both surgeons and machines working. Since the service is also an expensive thing to have, it needs to be kept busy if only to bring in surgical fees. But does it make sense to have all that expensive equipment? Hospital administrators will tell you it does because the hospital down the road has it. If Mass General is in competition with Beth Israel, and both compete with Mount Sinai, all of them need the most advanced machines. Then there are the HMOs, which have their accountants making medical decisions, effectively rationing health care. Both approaches are meant to maximize profit. What happens is that sometimes people get too much care, sometimes too little. But in both cases, our health is a side effect of the obsession with making money. The irrationality of the system extends even to the rich, who are overtreated. We kill nearly 200,000 people a year through improper medical interventions. Many more die due to misuse of heavily advertised prescription medications, over-the-counter remedies, and other preparations.

The third hypothesis is that the health care system is built on a foundation of inequality. Only some of us actually receive or have access to the health care we need, while most don't.

Finally, the fourth hypothesis is that we have created a sick society, even as we invest more and more to repair the damage. We are exposed to more pollution and increasing levels of stress and therefore exposed, ironically, to more opportunities to display our cardiac surgery skills. We make more people miserable, so we spend more on psychiatry and on psychotropic drugs. This is clearly evident in the public health situation in contemporary Russia, where the collapse of universal health coverage exposed the population to all the ills of incipient capitalism. They have had waves of epidemics, diphtheria, whooping cough, and the completely novel situation in modern times of declining life expectancy—from about sixty-four years to about fifty-nine years. Ours is a sick society that demands ever greater expenditure to repair the damage to public health that it has itself inflicted.

Responses to the Crisis

The condition of health care has not gone unnoticed; indeed, there is widespread and growing dissatisfaction. A number of responses have been made to address the situation:

Ecosystem Health. Ecologists looking at the problem have derived an approach they call Ecosystem Health. They posit that there are ecosystems under stress for multiple causes: from pollutants, contaminated food and water, high stress, and changes in the daily rhythm of life. For example, with nearly universal electrical light, people sleep less and our physiology changes. If we examine human biology as a socialized biology, we notice that there are things that appear as constants of human biology that really are not. For instance, it has long been conventional wisdom that, as a natural part of the aging process, blood pressure increases with age. But it turns out that among the !Kung Bushmen of the Kalahari, blood pressure increases with age only up through puberty and then levels off. Our blood pressure pattern is a function of the kind of society in which we live. We can see this in the pattern of stress-reaction hormones that vary with one's social location. Recent Harvard studies have shown that among groups of teenagers from high school, all of whom are doing equally well academically, working-class kids showed prolonged rises in cortisol under any kind of stress while upper-class kids showed a quick spike and then decline. The physiology of working-class youngsters was altered by their social location, whether or not they acknowledged their working status. Evidently your body knows your class position no matter how well you have been taught to deny it. Human physiology, then, is a socialized physiology and differing social locations create different relationships with the environment. This knowledge has led to the ecosystem health concept, bringing together environmentalists and public health people to examine questions about how we rate the health of the whole ecosystem.

The Environmental Justice Movement. This movement arose from the observation—by others—that the best way to find an incinerator or a toxic waste dump is to look for an African-American neighborhood. With lower real estate values in minority neighborhoods, it is cheaper to put the incinerator there. Zoning rules, made by the powerful, are more lax there. So the health risks from pollution and industrial waste become yet another facet of oppression. Exposure to pollutants doesn't affect everybody equally. Exposure to occupational health hazards—the exposure of somebody who makes a living sandblasting buildings, for example—is very different from someone who works at a desk totaling up actuarial tables. Exposure to environmental insult also varies with class and the condition of oppression. The environmental justice movement has been a response to this, fighting the dumping of pollutants and attempting to equalize the risks of an industrial society.

Social Determination of Health. This approach has been growing among epidemiologists, partly due to the rediscovery of what Rudolph Wirchow and Frederick Engels pointed out in the nineteenth century: capitalism can undermine health. This is important to keep in mind when conservative and reactionary commentators assert that there isn't any real poverty anymore. They argue that though some people make more money than others and can afford a bigger color television, the poor are not without their TVs. The car of a poor family is a little bit older or perhaps they do not eat in restaurants as often, but this inequality does not negate the real truth as the right-wing pundits see it: basically, there is no longer any poverty. Of course, an answer is easily found in the numerous studies that show that black people pay for racist oppression with life spans ten years shorter than that of whites. Poor and oppressed minorities have 25 percent fewer successful encounters with the health care system than more privileged groups. Meanwhile, the rate of death or other harmful outcomes increases with the level of poverty in illnesses like coronary heart disease, cancer of all forms, obesity, growth retardation in children, unplanned pregnancies, and maternal mortality.

Those interested in the social determination of health include some English scholars, such as Richard Wilkinson, who have looked at the life expectancies of different ranks in the English civil service. He found there was a difference even among those groups that are better off than those exposed to obvious dire need. He noticed that mere social hierarchy, social differentiation, makes your health worse everywhere, not only among those in extreme poverty. Now this can be interpreted in two opposing ways, but both of them are operative. One is to say that inequality per se, rather than the level of poverty, can make a person sick. Another is to say, quite literally, that it is all in your head. In support of the latter, baboon studies are cited that seem to indicate that those with higher rank in their troupe have better health. Their arteries are cleaner, they respond to stress like upper-class people; their cortisol level shoots up under stress and then comes right down again. Lower-ranking baboons tend to have the effects of stress lasting longer; their life expectancies are lower. But if you intervene in animal communities and alter their social hierarchy, within a few months the baboon's physiology will take on the characteristics of its new social location. This leads some people to say that it is how people perceive their situation in society—and therefore that people must be taught to cope with where they are, that after all we create our own realities. That's a common phrase in some of the growth and therapy movements: we create our own realities. It's not so much that you're underpaid and poor, but that you feel lousy about it. And so we have devised cheer-up pills: the cure for depression is not to get rid of the depressing situation, but to help people feel better about it. Another way to look at this so-called social determination of health is to see it not as a simple result of inadequate incomes that need to be raised but as a consequence of a profoundly stratified, class-based society. Those who emphasize the latter feel that it is a more radical position than simply talking about how absolute deprivation is bad for your health, because the remedy for that would seem to be to increase income. Instead, they say, you have to eliminate the inequalities of class. Since the same studies can give rise to opposite conclusions, we need to emphasize that inequality affects your health in many different ways. When rich people think about poverty, they think about it only in the sense of having a little bit less, without examining the underlying structure of impoverishment. Poverty affects people, first of all, as chronic deprivation, actually having less food or worse food. Kids who live in damp, moldy apartments have worse health than kids who live in dry apartments. There are many other ways in which chronic deprivation itself is a menace to health.

There are what we call low-frequency, high-intensity threats, meaning those experiences that do not happen to everybody, but that could, and therefore are a constant threat to a sense of well-being. Robert Fogel, a right-wing, Chicago-school economist, pointed out in his book *Time on the Cross: The Economics of Negro Slavery* that most slaves were not whipped. He went on to say that slavery was not what we would imagine from reading *Uncle Tom's*

Cabin, that it had a certain economic rationality. What he neglects to say is that physical abuse of slaves, even when not employed, was a constant threat. Most slaves, perhaps, weren't whipped but all of them witnessed or knew about beatings. Similarly, most kids in impoverished neighborhoods are not shot, but getting shot is a constant menace every time you go to the store or go outside. These are examples of medium- and low-frequency, but very high-intensity, threats.

There also are high-frequency, low-intensity insults, the daily harassment one can see, for instance, in African-American communities. There, one is constantly forced to make strategic decisions. Am I walking so slowly that the cop is going to think I'm loitering? Or, am I walking so fast that he or she will think that I'm running away from the scene of a crime? If I come onto campus at night to work in my laboratory, will I first be stopped by the police who think that I'm a thief? The resident commissioner of Puerto Rico was once stopped by police on the way to his office in Washington. They laughed when he said he was a member of Congress and the resident commissioner. Ramos Antonini was black.

We are learning now from the study of neurotransmitters that our brain is not the only locus of social experience. The cerebrum gathers social experience and transmits it through many branches of the nervous system into the neurotransmitters. The neurotransmitters are chemically similar to substances in our immune system, in the white blood cells. In a certain sense, we think with our whole bodies, we feel with our whole bodies, and so the whole body is the locus of social experience that comes with these patterns of chronic conditions, of low-frequency threats or high-frequency insult. There are many dimensions to the experience of deprivation, but they are often lost in the hands of the statisticians, who simply see poverty as a quantitative difference in income.

The Health Care for All Movement. This group champions a national health insurance system and has done much work comparing the American system to the Canadian system; many progressive physicians are active in this movement.

Alternative Medicine. The alternative health movement deals mostly with individual health. It stresses diet, exercise, homeopathy, chiropractic, and naturopathic remedies—areas where people feel that they have not been treated adequately by the established medical system. They draw on a holistic approach to health rather than the targeted, magic-bullet approach of traditional allopathic medicine. They seem to be particularly effective in dealing with long-term chronic conditions rather than acute emergencies. For instance, for those who need radiation and chemotherapy for cancer, alternative practices are helpful in modulating the negative side-effects. The strategy of modern medicine is that cancerous tissue is sufficiently fragile and can, in effect, be poisoned in the hope that the radiation or chemotherapy will kill the cancer more than it kills you. The approach employed by alternative therapies is not to attack the cancer directly, but to try to build up the body's defenses. So the two approaches complement each other. Alternative medicine is very attractive and very powerful, but its primary appeal is to people who have control over their lives and access to the resources and techniques of alternative health care. It is not a mass movement; the holism it advocates stops at the edge of your skin. It is not a societal holism. Nonetheless, it is a powerful antidote to those movements that simply demand health care for all, without asking what kind of health care.

A Radical Critique

A radical critique of medicine has to deal with the things that make people sick and the kind and quality of health care people get. A Marxist approach to health would attempt to integrate the insights of ecosystem health, environmental justice, the social determination of health, "health care for all," and alternative medicine. One aspect of my approach to the issues of health care comes from my background as an ecologist. I looked at variability in health across geographic locations, occupational groups, age groups, or other socially defined categories. Just how variable, I asked, is the outcome in health care in different states in the United States, different counties in Kansas, different provinces in Cuba, different health districts in a Brazilian state, or in a Canadian province? Interesting patterns emerged from that work. My colleagues and I examined the rate of infant mortality in each of these regions, both as an average and how, in each place, the rates varied, reflecting the quality of health care, among other factors, from the best to the worst. What we saw was that infant mortality rates in United States were more or less comparable to Cuba, that Kansas had a rate a little higher than the U.S. average, and Rio Grande do Sul in Brazil had a more typical, and much higher, third-world infant mortality rate. That Cuba scored so high was not very surprising.

However, when we viewed the same data from the perspective of the range from the best to the worst rates of infant mortality, that is, the variability within given populations, an effective measure of fairness, much more was revealed. The numbers for counties in Kansas showed the greatest variation, while the data that compared U.S. states showed somewhat less difference. The difference across health districts of Rio Grande was even less, and the least variation was in Cuba. Similar things happen when we look at all causes of death. Once again, we observed average rates as well as the disparity; we divided the variation, the difference between best and worst, by the average. For Kansas the range divided by the average is .85, but in Cuba it was .34. We saw that the cancer rates in Kansas and in Cuba are comparable, but the variability is higher in Kansas than in Cuba. When we examined Canadian data, we found that Saskatchewan was somewhere between Kansas and Cuba.

The reason we chose these places is that Brazil, Canada, and Kansas all have capitalist economies in which investment decisions are based on maximizing profit rather than any social imperative meant to equalize economic circumstance. Saskatchewan and Rio Grande do Sul along with Cuba have national health systems that provide fairly uniform coverage over a given geographic area. The Canadian and Brazilian regions have the advantage of a better and more just health care system but, unlike Cuba, have the disadvantages of capitalism, giving them an intermediate location in the variability of health outcomes.

This method can also be applied when comparing different diseases. One question we want to answer is whether variability will be greater across states and other large geographic regions, or across small areas like counties. There are good reasons why it might go either way. For example, weather could impact the data in large areas like states. But weather is not the only variable; others may vary greatly over smaller geographic units, only to be lost in the averages we develop for large areas. When we are able to look at smaller areas, like different neighborhoods within the city of Wichita, Kansas, we find a threefold variation in infant mortality. We also notice that unemployment in Kansas averages 9 or 10 percent in most Kansas counties but is 30 percent in northeast Wichita. Why? Because neighborhoods are not simply random pieces of environment. They're structured. Wherever there is a rich neighborhood, you need a poor neighborhood, like northeast Wichita, to serve it. And so whenever we can get data across neighborhoods, we see very large variations in social conditions and, as a consequence, in the quality and quantity of health care—clearly unnecessary from the point of view of any limitation in our medical knowledge or resources.

Another interesting case can be found in Mexico, where a study was conducted of several villages, ranking them according to how marginalized they were from Mexican life. Examined were such variables as whether there was running water or what proportion of the people spoke Spanish. The research showed that the more marginal communities had worse health outcomes. But, unexpectedly, the data also showed that there was tremendous difference among the outcomes in poor villages that you didn't get among the villages that were integrated into the Mexican economy.

It is an as-yet-unrecognized ecological principle in public health that when a community or an individual organism is stressed for any reason (low income, a very severe climate, for example), it will be extremely sensitive to other disparities. So, if people have very low income, changing seasonal temperatures become very important. In late autumn and early winter, emergency rooms have a lot of people coming in with burns from kerosene stoves, ovens, and other dangerous means used to compensate for inadequate heat in their houses. For such people, a small difference in temperature can have a big effect on their health—one that doesn't affect the more affluent. The same is true in relation to food. When people are unemployed, or if the prices go up, they cut back on food and other kinds of expenditures with an immediate impact on nutrition. If you are a superb shopper, and if you clip all the coupons and scrutinize the supermarket ads, you might just get by on the Department of Agriculture poverty-level basket; the people who dream up these baskets assume you are a wiz at finding bargains. But suppose you are not so good, or that you read the ads but cannot get away for two hours for comparison shopping. Or that you live in a neighborhood where the local supermarket was not as profitable as the national chain that owned it thought it should be, and is gone, and with it your opportunity to get quality food. Or suppose that you would love to eat organic food for lunch but what you have is a half-hour break to go down to the vending machines. Under those circumstances, individual differences in where you work, how much energy you have, whether you can have a babysitter available, can have a big impact on your health.

The Illusion of Choice

Poor health tends to cluster in poor communities. Conservatives will say, "Well, obviously poverty is not good for you, but after all, not all kids turn out badly. I made it, why can't you? Some people have become CEOs of corporations who came out of that neighborhood." What they miss is the notion of increased vulnerability. The apparently trivial difference in experience can have a vast effect on the health of someone who is marginal. Suppose a pupil is a bit nearsighted but, because she is tall, is seated at the back of the classroom. The teacher is overworked and does not notice that the student cannot see the blackboard. She fidgets; she gets into a fight with the kid at the next desk. Suddenly she has become someone with a "learning problem" and is transferred to a vocational course even though she might have been a great poet. In a more affluent community, where the classes are smaller and teachers pay attention, this kid would simply end up with glasses. Individual differences can come from anything, from personal experiences growing up, even from genetics. But even when genetics is responsible for a given human characteristic, it is only responsible within a particular context. For instance, in a factory emitting toxic fumes, people will develop cancer at a higher rate; those most likely to develop the cancer have livers that are not able to effectively process a particular chemical as well. This is a genetic variable and thus a genetic disease, but it occurs only with exposure to those fumes. The cancer is not a result of genetics alone; it is also caused by the environment.

Trivial biological differences can become the focus around which important life outcomes are located; the most obvious is pigmentation. The difference in melanin between Americans of African and European origin is, from the point of view of genetics and physiology, trivial. It is simply the way in which a pigment is deposited in the skin. Yet this difference can cost you ten years of life. So is this a lethal gene? Is this a gene for a higher spread of pigmentation—one that also makes you more vulnerable to arrest? A standard geneticist would look at family histories and determine that if your uncle was arrested, there would be a higher probability of you being arrested as well. Conclusion: the cause of criminality is genetic. Following the rules of genetics in this mechanistic way, he or she will have proved that crime is hereditary. This makes as much sense as the notion that black people get more tuberculosis because they have bad genes. Genetics is not an alternative explanation of social conditions; it is a component of an investigation of causal factors. There is an intimate interdependence among biological, genetic, environmental, and social factors.

Behavior is one of the areas where public health workers want to intervene, arguing that much that differentiates health outcomes in poor neighborhoods from rich ones can be associated with behavior, such as smoking, exercise, and diet. Conservatives, finally forced to concede that there are big differences in health outcomes between rich and poor, now say, "Yes, this is because the poor make unwise decisions. The appropriate remedy is education. We know that kids do better if their mothers have had more schooling, so what we need are education programs to teach people to make the best of their situation." But some health education programs are valuable. Safety orientation within factories does help people cope with unsafe conditions. So let us take a closer look at this question of choice. The Centers for Disease Control and others who deal with these issues say only some things can be chosen, while others are imposed by the environment. They would have us distinguish between disadvantages imposed on us, that may be unfair and/or can be eliminated, from those that were freely chosen and for which we can only blame ourselves. A Marxist confronted with choices among mutually exclusive categories like choice versus environment, heredity versus experience, biological versus social, knows that the categories themselves must be challenged. Choice also implies the lack of choice. Choices are always made from a set of alternatives that are presented to you by somebody else. We know this from elections and from shopping. We choose food, but only from the products a company has chosen to make available to us. The choice is distinguished by the lack of choice, that is, unchoice. The same is true with respect to the opportunity to exercise choice. There are always preconditions to the exercise of choice. If the conditions of life are very poor or oppressive, some of the things that are unwise choices under other circumstances become the lesser evil.

Public health people, like nearly everyone else, worry a great deal about teen pregnancies, which generally are not a good idea. Teen mothers are not experienced; they may have difficulty taking care of their babies; and the babies are more likely to be underweight. Nevertheless, it turns out that the health of a baby born to an African-American teenager is on average better than the health of a baby born to an African-American woman in her twenties. Why? The environment of racism erodes health to such an extent that it makes a certain amount of sense to have your babies early if you're going to have them. This is something that is not obvious when you simply say, "Teen pregnancy is a danger to people." We need to look at teen pregnancy in a much broader social context before we can think about making it simply a public health issue.

Smoking is another example. Smoking increases inversely with the degree of freedom one has at work. People who have few choices in life at least can make the choice to smoke. It is one of the few legitimate ways in some jobs to take a break and step outside. So there are people who choose: "Yes," they say, "it might give me cancer in twenty years, but it sure keeps me alive today." The unhealthy choices people make are not irrational choices. We have to see them as constrained rationality, making the best of a bad situation. Most of the apparently unwise decisions people make have a relative rationality to them when their circumstance is taken into account, so it is unlikely their behavior will change simply by lecturing to them. You have to change the context within which choice is made.

Yet another dimension of choice is found in the way we perceive time. When making a choice about health, we assume that something we do now will have an impact later on. That may seem obvious, but it is not the experience of everyone. Most people do not experience the kind or quality of freedom that gives them control over their own lives, that would allow them to say, "I will quit smoking now so that I won't get cancer in twenty years." Not everyone can organize their lives along an orderly annual timescale. In the inner city of San Juan, Puerto Rico, the life pattern is such that one can work unloading a ship for twenty-three hours a day for two days, then sleep for three days, then unexpectedly work in a restaurant for another two days because his or her cousin has to go to a funeral in the mountains. Time does not have the same structure when you can't make solid plans now for what is going to happen to you later.

In contrast, the lives of, say, academics are notable for the way time is organized. Students can and do choose courses of study that, in two or three years, will prepare them for a career. On a shorter timescale, a professor may conveniently order his or her teaching schedule around patterns of Monday, Wednesday, and Friday, or Tuesday and Thursday. Physicians decide when to see patients, when to be in the library, when to go to seminars. So some people can actually structure their lives in such a way that we can actually make predictions. Not absolute predictions, obviously. Things can come up; we can be hit by a car. But, basically, the more control you have over your life and your experience of life, the more it makes sense to make the kind of decisions that public health experts recommend, the more the possibility, then, of exercising choice. So the answer to those who talk about decision-making and choice is to tell them, first of all, to expand the range of choices. Secondly, they need to provide the tools for making those choices. Third, of course, people need to control their own lives, so that they can exercise all their faculties to make meaningful choices. In taking each of these steps, we directly challenge the false dichotomies that rule thinking about public health and constrain it within predetermined societal boundaries.

What Can Be Done?

At a recent meeting I attended, a paper was distributed that posed the following dilemma: Why, living in a democracy, where all citizens have the vote, do we permit policies that create inequalities that have such a negative impact on our health? How do we explain this? We have schemes to improve agriculture but they increase hunger. We create hospitals and they become the centers for the propagation of new diseases. We invest in engineering projects to control floods and they increase flood damage. What has gone wrong? One answer might be that we are just not smart enough. Or the problems are just too complicated, or we are selfish, or we have some defect. Or, after having failed to eliminate hunger, improve people's health, and do away with inequalities, and failed, perhaps we need to face facts and conclude that it just cannot be done. Or perhaps we're just the kind of species that is incapable of living a cooperative life in a sensible relation to nature.

We should reject any of these unduly pessimistic conclusions. The history of struggle is long and not without achievements. But struggle is also difficult. For example, it is easy to depend on the illusion of democracy and a beneficent government to solve our problems. But when we look at the policies that emerge from those institutions of democracy, we see that those ostensibly aimed at improving the people's lives are nearly always hobbled by some hidden side condition. I am sure that, on the whole, President Clinton would rather have people covered by health insurance than not. But that is subject to the side condition that insurance industry profitability must be protected. He probably would like medicines to be cheaper, but only if the pharmaceutical industry continues to make high profits. Abroad, the United States would like peasants to have land, but only if not expropriated from plantation owners. The basic reason that programs fail is not incompetence, ignorance, or stupidity, but because they are constrained by the interests of the powerful. Sometimes we discover that part of a program is carried out successfully, and part not. An enterprise zone might be established in an inner city that actually brings in investment, but there is no impact on poverty because the assumption that benefits would trickle down was an illusion. A reasonable return on investment was the goal of the developers. When that was achieved, nothing else mattered.

A good way to see how these hidden constraints, these systemic barriers, operate is in the delivery of health services elsewhere. Health care in the United States exists against the background of this country's unrestrained capitalism. We have described at length both the prospects and problems of that system. But, in Europe, social democrats historically have taken a different approach—one that acknowledges inequality as an obstacle. They have treated unemployment, for example, as a social problem rather than an inevitable byproduct of a vigorous market. A town council will address it by financing a center for the unemployed, with counselors to advise them of their right to unemployment insurance and other benefit programs. The center may even organize a support group where people can deal with their feelings about not being able to bring home an income to the family. Local governments can address other social concerns. In London, there is a program to break down the isolation of young mothers, where they can meet one another, share experiences, and provide support. Of course, none of these measures affects profitability or challenges the market. So the council cannot create employment. Even the most farsighted programs initiated by European social democratic governments do not challenge the capitalist order in any way. What they do is to try to make things more equitable—for instance, through progressive income taxes or generous unemployment insurance. In Sweden, transport workers demanded improved food to reduce heart disease among truck drivers. They organized to improve the quality of food in the roadside canteens and collaborated with restaurant owners and canteen owners and food was improved. In other places, unions have negotiated collective agreements to change shift work, hours of work, and working conditions. The unions recognized that health concerns were but another aspect of class relations.

In some cases improving on-the-job health is relatively cost-free. No employer will object to putting up a sign reminding workers to wear their hard hats on the construction site. But it begins to get a little tricky when you talk about the reorganization of work or the expenditure of money. If the expenditure of money comes from taxes, through government programs to improve health, we can expect the business class to object. And if, after each new expenditure, they perceive some interference with their competitive position, their opposition may take some political form—for example, the repeal of some aspect of health and safety regulation. When an expenditure has to come from the individual employer, perhaps by way of a union demand, they will be even more resistant. They will say that it is bad for competition and threaten to close down and move somewhere else. If the union's demands deal with the organization of work itself, management will see workers impinging on the very core of class prerogative. In that situation, only a powerful and well-organized labor movement will be able to impose changes.

When health policy is looked at from the point of view of which issues involve a direct confrontation of fundamental, ruling-class interest, which ones involve simply relative benefits to a class, and which are relatively neutral, we can predict which kinds of measures are possible. This highlights the lie in the notion that society is trying to improve health for everybody. We need to see health care in a more complex way. Health is part of the wage goods of a society, part of the value of labor power, and therefore a regular object of contention in class struggle. But health is also a consumer good, particularly for the affluent, who can buy improvements in health for themselves. Rather than improve water quality, they buy bottled water; rather than improve air quality, they employ oxygen tanks in their living rooms. Health is also a commodity invested in by the health industries including hospitals, HMOs, and pharmaceutical companies. They sell health care to as large a market as can afford to pay for it; they even push it on people who do not need it. Like any aggressive business, the health industries engage in public relations—the winning of hearts and minds. Some of the clinics that were established in Southeast Asia during the Vietnam War and earlier, during the Malayan insurrection, were for this purpose. Doctors, at great sacrifice, would go into the jungle and set up clinics and work very hard under very difficult conditions for low pay, seeing themselves either as bringing benefits to people who needed it or, more consciously, as trying to prevent communism. It was yet another reincarnation of the White Man's Burden that justified nineteenth-century imperialism.

If good health depends on your capacity to carry out those activities that are necessary and appropriate according to your station in life, it matters how that station is determined. Those who can determine for themselves what constitutes necessary and desirable activities are clearly different from the people who have that determination made for them. This distinction is clear when an employer negotiates health insurance for his or her employees; for the employer, the cost of the benefits package will always come before what employees may think they need. So health is always a point of contention in class struggle. So is medical and scientific research; knowledge and ignorance are determined, as in all scientific research, by who owns the research industry, who commands the production of knowledge production. There is class struggle in the debates around what kind of research ought to be done. Increasingly, research in the health field is dominated by the pharmaceutical and electronic industries.

There are intellectual concerns about how to analyze data, about how to think about disease, about how widely we need to look at the epidemiological, historical, and social questions they raise; there are also issues of health service and health policy. But these issues are all part of one integral system that has to be our battleground in the future. We have to take up health as a pervasive issue as we do with problems of the environment; they are aspects of class struggle, not an alternative to it.

28

Science and Progress:
Seven Developmentalist Myths in Agriculture

In the third world, the view of "science" as unqualified progress is expressed in developmentalism, the view that progress takes place along a single axis from less developed to more developed, and therefore the task of the revolutionary society is to proceed as quickly as possible along that axis of progress to overtake the advanced countries. The consequence of this view is the rapid reproduction of the worst features of world (capitalist) science and technology, the uncritical acceptance of the "modern." Developmentalism fails to recognize that the pattern of modern technology is not dictated by nature but is developed through the interaction of the capitalist need to control the labor force, the desired outcome of research in the form of commodities, the intellectual climate in which the scientists work, the pattern of knowledge and ignorance coming from previous work, and the nature of the scientific problems to be solved.

Here I examine the developmentalist ideology as it applies to agricultural technology, and contrast it with a more dialectical, political, and ecologically based approach. In the field of agricultural technology, developmentalism is supported by seven myths about what is "modern":

1. *Backward is labor-intensive, modern is capital-intensive agriculture.* From this it follows that criticism of high-tech agriculture is misinterpreted as an appeal to return to primitive hand cultivation, and the critics are accused of

"trying to deny to our people the advance that you have achieved." This reflects first of all a misunderstanding of development. Its view of technology is a very nineteenth-century one, a thermodynamic model in which vast amounts of energy are used to move vast amounts of material. It has proven to be the least efficient production system in terms of energy; harmful to its productive base through erosion, water depletion, salinization, destruction of organic matter and soil organisms; and it endangers the health of people and wildlife.

The alternative that some of us have been advocating is more analogous to physiology and electronics. Hormones are tiny quantities of matter that produce big effects; the nerve impulse is an insignificant amount of energy compared to the motion it is capable of initiating. The strategy of ecological agriculture is not the invention of a bigger bag of tricks allowing more kinds of interventions in crop production but the design of systems that require minimum intervention. This is achieved by detailed knowledge of the processes affecting soil fertility, the population dynamics of insects (both pests and useful), and microclimatology. It is not an anti-technology stance. My own experience as a farmer in the mountains of Puerto Rico preparing land with a hoe left me with no nostalgia for that most burdensome of tasks. But rather than seek to remove the physical toil of plowing from human muscle to giant machines, we look for ways of loosening soil structure and reducing the tillage requirements. Our claim is that *the evolution of agricultural technology should be from labor-intensive through capital-intensive to knowledge- and thought-intensive.*

2. *Diversity is backward, uniform monoculture is modern.* Once again, this myth is based on experience: the diversity of the *minifundia* has in many places been replaced by the uniformity of agribusiness's monoculture. But monoculture inevitably creates new and serious pest problems, prevents us from using the variability of soils and climate to our advantage, depletes the soil, and makes necessary the heavy use of costly inputs.

Agroecologists see the possibilities of using patterns of diversity to manipulate the microclimate; for example, a shelter belt of trees on a hillside can act as a dam holding back the flow of cooler air and creates a belt of warmer air for a distance of about ten times the height of the trees. This belt, broad enough to farm with suitable mechanization, allows cultivation of a crop that requires higher temperatures. Diversity helps pest control; for instance, by one crop's providing the nectar needed by wasps that parasitize pests of another crop or by interrupting the spread of an epidemic. Crop diversity would also allow better use of labor in the face of the uncertainty of nature since some crops such as tomatoes require rapid harvest as soon as ripening occurs, while others such as cassava can be left in the ground until needed. The advantages of socialism will show up not so much in the management of larger monocultures, but in the planning of diversity. The evolutionary sequence should be from the random heterogeneity of the peasant minifundia through the homogeneity of the capitalist agribusiness, to the planned heterogeneity of an ecologically rational agriculture.

3. *Small scale is backward, large scale is modern.* The economies of large scale are recognized, but the disadvantages must also be acknowledged. For example, in dairying it helps to get to know the individual cows in a herd in order to adjust nutrition and detect the first signs of disease. This usually means that herds larger than 50 or 100 animals are not as productive as smaller herds. In field crops, large scale prevents utilization of each piece of ground for its most suitable cropping pattern. Therefore, there is an optimum size of plot that is not the greatest possible but rather large enough to make use of the necessary mechanization, and small enough to permit the use of edge effects, mosaic patterns of diversity, and adaptation of crops to topography. The unit of planning must be large in order to take into account regional patterns of hydrology, pest migrations, labor supply, and consumption needs. But the unit of planning is not the same as the unit of production, which can be much smaller. This is not a "small is beautiful" position but rather the search for an optimal geographical scale for the needs of a revolutionary society.

4. *Backward is subjection to nature, modern implies increasingly complete control over everything that happens in the field or orchard or pasture.* However, nature is inherently variable. We can override part of the natural variability by major, costly inputs. But these often create a new vulnerability that replaces the old. For instance, irrigation works reduce the immediate dependence on rainfall but increase the vulnerability to variations in the price of oil. High-yielding crop varieties often depend on a complete technical package under optimal conditions, and lose their advantage when there is

unusually severe weather or the package is not available. Small differences in the weather can drastically change the synchrony between crops and their pests or pollinators, and between pests and their predators. And every change we make in nature changes the direction or intensity of natural selection, causing new directions of evolution in the many species that coexist with us. New pests adapt to our new crops or technologies, old ones acquire resistance to our control measures, and beneficial predators may lose interest in the prey we had in mind for them.

An ecologically rational strategy would not pretend to set up a final, fully controlled production system but would acknowledge and utilize the variability of nature in several ways: climatic monitoring for planting schemes that take into account the trends in temperature and moisture on a scale of years or decades (the years of poor crops and reduced rainfall in the USSR are roughly ten years apart and have some predictability); growing together crops with somewhat different environmental requirements to guarantee the food supply no matter what the weather (wet years in the U.S. Midwest are better for corn but worse for wheat, which suffers more from fungus diseases in wet years); the selection of plant varieties and of mixtures of varieties for their broad tolerance of the unexpected; the dependence on an array of natural pest control measures, with predators and parasites that have their own patterns of response to the environment; and systems of redistribution so that local failures and bumper crops can compensate for one another without major impacts on human welfare.

5. *Folk knowledge is backward, scientific knowledge is modern.* The struggle against superstition has been and continues to be an important part of the process of liberation. However, in recent times there has been a growing appreciation of folk knowledge, particularly in the healing arts and agriculture. The aggressive claims that science is the only way to knowledge have been used to justify a chauvinist, class-based, and sexist contempt for the intellectual achievements of third world peoples, workers, and women of all countries. These claims are false. All knowledge does indeed come from experience, direct or indirect, and the reflection on that experience with intellectual tools derived from previous knowledge and experience. Modern science is one way in which that experience has been organized and used consciously for the purpose of acquiring knowledge. But all peoples learn, experiment, and analyze. In

agriculture as in health, in industrial production, and indeed in all spheres, the best condition for the creation of new scientific knowledge requires the combination of the detailed, intimate, local, and particular understanding that people have of their own circumstances with the more general, theoretical, but abstract knowledge that science acquires only by distancing itself from the particular. This can only come about under conditions of equality between the scientists and the consumers of science and therefore is observed most often in the context of radical political upsurges and revolutionary reconstruction of society.

Mass participation in innovation is especially feasible in agriculture, where the experience in one place supplements rather than competes with knowledge created elsewhere, and where the objects of interest are usually on the scale of the objects of everyday life, unlike atoms or molecules. It also requires liberation from the cult of expertise and specialists. In agriculture, the adoption of gentle, ecologically rational technology must be highly site-specific and requires the joint development of knowledge by research stations and farmers. In Cuba, the Young People's Technical Brigades and the National Association of Innovators and Rationalizers, as well as the amateur botanists' groups and conferences of farm innovators, encourage this process of mass participation. Besides tapping the creative abilities of the people, these activities create an awareness of the nature of scientific practice so that farmers know what experiments are, can judge them critically, and meet the scientists as equals, as comrades in a common endeavor. And it helps to wean the scientists away from the temptations to look north for all wisdom.

6. *Specialists are modern, generalists backward.* The rational kernel in this view is that there *is* too much to know within every discipline for anyone to know everything. The history of European thought has been an increasing subdivision of knowledge from the days of the philosopher-scholar-theologian, through general "scientists," to the present multiplication of specialties within previously coherent fields of study. For example, genetics, a part of biology, now includes molecular genetics, cytogenetics, population genetics, quantitative genetics (for plant and animal breeding), as well as further breakdowns by kinds of organisms studied. Developing countries speak with pride of the numbers of specialists who graduate from their schools. Uncritical admirers of specialization propose that groups of specialists working as teams can solve problems related to the subdivision of knowledge within a field.

However, specialization prevents the researchers from seeing the whole picture, both because of the narrowness of their training and because the ideology of expertise makes it a matter of pride to consider only precise, quantitative information as real science while the rest is "philosophy" (a bad word among positivistic scientists) or "not my department." The training of specialists rather than the education of scientists encourages the combination of micro-creativity and docility that permits scientists to work on the most monstrous projects of destruction without attention to their consequences. The great failings in the application of science to human well-being have come about not because of the failure to know the details of the structure or workings of something but because of the failure to examine the system in its complexity. The strategy of the Green Revolution is solving many and difficult technical problems of plant breeding, but the geneticists did not anticipate problems of pest ecology, land tenure, or political economy, and as a result increases in production are sometimes associated with increases in misery. The Aswan Dam was an engineering success in that it retained the water it was intended to retain. But by stopping the seasonal flooding that provided renewed soil fertility, the dam made farmers dependent on imported chemical fertilizers; the reduced flow of water into the Mediterranean Sea increased salinity and adversely affected fisheries; the outflow of the Nile was reduced to the point that it could no longer offset the erosion of the coastline; the irrigation ditches became the habitat for snails that transmit liver flukes.

It is a common experience that in large programs of development the ministries of health and agriculture do not talk to each other; thus it comes as a surprise when the expansion of cotton production increases malaria. Cotton is very heavily sprayed. The natural enemies of the mosquitoes are killed, allowing the mosquitoes that transmit malaria to thrive in habitats created for them by the clearing of forest. The immigration of a labor force not previously adapted to malaria allows the parasites ideal susceptible hosts. There is a vast oral tradition of such cautionary tales. The point is that most of these "unexpected" outcomes are predictable, at least in principle. There is no longer any excuse for planners not to ask the obvious questions about a program, such as: what will it do to the position of women? New technologies are usually handed over to men, and traditional women's occupations are displaced. For instance, the use of herbicides displaces women from weeding. How will vegetational changes alter the biology of potential disease vectors? Will the new productive activity be compatible with the water needs of the people? Will the production of export crops make the food supply more vulnerable?

The outcome of short-sighted specialization is that each department takes as its starting point the products of the department next door. Crops are bred for their performance in monoculture because the machinery was designed for operations in pure stands of a single crop. The engineers design machinery for monoculture because the agronomists inform them that it can replicate what farmers do. The farmers plant monoculture because their varieties and machinery are suitable to monoculture. Each party is making rational decisions given the constraints imposed by the others, giving the whole trajectory of technological development the appearance of inevitability and necessity, while nobody looks out for the process as a whole.

7. *The smaller the object of study, the more modern.* This is a continuation of the old pecking order of the sciences introduced by Comte, which sees the study of atoms as superior to the study of molecules, which is superior to the study of cells, which is superior to the study of organisms, which is superior to the study of populations, etc. This approach also elevates laboratory experiments above field studies, investigations of collections of specimens, and theoretical work done in libraries. In its present insidious form, it uses the term "modern biology" to refer exclusively to molecular genetics, genetic engineering, and biotechnology. It implicitly relegates other branches of biology to lower ranks and has even threatened the existence of museum collections, which take decades to develop. It is important to recognize that a modern systematic biology, modern population genetics, modern community ecology or biogeography or epidemiology also exist. The reductionist bias implicit in this very narrow view of "modern" is especially harmful in developing countries, which often give overwhelming priority to the more expensive branches of modern biology while ignoring the other areas which are of equal practical and theoretical importance. A healthy biological science requires the combination of study at all levels of organization and in the laboratory, field, library, and museum.

The struggle against these developmentalist myths about modernization is not anti-scientific. Rather, it is a program for a different kind of science. We have to insist that existing "modern" high-technology agriculture is not generic progress but a particular form of technological development under capitalist political or intellectual domination. The alternative is the development of new technologies designed not to create new input commodities or control a reluctant labor force but to provide high, sustainable yields of necessary products with minimum use of resources and damage to the environment and to people, and in a work process that promotes health and creativity.

The arguments outlined here in relation to agricultural science and technology have their parallels in critiques of medicine, industrial design, urban planning, and, indeed, all of applied science. In each case, we must recognize that the needs of capitalism for profit and social control over labor set a general agenda for science; the recruitment and organization of scientists creates a scientific community that accepts that agenda; and the ideology of science generates the intellectual environment within which the prevailing directions seem to be self-evidently the only way to proceed.

29

The Maturing of Capitalist Agriculture: Farmer as Proletarian

We are all familiar with the classical story of how capitalism came to dominate industrial production and how capitalist relations of production swallowed up the individual artisanal producer. We recognize the power that the capitalist mode has to infiltrate and finally transform other forms of the organization of production and exchange. We sometimes think that the power of that transformation is so great that all of the significant action already occurred in the past, at least in Europe and North America, and was essentially over by the end of the nineteenth century. In the society we inhabit, it is a *fait accompli* whose dynamic we can only understand by reconstructing the past because it is not happening around us. On second thought, we realize that the transition was still in progress until very recently in a few skilled domains like medical care and entertainment, where individual artisans were able to ply their trade throughout most of this century, but these fossils of early capitalist relations seem exceptional because of their requirement of special talents or of necessary skills acquired by long training. But the view that the transition to mature capitalism is essentially over except at the margins of the main body of commodity production is clearly wrong, because it ignores an immense sector of basic essential commodity production, agriculture, which is still in the throes of the transition.

The penetration of capital into agriculture has been a long process of a different form than the classical case, of industrial production usually exemplified by cloth weaving in the eighteenth and nineteenth centuries. Indeed, on the surface agriculture would seem to have been resistant to capital. After all, despite a 72 percent drop in the number of individual farm enterprises in the United States from 6.7 million in 1930, there are still about 1.8 million independent farm producers today. This means that even though only 6 percent of these establishments account for 60 percent of the total value of farm production, there are over 100,000 separate enterprises producing more than half of all the value of the output. In the industrial manufacturing sector the four largest enterprises account for an average of 40 percent of value produced, and even in a highly differentiated product like clothing the top four companies produce over 15 percent of the value.

There has also been a major increase in the proportion of farmland that is leased to farmers who also own their own land. Roughly 55 percent of farmland is now operated by owner-renters who are, for the most part, small producers. Finally, despite the conventional wisdom that corporate farming is taking over, the proportion of farms and farmland operated by managers representing absentee owners has remained about 1 percent since the last century. Thus if we are to look for evidence of the capitalist transformation of agriculture, we will not find it in the classical industrial model. We do not find a concentration of more and more productive capacity in the hands of a very small number of farmers, employing a large wage labor force carrying out its tasks under close supervision and according to a tightly controlled schedule. There are, of course, some examples of a factory-like labor process in farming, especially in the harvesting of fruits and fresh vegetables, and these are often pointed to as evidence of a capitalist transformation to factory farming. However, the vast majority of farm enterprises do not employ a large labor force, but more typically have one or two hired laborers, usually for only part of the year.

In analyzing the process of the capitalist transformation of agriculture we must distinguish between farming and the agrifood system. Farming is the physical process of turning inputs like seed, feed, water, fertilizer, and pesticides into primary products like wheat, potatoes, and cattle on a specific site, the farm, using soil, labor, and machinery. The failure of classical capitalist concentration in farming arises from both financial and physical features of farm production. First, the ownership of farmland is unattractive to capital because it cannot be depreciated, and investment in farmland has very low liquidity as a consequence of the thin farm real estate market. Second, the labor process on very large farms is hard to control because farming operations are spatially extensive. Third, economies of scale are hard to achieve beyond what has already been realized by medium-scale enterprises. Fourth, risks from external natural events like weather, new diseases, and pests are hard to control. Finally, the cycle of reproduction of capital cannot be shortened because it is linked to an annual growth cycle in plants, or a fixed reproductive cycle in large animals. An important exception to this constraint has been in poultry, where there has been considerable success in shortening the reproductive cycle, and this has had important ramifications for the development of capitalist farming, as we will see. For all of these reasons we do not expect to see, and have not seen, the wholesale direct takeover of farm ownership by large corporate enterprises employing large, well-controlled labor forces.

The agrifood system, however, is not simply farming. It includes the farm operation, and the production, transportation, and marketing of the inputs to farming, as well as the transportation, processing, and marketing of the farm outputs. Whereas farming is a physically essential step in the entire chain of agricultural production, the provision of farm inputs and the transformation of farm outputs into consumer commodities have come to dominate the economy of agriculture. Farming itself now accounts for only about 10 percent of the value added in the agrifood system, with 25 percent of the food dollar paying for farm inputs and the remaining 65 percent gained by transportation, processing, and marketing that converts farm products into consumer commodities. At the beginning of the century the value added on the farm was around 40 percent of the total food dollar, and many of the inputs were produced directly on the farm in the form of seed, draught animals, feed for the animals, manure and green manure for fertilizer, and family labor. For the most part these inputs are now purchased in the form of commercial seed, tractors, fuel, refined or synthesized chemical fertilizers, and machinery and manufactured chemical substitutes for labor. Thus it is the production of farm inputs and the transformation of farm outputs that have provided an opportunity for industrial capital to capture profits in the agricultural sector.

Like any other industrial processes, the production of farm machinery, chemicals, and seeds, and the turning of threshed wheat into a box of breakfast cereal at the supermarket checkout counter are completely controlled by capital and its demands. The problem for capital, however, is that sitting in the middle of the transformation of petroleum into potato chips is an essential step, farming, in the hands of two million petty producers. They cannot be dispensed with, they own certain essential means of production whose ownership cannot be concentrated (land in particular), and, though economically rational, consume their surplus rather than turning it into capital. Agriculture is unique among all the sectors of capitalist production by possessing at its productive center an essential process organized around large numbers of independent petty producers. It is as if the spinning of yarn, the weaving of cloth, and the sewing of garments were in the hands of a few large capital enterprises (as they are), but the dyeing and finishing of the raw woven material were unavoidably the exclusive province of hundreds of thousands of home producers who bought the unfinished cloth and sold their product to clothing factories.

Farm producers have historically been in possession of two powers that stood in the way of the development of capital in agriculture. First, farmers could make choices about the physical process of farm production, including what was grown and how much, and what inputs were to be used. These choices were always constrained, partly because of local conditions of climate and soil, and partly because of the local nature of markets for farm products. Second, farmers were themselves traditionally potential competitors with the commercial providers of inputs, because they could choose to produce seed, traction power, and fertilizer themselves. The problem for industrial capital, then, has been to wrest control of the choices from the farmers, forcing them into a farming process that uses a package of inputs of maximum value to the producers of those inputs, and tailoring the nature of farm products to match the demands of a few major purchasers of farm outputs who have the power to determine the price paid. Whatever production risks remain are, of course, retained by the farmer. As the farmer loses any power to choose the actual nature and tempo of the production process in which he or she is engaged, while at the same time losing any ability to sell the product in an open market, the farmer becomes a mere operative in a determined chain whose product is alienated from the producer. That is, the farmer becomes proletarianized. It is of little import that the farmer retains legal title to the land and buildings and thus, in some literal sense, is the owner of some of the means of production. There is no alternative economic use for these means. The essence of proletarianization is in the loss of control over one's labor process and the alienation of the product of that labor.

How has this transformation of farming been accomplished? In the first stages, in the century between the invention of reaping machines and the end of the Second World War, innovations in farming directly addressed that problem of the availability, cost, and control of farm labor through mechanization. No farmer could resist the arrival of the tractor, nor could one be home-built. After the Second World War, refined and synthetic chemical treatments became the chief purchased inputs in the form of fertilizer, insecticides, and labor-saving herbicides. Again, these purchased inputs could not be resisted because of the large increase in yields and the reduction in labor. Herbicides, in particular, also reduced the requirement for tillage machinery, insecticides reduced the uncertainty of a successful crop, hormone sprays allowed for a close control of ripening time in fruit crops, and antibiotics prevented animal diseases. Once again there could be no competition between these industrial products and self-produced farm inputs.

The analysis of the growing role of capital inputs cannot be made, however, if we lose sight of a central feature of the productive process: *the concrete use of all these inputs is to produce living organisms.* The steps of mechanization and the use of chemicals were not possible in isolation from the nature of the organisms being produced. In agriculture, unlike in other sectors of production, living organisms are at the nexus of all input streams and are the primary sources of all output transformations. But living organisms are mortal, so their *production* requires their *reproduction.* That is, every cycle of farm production begins with seeds or immature animals to which value is added by on-the-farm operations, so seed (or the "seed" animal) is the central input into farming. The control of the biological nature of these seed organisms is a critical element in the control of the entire process of agricultural production, which puts the provider of this input in a unique position to valorize other inputs. For example, while a dramatic drop in the price of nitrogen fertilizer at the end of the Second World War made it economically possible for farmers to use this input in large quantities, for this input to be useful it was necessary to breed plants, corn hybrids in particular, that could turn a massive nitrogen application into crop yield. The successful mechanization of tomato harvesting was only possible by a close cooperation of the machine designers with plant breeders. The breeders completely remade the biology of the tomato plant, turning a loosely branched plant that flowers and sets easily bruised fruit continuously over the growing season into a short, stout, Christmas tree-like plant whose tough fruits all ripen at about the same time.

The consequence of the central position of seed input in the production process is that seed companies are potentially in an extraordinarily powerful position to appropriate a large fraction of the surplus in agriculture. There is a barrier to this realization, however. The seed of a desirable variety, when planted by the farmer, produces plants that produce yet more seed of the variety. Thus the seed company has provided the farmer with a free good, the genetic information contained in the seed, that is reproduced by the farmer over and over again in the very act of farming. Some way must be found to prevent the farmer from reproducing the seed for next year's crop. The historical answer to this problem was the development of the inbred/hybrid method of breeding, using hybrid crosses between inbred lines, which makes it possible to sell seed that will produce hybrid plants, but which do not reproduce hybrids. Because the second generation would not be true hybrids and thus would lose yield and be more variable, the farmer must go back to the seed company every year to buy new seed. As a result of the immense profits made by seed companies selling hybrid maize seed, the method was spread into other organisms such as tomatoes and chickens. Moreover, major commercial hybrid seed and chicken breeders like Dekalb, Funk, and Northrup-King were at one time acquired by pharmaceutical and chemical companies like CibaGeigy, Monsanto, and Dow, although subsequently there have been divestments and realignments. Only the largest hybrid seed company, Pioneer Hy-Bred, remained obdurately independent until, in 1997, 20 percent of its equity and two seats on its board were purchased by DuPont.

Generally, the ability of commercial seed companies to control seed inputs by the inbred/hybrid method was severely limited. First, the method cannot be made economically workable in many important crops like soybeans and wheat, or in large animals. Second, though the inbred/hybrid method was successful for general yield increase, large numbers of important specific characters, such as resistance to particular diseases or resistance to herbicides, or increases in oil content in oil seeds, do not show hybrid vigor and must be introduced by other breeding methods. Third, there are characteristics that would be desirable to introduce into an agronomically important species but which are present in other organisms that cannot breed with the species under cultivation. The most famous example was the desire to make corn plants able to fix nitrogen from the atmosphere, as legumes are able to do, by making their roots hospitable to nitrogen-fixing bacteria. Although this would reduce the market for nitrogenous fertilizers, it would place the provision of nitrogen in the hands of seed companies!

The limitations on what changes could be made to agronomic species that would be profitable to seed companies and their chemical company partners or owners meant that the penetration of capital into agriculture had reached its apparent limits in the 1970s. The introduction of major new forms of mechanization into farm production had come to an end, partly because of the dramatic change in the cost of fuel and partly because a steady supply of immigrant labor that could be deported stalled progress in agricultural labor organizing. The growing public consciousness of the polluting effects of fertilizers and pesticides and the development of OSHA regulations to protect farmworkers against the deleterious effects of insecticide and herbicide sprays discouraged radical changes in the uses of chemicals or even continued growth in the use of older materials. In addition, these fertilizers and pesticides were being used at very high rates, probably higher than could be economically justified by the farmer. There was, for example, no growth in fertilizers after 1975 or in synthetic pesticide application rates beginning in about 1980. Any further possibility for input providers and output purchasers to increase their appropriation of the surplus in agriculture depended on 1) making some radical changes in the biology of agronomic species; and 2) guaranteeing that such changed biological systems would remain within their ownership and control. Moreover, that appropriation could be greatly increased by a greater consolidation of both input and post-farm production sectors (purchasing, processing, and distribution), to provide near monopoly control. Enter biotechnology.

Biotechnology and the Control of Property

The purpose of the commercial use of biotechnology is to extend the control of capital over agricultural production. To accomplish this purpose biotechnological innovation should meet three criteria. First, the time and cost of its development must be within the limits set by capital investments in research. Thus the attempt to introduce nitrogen fixation into non-leguminous plants has been largely abandoned by Agricetus, Agrigenetica, Biotechnica, and other biotechnology enterprises after spending over $75 million on the problem over more than ten years, despite the evidence that it ought to be possible, and despite the immense profits that could be made if it were successful. Second, the development must not provoke a significant challenge from politically effective forces concerned with health and environmental issues. All biotechnological innovations have been challenged on the basis of environmental and health risks, and this contributed significantly to the demise of at least one early biotech project. An important impetus for introducing biotechnology is that the resistance to further applications of fertilizer and pesticides was impeding further increase in the appropriation of the surplus in agriculture by input producers. Third, *ownership and control of the product of biotechnology must not pass into the hands of the farmer but must remain with the commercial provider of the input.*

The requirement that the biotechnological innovator maintain ownership and control over the altered variety creates a contradiction. As previously discussed, the farmer acquires a free good, the genetic information contained in the seed, when he or she purchases a new variety, and the breeder loses its ownership. The property rights protection offered by the inbred/hybrid method is limited to a few organisms and a few agronomic characteristics, and biotechnology has been introduced in precisely those instances where the inbred/hybrid method does not apply. How, then, can breeders appropriate a greater share of the surplus when they are giving away the critical material, the genes? The answer has been provided by a combination of legal and biological weapons in the hands of the breeders. These weapons are legal rights granted to breeders by the Plant Variety Protection Act and subsequent court decisions, in combination with the use of standard DNA "fingerprinting" that allows an unambiguous determination of the source of farm products. It is now standard that a farmer who wishes to purchase a bioengineered seed must sign a contract with the seed producer giving away all property rights in the next generation of seed produced by the crop. Not only does the farmer undertake not to sell seed from the crop to other farmers (brown bagging), but, more revolutionary, *the farmer is prohibited from using the next generation of seed to produce next year's crop on his or her own farm.* All farmers who buy seed of Monsanto's Roundup Ready soybeans, or that company's seed potatoes for a special variety that makes "light" potato chips with low oil retention, must, by the terms of the contract, return to Monsanto in the next season if they wish to continue production of those varieties. (Monsanto is the producer of Roundup, a potent herbicide that kills all plants including soybeans. "Roundup Ready" soybeans, produced by genetic engineering, can be grown in fields heavily treated with Roundup without killing them and, presumably, without materially affecting their yield.) The enforcement of such a contract depends on the ability of Monsanto to identify a crop, and this can easily be done from a single plant or even a single seed because the DNA of the engineered variety contains certain characteristic sequences, placed there deliberately by the genetic engineers, that are unique to the variety. The assay of crops for such labeled sequences is called "genome control" by the biotechnology laboratories of seed producers, and a considerable laboratory effort has been put into developing these detection techniques. Nevertheless some brown bagging and replanting has been taking place. In reaction, Monsanto has placed full-page advertisements in magazines read by farmers, threatening and cajoling:

> When a farmer saves and replants Monsanto patented biotech seed, he understands that what he is doing is wrong. And that, even if he did not sign an agreement at the time he acquired the seed [that is, replanted or bought "brown bag" seed from a neighbor], he is committing an act of piracy. . . . Furthermore, seed piracy could cost a farmer hundreds of dollars per acre in cash settlements and legal fees, plus multiple years of on-farm and business records inspection.

It only takes a few widely publicized legal judgments to keep the rest in line. But the story of property rights has yet one more chapter. The inbred/hybrid method only applies to a few organisms, and the contract system requires threats, monitoring, and litigation to make it work. It is biotechnology that has now perfected the solution to ownership in seed crops. It was announced on March 3, 1998, that a patent had been granted for a genetic manipulation that would allow plants to set seed and therefore make a crop, *but would render those seeds unable to germinate.* Thus, at one blow, the problem of capitalist seed production, first addressed by the invention of the inbred/hybrid method at the beginning of this century, has been solved for all seed crops. As the inventors point out, there is still development to be carried out before this bit of biotechnology becomes a commercial reality, but there seems no bar to its transfer to any crop. And who are the inventors and owners of this patent? They are the Delta and Pine Land Company, a leading breeder and producer of cottonseed and soybean seed, and the Agricultural Research Service of the United States Department of Agriculture. Yet there is no suggestion that this development will be of any benefit to farmers or consumers. We could hardly ask for a more blatant case of state support of private property interests to the exclusion of any public benefit.

The use of the contract to enforce breeders' property rights allows us to make some predictions about the limitations of genetic engineering. At the present time the hormone BST, which causes dairy cows to direct more of their metabolism to milk production, is produced commercially by Monsanto in fermenters using genetically altered bacteria. But cattle normally produce their own BST, and there is no reason that the regulatory DNA that controls production of this protein in cows could not be altered to increase the amount. This would then make the purchase and administration of commercial BST unnecessary. We can predict, however, that this is unlikely. First, dairy herds have always been largely self-reproducing on small to medium-sized enterprises, and there are no major commercial dairy herd breeders equivalent to major seed companies. Second, enforcement would be very difficult. It is easy for a representative of Monsanto to "acquire" a single potato or a few seeds from any farmer's field, or from a local elevator. It is considerably more intrusive to take the blood or tissue sample from a farmer's dairy herd that would be needed for "genome control." Moreover, since dairy herds are not all reproduced at one time but have overlapping generations, it would be impossible to say, except after a number of years, whether a cow was one of the originally purchased stock, or an offspring.

Production Contracts, Biotechnology, and the Control of Farming

If the only effect of biotechnology and the contract system of guaranteeing property rights were to extend the domain of manufactured inputs into farming, nothing very revolutionary would have occurred. Farmers for a long time have been the purchasers of manufactured inputs. The major structural changes that are occurring in agriculture arise from a vertical integration of farm production in such a way that the purchasers of farm outputs take control of the entire production process. This vertical integration is made possible by 1) a technical linking of the inputs and outputs; 2) the dual function of a single capital enterprise as both the monopsonist (near monopoly) purchaser of outputs and the provider of critical inputs; and 3) a contract mechanism that links farmers into the loop of inputs and outputs. The use of such contracts predates biotechnology. Wherever the purchaser of farm outputs is also the processor of those outputs for the market, the possibility of vertical integration has existed. Contract farming has been a common feature of vegetable production for canning. Tomato canneries in Ohio were built in a location central to the farms, the canning company provided the seed and chemical inputs and collected the ripe tomatoes. The farmer provided the land and labor. But the system has evolved greatly since the first canning contracts. The critical role played by biotechnology has been in the material linking of inputs and outputs. In order to guarantee an efficient integrated system of production, the biological inputs into the chain of production, the organisms being raised, are engineered to fit the package of other inputs, the mechanics of the farming process, and the qualities that the final output is to have for the market. Whereas some of these aims can be accomplished by conventional methods of breeding organisms, many of the needed qualities, such as specific disease resistance or qualitative changes in the composition of the organism, are best produced by biotechnological manipulations. Moreover, various cloning and cell culture techniques make it possible to reduplicate large numbers of input organisms with desired heritable qualities, no matter how those qualities were originally produced.

An example of the nature of contract farming is in the production of broilers (chickens raised as meat) where the system is especially entrenched. A major supplier of chickens to supermarkets and fast-food restaurants is Tyson Farms of South Carolina. Tyson chickens are produced, not by Tyson "Farms," but by small farmers, owning about 100 acres, producing an average of 250,000 chickens per year, with a gross income of about $65,000 and a net of around $12,000.

This production is under a four-year contract with Tyson (or other similar regional firms), a contract that makes Tyson the sole provider of the chicks to be raised, the feed, and veterinary services. The company is also the sole determiner of the number, frequency, and type of chicks provided. Tyson then collects the mature birds after seven weeks, at a date and time of their own determination, providing the scales on which the birds are weighed and the trucks to take them away. The farmer provides the labor, the buildings in which the chicks are raised, and the land on which the buildings stand. The detailed control of inputs and farming practices are entirely in the hands of Tyson. So, "the Producer (farmer) warrants that he will not use or allows to be used . . . any feed, medication, herbicides, pesticides, rodenticides, insecticides or any other items except as supplied or approved in writing by the Company." In addition, the farmer must adhere to the Company's "Broiler Growing Guide" and a failure to do so puts the farmer into "Intensified Management" status under the direct supervision of the Company's "Broiler Management and Technical Advisor."

The chicken farmer has ceased to be an independent artisan, buying materials, transforming them by his or her labor, and selling the product on a market. The contract farmer buys nothing, sells nothing, nor makes any decisions about the physical process of transformation. The farmer does own some of the means of production, land and buildings, but has no control over the labor process or over the alienated product. The farmer has then become the typical "putting out" worker characteristic of the first stages of capitalist production in the seventeenth and eighteenth centuries. What the farmer has gained is a more stable source of income, at the price of becoming an operative in an assembly line. The change in the farmer's position from an independent producer, selling in a market with many buyers, into a proletarian without options, is reflected in the nature of the recommendation in the 1998 report of the National Commission on Small Farms:

> Congress should amend the AFPA [Agricultural Fair Practices Act] to provide the USDA with administrative enforcement and civil penalty authority that will, in turn, *enable growers to organize associations and bargain collectively without fear of discrimination or reprisal.* [emphasis added]

The combination of biotechnological manipulation and contract farming can also have a catastrophic effect on third world economies. Much of the import of agricultural products from the third world consists in qualitatively unique materials like coffee, flavorings, essences, and food oils with special properties. Moreover, the production of these materials is at a low technological level with high labor inputs, in countries with unstable political and economic regimes. As a result, the price and availability of, say, palm oil from the Philippines are unstable. The characteristics make such agricultural products prime targets for gene transfer into domestic species which will then be grown as specialty crops under contract to processors. Calgene has engineered a high lauric acid canola (rapeseed) strain for oils that are used for soaps, shampoos, cosmetics, and food products that formerly required imported palm oils. These special canola strains are now produced in the Midwest under contract, displacing Philippine production on which a large fraction of the rural population depends economically. And the genes for the biosynthesis of caffeine have been successfully transferred to soybeans. If the essential oil genes for coffee flavor can also be transferred, then Central and South America and Africa will lose their market for beans destined for powdered coffee.

It would be a mistake to think that agriculture has followed the classical picture of the spread of capitalism. Unlike in industrial production, the first step in the capture of agriculture by capital was the immense flowering of input industries and output processors, who appropriated the surplus in agriculture by selling the petty entrepreneurial farmer what he needed and buying what he produced. No parallel exists in the industrial sphere. It is only with the saturation of that possibility of appropriation that wholly new techniques have come into play. By concentrating on the central material link in farm production, the living organism, which at the same time was the most resistant to capitalization, biotechnology has accomplished two steps in the penetration of capital. First, it widened the sphere of input commodity production by including a wide array of organisms that had previously escaped. Second, and more profound, it made vertical integration possible with the accompanying proletarianization of the farmer. It is this second stage that is the capitalist agriculture of the future, because the physical nature of farm production, inevitably tied to the land, is such as to maintain its unique organization as a productive process.

30

How Cuba Is Going Ecological

The question I will try to answer is: How is Cuba doing it?[1]

While the world environmental problems continue to worsen despite intensive research and rhetoric, how come a poor third world country besieged by a hostile neighbor has been able to embark on an ecological pathway of development that combines sustainability, equity, and quality-of-life goals? How did it achieve a commitment to an integral program of protected areas, ecological and organic agriculture, public health levels behind only the Scandinavian countries, environmental education, occupational health, urban planning and economic development compatible with environmental protection, and compliance with the major world treaties on the environment?[2]

Although the commitment to agroecology and ecological development is relatively new, it is not, as is often misrepresented, an improvised emergency response to the Special Period, the economic crisis brought on by the collapse of Cuba's trade relations with the Soviet Union and the tightening of U.S. economic warfare. Rather, it has its roots in a complex history of colonial science, anti-imperialism, the emergence of a self-conscious community of ecologists, and the transformations of Cuban society since 1959.

Scholars of different disciplines tend to prefer different analytic frameworks, to explain observed processes. Historians may trace the sequence of steps from the early literacy campaigns and plans for botanical gardens through the present national environmental plan. Sociologists might point to the institutional frameworks, the role of the United Nations, government departments, research institutions, and NGOs that analyze and propose. Policy analysts might focus on the legal framework and the when and how particular decisions were made, or look at the unique events, the right person in the right place at the right time to achieve the ecological commitment. Intellectual historians might show the unfolding and deepening of environmental awareness and concern, the conflicts about pesticides, and the philosophical underpinnings that made the outcome almost inevitable. Economists might point to the Special Period and show how the urgencies of scarcity forced a rethinking of industrial and agricultural strategy.[3]

As a Marxist, I view these approaches as different modes of abstraction applied to the same complex, multilevel reality, the whole that is the full explanation. Therefore, I will try to place these various descriptions and interpretations in the context of an evolving, very Cuban socialism.[4] A complex, nuanced explanation of a phenomenon is not the antithesis of theory and generalization. Rather, it demands a theory of complexity and process.[5]

My primary concern is not a description of the state of the Cuban environment or a catalog of successes and failures but rather the trajectory of Cuban society's evolving relation with the rest of nature. My thesis is that each kind of society develops its own relations with the rest of nature, and that an ecological pathway of development is at least latent in socialist development, coequal with equity and participation. Despite all the zigzags, vacillations and disputes, it emerges as an increasingly central characteristic. And this is imperative, for socialism cannot succeed without committing to an ecological pathway. Indeed, the failure of the Soviet Union and Eastern Europe to do so was one symptom of the disintegration of the European socialist project.

I will start with Cuban science in general, then environmental science and policy, and finally the specific case of agriculture.

Science

José Martí's modernist valuing of learning was joined with the traditional socialist appreciation of science to encourage the young revolutionaries to give a high priority to science from the earliest days of the revolution.[6] The traditional socialist view was that scientific knowledge had been produced out of the wealth created by working people but was monopolized by the rich to be used for profit and to build the instruments of power. Therefore the recapture of scientific knowledge for the people was a common goal of radicals throughout the world, and any scientific learning was considered a victory. Further, scientific literacy was seen as liberation from religious obscurantism and bigotry. Scientific news or controversies frequently appeared in the socialist and communist publications. Public lectures in England, the United States, and Russia contributed to this goal. In pre-revolutionary Cuba the *lectores* in the tobacco factories were hired by the workers to read from world classics and scientific literature while they worked.

Thus it was natural for Cuban revolutionaries to look toward science for economic development and as part of the necessary culture for a free people. In 1960, Fidel Castro was invited to speak to the Cuban Speleological Society.[7] In that talk he proposed that "The future of our country will be a future of men of science." (In Silvia Martinez's 2003 book, she corrects the sexism and paraphrases it as "men and women of science.")[8]

The preconditions for today's modern Cuban science were laid in the early years of the revolution with the literacy campaigns, starting with the battle for the sixth grade. The enemies of the revolution realized the significance of education, and CIA-sponsored counterrevolutionary bands murdered two young alphabetizers, Conrado Benitez Garcia and Manuel Ascunce Domenech. But the country became fully literate and has continued to extend mass education through secondary and, increasingly, university level, with university centers in every municipality and special programs for seniors, dropouts, people with disabilities, and workers displaced by the downsizing of the sugar industry. Now Cuba, with only 2 percent of the population of Latin America, has 11 percent of its scientists, a large fraction of them women. More than 1.3 percent of the population works in science, a level comparable to the most developed countries. There are more than 100 major research centers as well as research sections of institutions dedicated to other goals. Physics alone has forty laboratory groups with some five hundred researchers concentrated in solid state and nuclear physics, optics, space and geophysics, mathematical physics and medical physics. There are some eighty centers doing social science research, studying such topics as marginality, social dysfunction, issues of race and gender, and residual and newly arisen inequality. Cuban science has been outstanding in the areas of public health and medicine, agriculture, electronics, and pedagogy.

Some of the major achievements of Cuban science and technology:

- 271 new medications
- 24 diagnostic systems
- SUMA (ultramicroanalytic system for HIV detection)
- Production of 90 percent of the medicines needed
- Melagenina (84 percent effective against vitiligo)
- Meningococcus B vaccine
- Hepatitis B vaccine
- *Haemphilus influenzae* vaccine
- Skin-growth factor for treatment of burns
- Monoclonal antibody HB3 for epithelial tumors, especially of the head and neck
- PPG anti-cholesterol agent
- Anti-retroviral medications against AIDS
- Control of HIV/AIDS to .03 percent of at-risk population through detection, quarantine, treatment, and education
- Infant mortality at around 6.5 per 1,000 live births (tied with Canada for best in hemisphere)
- Elimination of poliomyelitis, malaria, and infant AIDS
- Integrated program for the treatment of retinosa pigmentosa
- Neuro-rehabilitation center
- Orthopedics: development of external fixators
- Psychiatry: emphasis on outpatient care, occupational therapy, and integration into the community
- 100 percent HIV-free blood bank
- Sanitary seed production *(vitroplamas)*
- Sugarcane derivatives for coolants, medicines, energy, and paper production
- Biological methods for preserving and enhancing soil fertility
- Systems of biological control of pests through release of parasites
- Reforestation and protected areas

For any third world country the problem arises of how it can create a science that is at the same time international, in the sense of being linked to the advanced communities of world science, and yet has its own agenda directed at the needs of its society. And how can a country do this on a low budget? In Cuba this comes to some 10,000 pesos per scientific worker. The peso is currently 26 to the U.S. dollar, but this grossly underestimates its real value within Cuba. In the United States, the expenditure is about $200,000 per scientist.

The lack of resources makes strict prioritizing a necessity. This produces a glaring unevenness of availability. While everybody who needs kidney dialysis receives it, reagents are often not available for student chemistry laboratory exercises. As of this writing, a group of my colleagues studying mosquito development in breeding sites in Havana are unable to measure the water temperature regularly, because they do not have sufficient thermometers.

Cuban science is right up there with the best of world science, but it also has its own special features. First, it is publicly owned science. With the recent program to establish university centers in all 149 municipalities and to make all centers of learning "microuniversities," research is even more broadly diffused.

Public ownership makes it possible to plan science, to incorporate science into national plans, and to have policies linking the recruitment and training of scientists to research directions. The mass literacy campaigns of the 1960s enlarged the pool of potential scientists, while the struggles for women's equality and against racism opened up new sources of talent. Today, women, many in positions of leadership, make up 52 percent of the scientific work force,[9] and 65 percent of the scientific and technical work force.[10] Among the people I have worked with, the Minister of Science, Technology and the Environment, is a woman. The dean of the mathematics faculty at the University of Havana is a woman. So is the director of the Institute of Fruit Research, the head of the Center for Animal and Plant Protection, its laboratory of phytopathology, the Center for Mathematics and Cybernetics Applied to Medicine, the Carlos Finlay Institute, and two of the four department heads at the Institute of Ecology and Systematics. A growing number of the scientists are Afro-Cubans. Education is free of charge. Study is regarded as productive labor, the task of producing a skilled and well-informed citizen, rather than as an investment in future high income. Therefore, there are no economic barriers to study.

When the British Marxist J. D. Bernal first posed the need for the planning of collective scientific work in the 1930s, the very idea was met with derision and hostility.[11] It was denounced as the totalitarian repression of free, individualistic science. But now scientific strategies are an accepted part of government policy throughout the world.

Scientific planning confronts a major contradiction: How do you plan for the still unknown? We can assess national needs and establish priorities. But surprise is inevitable in science, and therefore it is necessary to be able to follow new pathways that present themselves during the lifetime of the plan. But how do you recognize surprises that should be followed up? Exciting new directions and innovations are not yet the consensus of a scientific community, but rather the initiative of a few individuals. However, the agreed upon priorities represent the consensus of the leaders of science—those who created the field the way it is—and therefore who are less likely to be critical of its direction. This makes it necessary to have a very flexible form of planning with leeway for departures from the plan.

Cuban scientific planning sets general goals. At a national level, some general research goals are proposed as priority areas. Provincial governments have their own priorities, and so do the various ministries and institutions. The individual institutions join in the projects that fall within their areas, and there is leeway for individuals to carry on their own work where resources permit. The projects that come down from above have actually gotten there at the initiative of the researchers, so that the process of planning moves up and down many times in the formal structure before a plan is adopted. There are frequent progress reports during a scientific investigation. This includes discussions with peers and the public that will be affected so that the research is an even more collective process than the interdisciplinary teams would suggest.[12]

In 2001, 3,093 formal research results were reported, of which 403 came from national programs, 1,584 from ministries and branches of the society, and 1,077 from territorial authorities. From the nonprofessional end, the National Association of Innovators and Rationalizers (ANIR) presents some tens of thousands of innovations each year, which indicates the breadth of mass participation in Cuban innovation. (ANIR has more than half a million members who have come up with over 100,000 solutions to mostly technical problems.) This experience refutes the notion that innovation would stagnate in the absence of opportunities to get rich from inventions. Amateur groups in computing, botany, and other fields supplement professional research.

The public nature of science makes it open science. There is no hiding of information for proprietary reasons, as is increasingly common in the United States and other capitalist societies. Inventors can receive economic rewards from their inventions, but they do not have the authority to suppress them or restrict their use. There is little duplication of effort by competing entities. This makes it possible for Cuban scientists to collaborate across institutional boundaries. The development of the recent completely synthetic vaccine against *Hemophilus influenzae* was the work of many research centers concerned with the biochemistry, immunology, clinical, and industrial aspects of this innovative project. The National Study of Biodiversity was prepared through the collaboration of the Ministry of Science, Technology and the Environment (CITMA), Ministry of Higher Education, Ministry of Agriculture, Ministry of Public Health, Ministry of the Economy and Planning, and many institutes within these ministries. The same was true for the National Atlas and other major efforts; all involved broad collaboration. In the absence of the obscene race for patents, Cuba has the capacity to wait and see what unexpected consequences an innovation may have. Cubans have been working with genetically modified organisms for more than seventeen years but have not released any GMO plant varieties, because they are still exploring the possible risks to the environment.[13]

A major obstacle to the serious evaluation of research and policy efforts in the United States is that the annual reports of grant holders and agencies to their sponsors are a mixture of real assessment and self praise, a pitch for more funding that downplays difficulties. The large scale of many projects makes it impossible to replicate or for outside examiners to know them well enough to evaluate critically. CITMA is able to look more objectively at the state of the environment and identify weaknesses. In January 1997, CITMA convened a workshop on the environment as a national consultation, "Rio + 5," to evaluate Cuban compliance with the agreements that came from the U.N. Conference on Environment and Development, the Earth Summit, that had been held in Rio de Janeiro in 1992.[14] Invitations were extended to various government departments, agencies, NGOs, and delegations from nearby provinces. Participants considered each of the categories discussed in Agenda 21, the declaration and plan of action that emerged from the Rio Earth Summit, and also added several of their own. For each problem area, they described achievements and also the major difficulties. For instance, in the chapter on agriculture, the list of achievements includes the system for plant disease prediction, the establishment of centers for the reproduction of natural enemies of pests and diseases, and the expansion of polyculture. Among the deficiencies listed are insufficient and unstable extension work, lack of studies of the environmental impacts of new production systems, and, especially interesting, "the existence at various levels of the opinion that the practice of sustainable agriculture is only a consequence of the Special Period destined to disappear when the present limitations make it [possible] and there will be a regression to high inputs of fertilizers, pesticides, mechanization, etc."[15]

Later that year, the National Environmental Strategy evaluated recent experience (up to 1997). After listing their achievements, they report (my editing of the Cuban version in English):

> Parallel to these gains there have been mistakes and shortcomings, due mainly to insufficient environmental awareness, knowledge and education, the lack of better management, limited introduction and generalization of scientific and technological achievements, still insufficient incorporation of the environmental dimension in the policies, developmental plans and programs, and the absence of a sufficiently integrative and coherent juridical system. Moreover, the scarcity of material and financial resources has prevented us from attaining higher levels of environmental protection, which has worsened in the last few years due to the economic situation in which the country has been immersed because of the loss of commercial ties with the former socialist bloc and the continued and intensified economic blockade by the United States of America.

Although it is often asserted in the United States that the Cuban government blames the United States for all its troubles, here we see a nuanced analysis that includes the economic warfare against Cuba as only one factor among many that influence the situation.

One common misconception is that the science of a developing country should concentrate on the applications of the achievements of world science and that fundamental research is a luxury of the rich. However, this condemns a country to dependence on the basic research done in other countries for other reasons. A coherent scientific community has to develop its own underpinnings for its work and the education and morale of its participants. In the National Environmental and Development Program, established by CITMA in 1995, there is a group working on protection of the atmosphere. It includes such immediately practical tasks as the monitoring of air pollution on different temporal and spatial scales and linking this to morbidity and mortality data. It studies the chemistry of rain and the ocean/atmosphere exchange.

The group also includes neurobiologists who work with problems of autism, trauma, and the neurological correlates of emotion, investigating potential links between air pollution and these issues.

In agriculture, a doctoral thesis has to include a section on the contribution of the work to practice and how it enriches science. However, Cuban science includes themes that are not directly related to practice, such as underwater archaeology and the theory of complex systems. For example, a recent international symposium on complexity included presentations by Cubans such as Entropy and Complexity: The Problem of Irreversibility; Contingency and Causality in Natural Disasters; Complexity and Morphogenesis: From the Properties of Systems to the Very Existence of Systems; Construction of a Critical-Analytical Model for the Study of Cultural Identities in the Social Complexity; Esthetics and Reasoning about Complexity: An Epistemological and Methodological Approach; Evidence for the Mind as a Dynamic Attractor; The Treatment of Attention Deficit as a Non-Equilibrium State; Transformations of a Citrus Agro-Ecosystem in Conversion to Organic; and The Complexity Barrier: The Next Challenge for Immunology.[16] As is increasingly common in Cuba, the symposium included musical performances as part of the plenary sessions.

Though Cuban science has a special style, it is very much influenced by the Marxist dialectical philosophy of science, with its emphasis on historicity, social determination of science, wholeness, connectedness, integrated levels of phenomena, and prioritizing of processes over things. All doctoral candidates must study Social Problems of Science and Technology, which emerged in the 1990s as a distinct field of study.

The impact of this preparation can be observed in the self-conscious view of the development of science as a social process, with the organization, recruitment, priorities, preferred approaches, and tools of investigation all being recognized as products of the social relations that promote, support, apply, and reward scientific endeavor. This allows for a critical examination of the state of a field internationally and the capacity to make active choices about what to concentrate on. For instance, by recognizing that the pharmaceutical industries develop only those medications for which there are large and lucrative markets, the Cubans have been able to select areas of research that are ignored, because the knowledge is not easily turned into commodities, or because the disease is uncommon among the wealthy. Thus Cuba has been in the forefront in work on retinosa pigmentosa, vitiligo, and malaria. It also encourages a view of science that combines its contributions to the economy and to the general culture of society with an awareness of its own internal needs for balance among the disciplines, the integration of practical and theoretical concerns, and the cooperative organization of research.

A major characteristic in the Marxist dialectical perspective is wholeness and the critique of reductionism. A recurrent theme in all of Cuban science is the breadth with which problems are approached and the willingness to span levels of organization. Agostín Lage, immunologist and director of the Center for Molecular Immunology, has been an outspoken critic of molecular and genetic reductionism. He sees the immune system as "a system of recognition and control of the composition of its own organism, whose regulation depends not only on the presence or absence of specific cellular clones but also on the interaction of these clones among themselves (supra-clonal properties)."[17] He sees the future of immunology as including interaction with neurobiology and calls for a synthesis of the high-tech molecular sciences with social medicine. Lage also raises ethical issues in science, particularly the question of whether science is used to increase or remove inequality in the world.

This multilevel approach pervades much of Cuban science. In medicine, modern technical tools coexist with herbal medicine (the "green pharmacy"), social epidemiology, and various kinds of alternative medicine. These are not seen as in opposition: the Carlos Finlay Institute, which has pioneered in the development of molecular biology to produce vaccines, antibiotics, and anti-cholesterol agents, also has a controlled experimental program for testing the macrobiotic diet. The very successful Cuban response to HIV/AIDS has combined chemotherapy (everyone who needs retroviral drugs receives them) with population-level interventions that include temporary quarantine and community education. Work in rehabilitation combines advanced neuro-science and occupational therapy. The prevalence of HIV/AIDS in Cuba is currently about 0.035 percent, and there have been no cases of infant infection since 1997. As a result of its strong commitment and broad approach, Cuba is the healthiest third world country and is tied with Canada for the lowest infant mortality in the hemisphere, making it a global health leader.

Another dialectical theme is the priority given to processes over things. Nilda Perez, a leading Cuban agricultural ecologist, poses the direction of change as from an agriculture of inputs to an agriculture of processes.

Science is defined more broadly in Cuba than in the United States. The recent colloquium that produced the book *Cuba, Dawn of the Third Millennium: Science, Society and Technology* included participants from the usual fields as well as economics; pedagogy; science, society and technology; and communications and audiovisual media. The discussions started with each participant describing the state of his or her own field and perspectives for the future and then entered into open discussion around general questions. Ethical issues were recurrent themes.

Thus Cuban science made possible the effective commitment to an ecological pathway of development by being publicly owned, planned, collaborative, holistic, multilevel, integral to the education of all Cubans, and committed to meeting the material and cultural needs of the people.

The above description stresses the directions in which Cuban science is different from science in capitalist societies. In Cuba, not all institutions work the way they are supposed to, not everybody thinks dialectically, and we can always find examples of narrowness and parochial interest. But what is significant in Cuba is the overall direction of change, the pathway that is being built.

Development of an Environmental Program

The political program of the Cuban Revolution did not have an explicitly ecological perspective at the beginning. The urgent concerns of the new government were eliminating extreme poverty, providing water and sanitation, housing, and literacy. But even before the triumph, the commander of the 26th of July Movement in Matanzas, Onaney Muñiz, was planning for a botanical garden. The destruction of Cuba's forests, the erosion caused by monoculture and the sugarcane economy, the prevalence of infectious diseases that could be prevented, and the need to develop the resources of the country to eliminate poverty all led to the creation of separate programs that later nourished ecological development as a conscious goal.

The literacy campaign of the 1960s made possible the massive commitment to science, which in turn laid the foundations for environmental sciences. As soon as Rachel Carson's *Silent Spring* reached Cuba, Fidel Castro was circulating it among his associates, and environmental consciousness began to spread. Already in the 1960s there were programs for reforestation, the Voisin system of rotational grazing, the digging of thousands of ponds as micro-reservoirs, cleaning up of foci of infection, and massive immunization campaigns. The Institute for Physical Planning, a new discipline for Cuba, undertook the first environmental studies for the selection of sites for development.[18] El Grupo, the Center for the Integral Development of the Capital, is one of the catalysts for innovative participatory neighborhood development.

The priority goals in agriculture were a stable food supply, income and safety for the rural population, sugar for export, and inputs for industry. The dangers in the use of pesticides were recognized and confronted, at first, mostly by measures to protect agricultural workers. But agricultural development was still largely within the paradigm of the Green Revolution, which depended on high-yielding plant varieties and massive infusions of mechanical and chemical inputs, most of which had to be imported.

There was not yet any organized field of ecology. Cuban biology was typical of colonial biology throughout the tropics: biology applied to medicine and agriculture, and systematic botany and zoology, with systematists making ecological observations. At the University of Havana the zoology curriculum started with two years of mostly descriptive zoology, surveying the major families of animal life. In discussions within the university in 1968, the administration and the students supported proposals for electives in ecological topics, but many in the faculty thought this would force the omission of whole families of animals from the curriculum, "thus losing the whole picture of evolution," as if evolution were a catalog of its results. In any case, the faculty was not yet itself prepared to teach those subjects. But there was already experimentation with Voisin's rotational grazing system, polyculture, and biological pest control.

Table 30.1 shows some major events in the evolution of an environmental perspective and commitment. In general, the 1960s were a period of laying the foundations for later development. The public health system was able to eliminate polio by 1963, malaria by 1968, and diphtheria in 1971. The first law of agrarian reform in 1959 made national land available for development programs. Within three years, illiteracy was almost eradicated. The abolition of legal racism, the recognition of equal rights for women, and the expansion of free education and scholarships widened the pool of potential scientists. Cuba sent thousands of students to study abroad, mostly in the East European countries. Maps were prepared of soil, water resources, and endangered species.

Table 30.1
Some milestones in the development of an ecological pathway

1960s • Founding of the Institute of Physical Planning and the Group for
the Integrated Development of the Capital.
• Introduction of the Voisin system of rotational grazing.

• The beginning of restoration of open pit mining areas.

• Construction of some 1,400 micro reservoirs for energy, water
resources, recreation, and fish production.

1970s • Transition toward low-input agriculture.
• The creation of botanical gardens.

1972–3 • National atlas.

1974 • Cuba joins UNESCO's Man and the Biosphere program and selects
the Sierra del Rosario montane rainforest as its study area.

1975 • Zoning of Havana.
• First Congress of the Communist Party adopts thesis on environment.

• Waste treatment facilities are required for all new plants.

1976 • Constitution adopted. Article 27 links environmental protection
to sustainable economic and social development and recognizes the
obligations of the state and citizens to protect the environment.
• COMARNA (Commission on Natural Resources and the Environment) established.

1978 • Law passes allowing veto of development projects that would
harm the environment.

1980s • Experiments with ecological agriculture and *organopónicos*.
• Establishment of the Centers for the Reproduction of Entomopara-
sites and Entomopathogens (CREE) for biological pest control.

• Legal and institutional structures for environmental inspection and
licensing of development projects are created.

- Plan Turquino-Manatí for the sustainable development of the mountains is implemented.

- Widespread adoption of urban *organopónicos*.

1990s   • National Environmental Strategy is implemented.
         • Legal instruments for environmental protection, inspection, and enforcement are developed.

- National Survey of Biodiversity is established.

- Antonio Nuñez Jimenez Foundation for Nature and Humanity vision statement: "a Cuban society with a developed environmental consciousness that recognizes nature as part of its identity, and an institution active in the development of environmental and cultural values in Cuba and the world."

- Network of protected areas established.

- Forest cover reaches 23 percent of land area.

- System of Protected Areas.

In the 1970s, Carlos Rafael Rodriguez introduced his argument differentiating development from growth and arguing for integral development, laying the groundwork for a goal of harmonious development of the economy and social relations with nature. This implied a rejection of Stalin's approach, the view popular among East European Communist governments that production decided everything, and that only after abundance was achieved would society be able to confront the task of bringing social relations into harmony with the economy. Despite disagreements with Rodriguez on how the economy should be organized, until Ché Guevara left for Bolivia in 1967 he had already been stressing that social relations and economic development must evolve together. At the same time, UNESCO initiated its International Biological Program, a ten-year international program of biological studies that concentrated on the productivity of biological resources and human adaptation to environmental change. Cuba joined in and selected the montane rain forest of the Sierra del Rosario as its area of study. There, sitting on the sopping litter under an endless tropical rain, zoologists and botanists came together to begin to think of themselves as ecologists.

At the first national meeting on ecology in 1981, which I attended with the plant ecologists, representatives of research groups in botany, zoology, agriculture, oceanography, and from the tourism and food-processing industries gathered to debate pesticides and consider what can be done with industrial wastes. The food processors called our attention to the pollution they were causing and asked what they could do with the mountains of rice husks and mango pits they were accumulating. The tourism institute asked how to develop environmentally friendly facilities. We ended the meeting with a resolution calling for the Commission on the Environment to have enforcement powers. This was soon accomplished, and the commission was elevated to cabinet status, the present Ministry of Science, Technology and the Environment (CITMA). Its formation "resolved a contradiction in the old structure of leadership of environmental activity in the country in which ministries were in charge of environmental matters for the same resource that they exploited for productive purposes, making them both 'judge' and 'interested party' of the same activity."[19] The sugar industry was responsible for some 47 percent of the polluting outflow load on coastal ecosystems. But that industry also pioneered recycling systems, using residues for energy production and an almost closed-system production design.

A decade later, the Institutes of Zoology and Botany finally merged into the present Institute of Ecology and Systematics. This group has led the way in developing programs in biodiversity, protected areas, and protection of the coastlines and forests.

One issue that has not been fully resolved is nuclear power. To a country dependent on fuel imports, a nuclear power plant seemed very attractive. Soviet technical and economic help encouraged Cuba to begin the construction of a nuclear power station at Juraguá, near Cienfuegos. Misgivings arose: Would Cuba be safe in the event of a disastrous meltdown? In a small country, a major radioactive release would be even more devastating than in Russia. Would even normal operations, without a catastrophic event, poison the surroundings with radioactivity? Would the plant demand too much water? Could they be sure of finding a safe and secure home to put the deadly waste products? But while Cubans pondered these questions, the Soviet Union collapsed, no further aid was forthcoming, and alternative forms of energy generation advanced. The half-plant still sits there, and the issue has been put on hold. According to the engineer Jose Luis Garcia, Cuba has, in practice, renounced the electro-nuclear path, partly because alternatives have appeared in the short run based on national oil and gas. "But doubtless," he says, "in strategic terms, we cannot rule out the possibility that at some time we might opt for electro-nuclear energy."[20] In September 2004, the principal turbine at the Juraguá plant was removed to replace a damaged turbine at the Guiteras thermoelectric plant.

As in other fields, Cubans take a very broad view of the environment. The conception of an ecological pathway of development is emerging from the perspectives of conservation of natural areas, agriculture, public health, urban planning, alternative energy, clean production and waste disposal, community participation, environmental education, and issues involving different sectors of society, particularly vulnerable habitats. Workplace and neighborhood pollution problems are included within the same framework.

The National Environment Plan of 1995 integrates a vast array of problems and proposals and is executed by the coordinated efforts of government agencies, NGOs, and community participation.[21]

Agriculture

One of the outstanding achievements of Cuba's advance toward ecological development is the acceptance of agroecology as a national strategy. Agricultural development was at first dominated by the high-tech Green Revolution perspective of the international development community. But soon Cubans in many institutions began a critical reevaluation of the economic structure of agriculture, geography of production, farm organization, pest management, soil fertility and mechanization.

This came about through the convergence of several different initiatives. Within agriculture, people such as Nilda Perez, Luis Ovies, and Tenelfe Perez in plant protection, Miriam Fernandez in entomology, Magda Montes in citrus research, Rafael Martinez Viera and Antonio Castañeiras in the Alexander Humboldt Institute for Fundamental Research in Tropical Agriculture, and Ricardo Herrera in the Institute of Ecology and Systematics carried out projects in polyculture, soil microbiology, and biological pest control.[22] Ecologists began to speak up against the pesticide treadmill.

There was fierce debate about pesticides and the ecological pathway. The traditional progressivist viewpoint of European socialism was that there is an inevitable progression from "backward" to "modern." Capitalism inhibited the full development of the "modern" and monopolized its benefits while unloading the costs on the workers and peasants. Therefore, the task of a liberated country was to proceed as quickly as possible along that pathway of "progress," avoiding the barriers inherent under capitalist governance. Features of agricultural modernization included the transition from labor-intensive to capital-intensive, from small-scale to the economies of large-scale, from the patchwork heterogeneity of peasant production to the rationalized homogeneity of agribusiness and specialized state farms, from subjection to nature to the conquest of nature, from superstition to scientific knowledge. Advocates of this approach saw themselves as rigorous materialists and mocked the ecological viewpoint as "idealist," sentimental nostalgia for some golden past that never really existed.

As advocates of ecological socialism, we fought back with the argument that it was the height of idealism to expect that we could pass resolutions about production and have nature obey. We proposed that development was a branching process in which technical choices were not socially neutral, and that each kind of society had to find its own pattern of relating to the rest of nature. Accumulating experience was showing that agroecology was productive, economical, and safer than chemical means.[23] In particular, we argued that beyond the dichotomy of labor-intensive versus capital-intensive was a knowledge- and thought-intensive agriculture. Instead of mobilizing vast amounts of energy to move large masses of materials, we sought the design of systems that were as self-operating as possible. Mechanization was sometimes very important but at other times destructive of the soil, inefficient in very wet soils, too expensive, and a constraint on other agronomic practices. A combination of tractors and animal traction according to circumstances seemed a better choice.

Instead of having to decide between large-scale industrial type production and a "small is beautiful" approach *a priori*, we saw the scale of agriculture as dependent on natural and social conditions, with the units of planning embracing many units of production. Different scales of farming would be adjusted to the watershed, climatic zones and topography, population density, distribution of available resources, and the mobility of pests and their enemies.

The random patchwork of peasant agriculture, constrained by land tenure, and the harsh destructive landscapes of industrial farming would both be replaced by a planned mosaic of land uses in which each patch contributes its own products but also assists the production of other patches: forests give lumber, fuel, fruit, nuts, and honey but also regulate the flow of water, modulate the climate to a distance of about ten times the height of the trees, create a special microclimate downwind from the edge, offer shade for livestock and the workers, and provide a home to the natural enemies of pests and the pollinators of crops. There would no longer be specialized farms producing only one thing. Mixed enterprises would allow for recycling, a more diverse diet for the farmers, and a hedge against climatic surprises. It would have a more uniform demand for labor throughout the year. One example is the "El Carmen" UBPC, a new cooperative in Ciego de Avila. It has assumed national leadership in designing the transition from the previous conventional citrus monoculture to mixed production of fruit, annual crops, and livestock products.

The arrogant presumption of the conquest of nature had to be replaced with a strategy of nudging nature here and there while respecting its autonomy and complexity. Traditional knowledge could not be dismissed as superstition but must be understood as a pattern of insights and blindnesses, just like modern science. Our task was to look at both of them critically in order to integrate the detailed, particular, and nuanced peasant knowledge with the more general and comparative but abstracted knowledge of agricultural science, an integration that depended on scientists and farmers meeting as equals in a common enterprise. This is made easier by the fact that so many agricultural scientists come from peasant families. More recently, the Australian system of permaculture is being spread in Cuba by solidarity groups from New Zealand and Pro Naturaleza, an NGO organized by the Antonio Nuñez Jimenez Foundation.

The Special Period, with critical shortages of fuel, chemicals, and feed, revealed the fragility of high-tech agriculture and encouraged the adoption of ecological agriculture. But it also reduced the capacity to carry out measures already adopted. Environmental inspections lapsed for lack of monitoring supplies and fuel to get to the sites to be inspected. Tough, thorny weed trees invaded fields abandoned for lack of tractors. Badly polluting buses were kept in service for lack of spare parts or replacements. Economic urgencies encouraged ignoring some protective regulations. We had the paradoxical situation in which environmental conditions worsened while environmental consciousness deepened. When sound measures were introduced, some producers were convinced of their value only as an emergency measure. Our task became to convert these ecologists by necessity into ecologists by conviction before the emergency ended and they could resume the peaceful destruction of Cuba. This conversion is being carried out by education at all levels, the training of ecologically oriented agronomists, and ongoing debate.

Meanwhile in the 1970s and 1980s the Ministry of Defense developed a new doctrine of defense that assumed the possibility of Cuba being partly occupied by a hostile power. The Cuban response would be a war waged by all of its citizens. But this required local self-sufficiency in the absence of central organization and exchange. A civil defense manual from that time had chapters devoted to first aid, herbal medicine, organization of schools, and food production. This led to military experiments in low-input agriculture. And in 1987 Raoul Castro called for the widespread introduction of *organopónicos,* raised beds of enriched and composted soils where crops could be grown in small areas with no dependence on outside resources. The first pilot *organopónico* in Havana on Fifth Avenue and 44th Street in Playa was organized by the armed forces and is still a showcase of urban agriculture. Today, agriculture is evolving in the direction of agronomically and socially sustainable production that emphasizes combining rural, suburban, and urban farming; diversification; and biological and natural pest control.

Combining Rural, Suburban, and Urban Farming

Urban agriculture now takes place on some 30,000 hectares producing more than three million tons of fresh vegetables per year for eleven million people. Like most Cuban programs, it serves multiple purposes. It provides abundant, diverse fresh vegetables throughout the year for consumers. This has transformed the Cuban diet in the communities, schools, and workplaces and encouraged the spread of vegetarian restaurants. It lowers the costs of transportation and storage by selling directly to consumers. It provides employment for some 300,000 people at a time when capital is not available to invest in more industrial employment. This comes to about ten people per hectare, a labor-intensive system that would be regarded as highly inefficient in the United States, though each worker is producing ample vegetables to feed thirty-six people. In the context of the unemployment that appeared with the Special Period, it is socially efficient. Urban agriculture increases the green area of cities, detoxifies the air, and provides foci of neighborhood social integration.

Diversification

Geographic diversification is a protection against regional disasters such as hurricanes, which may have zones of destruction 200 miles across. More locally, instead of having large monoculture farms, enterprises are converting to mixed production of fruits, vegetables, basic grains, livestock, and fish. This results in a mosaic land use that makes better use of the topography and microclimates and permits recycling within the farm. Each patch of the mosaic has its own products but also contributes to the whole. As mentioned above, woodlands provide a wide range of forest products and ecological services. Pastures support livestock for meat and dairy products, manure for composting, combat erosion, and serve as nectar sources for honey and beneficial wasps. Oxen are integrated with tractors in a complex traction strategy, and horses and other animals are helpful in weed control. Diversification is a hedge against natural disasters that affect particular crops, and it more uniformly spreads the need for labor and assures local diversity of food supplies. By combining constant evaluation and decision making with the hard physical labor of farming, it also raises the technical level of agricultural labor for a rural population too educated to aspire to a life limited to wielding a machete. Soil fertility is maintained by composting, crop rotation, the use of nitrogen-fixing bacteria, fungi that mobilize potassium, phosphorus and other minerals, as well as the cultivation of earthworms.

Biological and Natural Pest Control

Biological and natural methods of pest control are proving more effective than chemical control, more economical, and protective of people's health and the environment. Here's just one example: sweet potatoes grown using integrated pest management yielded 8.9 metric tons per hectare at a net per hectare value of $904.70 compared to 7.8 tons worth $818.60 per hectare for sweet potatoes grown with conventional methods using pesticides. Ecological protection makes use of polyculture and the spatial arrangement of crops, rotation, encouragement of predators, introduction of parasitic wasps and fungi, and, finally, biological products such as neem. All urban agriculture is now organic, and much of the rest of Cuban agriculture is advancing in that direction. The Antonio Nuñez Jimenez Foundation for Nature and Humanity is a major

NGO in Cuba. It has developed strategic documents for urban sustainability and has been the leading group in the promotion of permaculture.[24]

The ecologists eventually won the struggle over developmentalism. It took a long time, and the debate was fierce at times, but it had a very different flavor from similar debates in the United States. All parties were looking for ways of meeting the needs of society so that the disagreements were just disagreements, not surrogates for clashing interests. Nobody was pushing pesticides or mechanization to make profit. And agroecology proponents were not demonized as Luddites, or worse.

Cuban Socialism and the Environment

We can now return to the original question: How is Cuba doing it? At the most abstract level, the short answer is socialism. That is, socialist social arrangements and ideological priorities made ecological development an almost "natural" correlate of the economic and social development and of the commitment to improving the quality of life as the primary goal of development. But an abstraction does not mobilize resources or change minds. Change occurs through the actions of particular people, through the decisions they make. And decisions are made in response to the questions that are posed, the social setting in which answers are sought, the tools available for providing answers, and the criteria for judging whether a solution is satisfactory.

The logic of decision making under Cuban socialism starts with the priority given to human need. Therefore questions about the environment arise more or less independently in areas such as urban development, health, agriculture, defense, conservation, and economics. There are no externalities such as environmental harm that can be thrust upon the society as a whole while responsibility is denied.

In each of these spheres, the general ideological commitment is reinforced by law. If one sphere lags, the others advance, and the convergence of ecological imperatives from differing sources creates the direction of movement.

The feedback mechanisms in the society favor ecology. Each success encourages further extension of the ecological commitment by showing that it is possible to develop in a way that departs from the conventional developmentalist wisdom. This is positive feedback. When ecological rationality is subordinated to expediency and destructive decisions are made, the mistakes are visible, and there is a collective incentive to correct the error. A case in point is the stone causeway built from the Cuban coast to the tourist center on Cayo Coco. Ecologists had warned that this would disrupt the hydrology of the area and harm the mangroves. But economic urgency prevailed. The causeway was built in spite of the warnings, and the mangroves began to die. But when this was observed, sections of the causeway were removed and replaced by bridge spans to permit the flow of the water.

In contrast, in capitalist society, each victory restricting the free destruction of our biosphere by business intensifies corporate resistance with an urgency to defend not only their profits but their property rights. Thus there is a powerful anti-ecological negative feedback expressed as "backlash," amid claims that environmentalists are "going too far." And environmental victories do not necessarily encourage further struggle.[25] When they don't unleash an environmental backlash, they are often co-opted as examples of "partnership." Of course, under capitalism there are also positive feedbacks, which are important in movement building.

In spite of the incentives and commitments to an ecological pathway, Cubans could have decided otherwise. In fact, they did so at the beginning when, in the absence of ecological consciousness, the urgency to meet the needs of the people led to harmful decisions. But when the first Green Revolution, developmentalist approach turned out to be destructive of productive capacity and poisoned people and nature, this was sufficient reason to reexamine the strategy. There were no greedy institutions committed to defending the harmful course with lobbyists, public relations firms, lawyers, and hired witnesses. It meant that Cuban scientific and political leadership, which is strongly committed to a broad, dynamic, and integral approach, was able to recognize the origins of the different developmental strategies in the world political economy and the implications of alternative choices. It meant that there were scientists prepared to argue the case for ecological development, receptive ears in the leadership and public to receive the arguments sympathetically, and a logic of decision making that made an ecological pathway of development along with equity and collectivity an essential part of Cuban socialism. That's how they are doing it.

31

Living the 11th Thesis

Philosophers have sought to understand the world. The point, however, is to
change it.

— Karl Marx, 11th thesis on Ludwig Feuerbach

When I was a boy I always assumed that I would grow up to be both a scientist and a Red. Rather than face a problem of combining activism and scholarship, I would have had a very difficult time trying to separate them.

Before I could read, my grandfather read to me from Bad Bishop Brown's *Science and History for Girls and Boys.*[1] My grandfather believed that at a minimum every socialist worker should be familiar with cosmology, evolution, and history. I never separated history, in which we are active participants, from science, the finding out how things are. My family had broken with organized religion five generations back, but my father sat me down for Bible study every Friday evening because it was an important part of the surrounding culture and important to many people, a fascinating account of how ideas develop in changing conditions, and because every atheist should know it as well as believers do.

On my first day of primary school, my grandmother urged me to learn everything they could teach me—but not to believe it all. She was all too aware of the "racial science" of 1930s Germany and the justifications for eugenics and male supremacy that were popular in our own country. Her attitude came from her knowledge of the uses of science for power and profit and from a worker's generic distrust of the rulers. Her advice formed my stance in academic life: consciously in, but not of, the university.

I grew up in a left-wing neighborhood of Brooklyn where the schools were empty on May Day and where I met my first Republican at age twelve. Issues of science, politics, and culture were debated in permanent clusters on the Brighton Beach boardwalk and were the bread and butter of mealtime conversation. Political commitment was assumed, how to act on that commitment was a matter of fierce debate.

As a teenager I became interested in genetics through my fascination with the work of the Soviet scientist Lysenko. He turned out to be dreadfully wrong, especially in trying to reach biological conclusions from philosophical principles. However, his criticism of the genetics of his time turned me toward the work of Waddington and Schmalhausen and others who would not simply dismiss him out of hand in Cold War fashion but had to respond to his challenge by developing a deeper view of the organism–environment interaction.

My wife, Rosario Morales, introduced me to Puerto Rico in 1951, and my eleven years there gave a Latin American perspective to my politics. The various left-wing victories in South America were a source of optimism even in those grim times. FBI surveillance in Puerto Rico blocked me from the jobs I was looking for and I ended up doing vegetable farming for a living on the island's western mountains.

As an undergraduate at Cornell University's School of Agriculture, I had been taught that the prime agricultural problem of the United States was the disposal of the farm surplus. But as a farmer in a poor region of Puerto Rico, I saw the significance of agriculture for people's lives. That experience introduced me to the realities of poverty as it undermines health, shortens lives, closes options, and stultifies personal growth, and to the specific forms that sexism takes among the rural poor. Direct labor organizing on the coffee plantations was combined with study. Rosario and I wrote the agrarian program of the Puerto Rican Communist Party in which we combined rather amateurish economic and social analysis with some firsthand insights into ecological production methods, diversification, conservation, and cooperatives.

I first went to Cuba in 1964 to help develop their population genetics and get a look at the Cuban Revolution. Over the years I became involved in the ongoing Cuban struggle for ecological agriculture and an ecological pathway of economic development that was just, egalitarian, and sustainable. Progressivist thinking, so powerful in the socialist tradition, expected that developing countries had to catch up with advanced countries along the single pathway of modernization. It dismissed critics of the high-tech pathway of industrial agriculture as "idealists," urban sentimentalists nostalgic for a bucolic rural golden age that never really existed. But there was another view, that each society creates its own ways of relating to the rest of nature, its own pattern of land use, its own appropriate technology, and its own criteria of efficiency. This discussion raged in Cuba in the 1970s and by the 1980s the ecological model had basically won although implementation was still a long process. The Special Period, that time of economic crisis after the collapse of the Soviet Union when the materials for high-tech became unavailable, allowed ecologists by conviction to recruit the ecologists by necessity. This was possible only because the ecologists by conviction had prepared the way.

I first met dialectical materialism in my early teens through the writings of the British Marxist scientists J. B. S. Haldane, J. D. Bernal, Joseph Needham, and others, and then on to Marx and Engels. It immediately grabbed me both intellectually and aesthetically. A dialectical view of nature and society has been a major theme of my research since. I have delighted in the dialectical emphasis on wholeness, connection and context, change, historicity, contradiction, irregularity, asymmetry, and the multiplicity of levels of phenomena, a refreshing counterweight to the prevalent reductionism then and now.

An example: after Rosario suggested I look at *Drosophila* in nature—not just in bottles in the laboratory—I started to work with the *Drosophila* in the neighborhood of our home in Puerto Rico. My question was: How do *Drosophila* species cope with the temporal and spatial gradients of their environments? I began examining the multiple ways that different *Drosophila* species responded to similar environmental challenges. I could collect *Drosophila* in a single day in the deserts of Gúanica and in the rain forest around our farm at the crest of the cordillera. It turned out that some species adapt physiologically to high temperature in two to three days, and show relatively little genetic differences in heat tolerance along a 3,000-foot altitude gradient (about twenty miles). Others had distinct genetic sub-populations in the different habitats. Still others adapted to and inhabited only a part of the available environmental range. One of the desert species was not any better at tolerating heat than some *Drosophila* from the rain forest, but were much better at finding the cool moist microsites and hiding in them after about 8 a.m. These findings led me to describe the concepts of co-gradient selection, where the direct impact of the environment enhances genetic differences among populations, and counter-gradient selection where genetic differences offset the direct impact of the environment. Since on my transect the high temperature was associated with dry conditions, natural selection acted to increase the size of the flies at Guánica while the effect of temperature on development made them smaller. The outcome turned out to be that the flies from the sea-level desert and the rain forest were of about the same size in their own habitats, but that Guánica flies were bigger when raised at the same temperature as rain forest flies.

In this work I questioned the prevailing reductionist bias in biology by insisting that phenomena take place on different levels, each with their own laws, but also connected. My bias was dialectical: the interaction among adaptations on the physiological, behavioral, and genetic levels. My preference for process, variability, and change set the agenda for my thesis.

The problem was how species can adapt to an environment when the environment wasn't always the same. When I began thesis work I was puzzled by the facile assumption that, faced with opposing demands, for example when the environment favors small size some of the time and large size the rest of the time, an organism would have to adopt some intermediate state as a compromise. But this is an unthinking application of the liberal bromide that when there are opposing views the truth lies somewhere in the middle. In my dissertation, the study of fitness sets was an attempt to examine when an intermediate position is truly an optimum and when is it the worst possible choice. The short answer turned out to be that when the alternatives are not too different, an intermediate position is indeed optimal, but when they are very different compared to the range of tolerance of the species, then one extreme alone or in some cases a mixture of extremes is preferable.

Work in natural selection within population genetics almost always assumed a constant environment, but I was interested in its inconstancy. I proposed that "environmental variation" must be an answer to many questions of evolutionary ecology and that organisms adapt not only to specific environmental features such as high temperature or alkaline soils but also to the pattern of the environment—its variability, its uncertainty, the grain of its patchiness, the correlations among different aspects of the environment. Moreover, these patterns of environment are not simply given, external to the organism: organisms select, transform, and define their own environments.

Regardless of the particular matter of an investigation (evolutionary ecology, agriculture, or more recently, public health), my core interest has always been the understanding of the dynamics of complex systems. Also, my political commitment requires that I question the relevance of my work. In one of Brecht's poems he says, "Truly we live in a terrible time . . . when to talk about trees is almost a crime because it is a kind of silence about injustice." Brecht was of course wrong about trees: nowadays when we talk of trees we are not ignoring injustice. But he was also right that scholarship that is indifferent to human suffering is immoral.

Poverty and oppression cost years of life and health, shrinks the horizons, and cuts off potential talents before they can flourish. My commitment to support the struggles of the poor and oppressed and my interest in variability combined to focus my attention on the physiological and social vulnerabilities of people.

I have been studying the body's capacity to restore itself after it is stressed by malnutrition, pollution, insecurity, and inadequate health care. Continual stress undermines the stabilizing mechanisms in the bodies of oppressed populations making them more vulnerable to anything that happens, to small differences in their environments. This shows up in increased variability in measures of blood pressure, body mass index, and life expectancy as compared to more uniform results in comfortable populations.

In examining the effects of poverty, it is not enough to examine the prevalence of separate diseases in different populations. Whereas specific pathogens or pollutants may precipitate specific named diseases, social conditions create more diffuse vulnerability that links medically unrelated diseases. For instance, malnutrition, infection, or pollution can breach the protective barriers of the intestine. But once breached for any of these reasons it becomes a locus of invasion by pollutants, microbes, or allergens. Therefore nutritional problems, infectious diseases, stress, and toxicities cause a great variety of seemingly unrelated diseases.

The prevailing notion since the 1960s had been that infectious disease would disappear with economic development. In the 1990s I helped form the Harvard Group on New and Resurgent Disease to reject that idea. Our argument was partly ecological: the rapid adaptation of vectors to changing habitats—to deforestation, irrigation projects, and population displacement by war and famine. We also focused on the equally rapid adaptation of pathogens to pesticides and antibiotics. But we also criticized the physical, institutional, and intellectual isolation of medical research from plant pathology and veterinary studies which could have shown sooner the broad pattern of upsurge of not only malaria, cholera, and AIDS, but also African swine fever, feline leukemia, tristeza disease of citrus, and bean golden mosaic virus. We have to expect epidemiological changes with growing economic disparities and with changes in land use, economic development, human settlement, and demography. The faith in the efficacy of antibiotics, vaccines, and pesticides against plant, animal, and human pathogens is naïve in the light of adaptive evolution. And the developmentalist expectation that economic growth will lead the rest of the world to affluence and to the elimination of infectious disease is being proved wrong by events.

The resurgence of infectious disease is but one manifestation of a more general crisis: the eco-social distress syndrome—the pervasive multilevel crisis of dysfunctional relations within our species and between it and the rest of nature. It includes in one network of actions and reactions patterns of disease, relations of production and reproduction, demography, our depletion and wanton destruction of natural resources, changing land use and settlement, and planetary climate change. It is more profound than previous crises, reaching higher into the atmosphere, deeper into the earth, more widespread in space, and more long lasting, penetrating more corners of our lives. It is both a generic crisis of the human species and a specific crisis of world capitalism. Therefore it is a primary concern of both my science and my politics.

The complexity of this whole world syndrome can be overwhelming, and yet to evade the complexity by taking the system apart to treat the problems one at a time can produce disasters. The great failings of scientific technology have come from posing problems in too small a way. Agricultural scientists who proposed the Green Revolution without taking pest evolution and insect ecology into account, and therefore expecting pesticides would control pests, have been surprised that pest problems increased with spraying. Similarly, antibiotics create new pathogens, economic development creates hunger, and flood control promotes floods. Problems have to be solved in their rich complexity; the study of complexity itself becomes an urgent practical as well as theoretical problem.

These interests inform my political work: within the left, my task has been to argue that our relations with the rest of nature cannot be separated from a global struggle for human liberation, and within the ecology movement my task has been to challenge the "harmony of nature" idealism of early environmentalism and to insist on identifying the social relations that lead to the present dysfunction. At the same time my politics have determined my scientific ethics. I believe that all theories are wrong that promote, justify, or tolerate injustice.

A leftist critique of the structure of intellectual life is a counterweight to the culture of the universities and foundations. The antiwar movement of the 1960s and 1970s took up the issues of the nature of the university as an organ of class rule and made the intellectual community itself an object of theoretical as well as practical interest. I joined Science for the People, an organization that started with a research strike at MIT in 1967 as a protest against military research on campus. As a member I helped in the challenge to the Green Revolution and genetic determinism. Antiwar activism also took me to Vietnam to investigate war crimes (especially the use of defoliants) and from there to organizing Science for Vietnam. We denounced the use of Agent Orange (used as defoliant in the Vietnamese jungle) that was causing birth defects among Vietnamese peasants. Agent Orange was one of the worst uses of chemical herbicides.

The Puerto Rican independence movement gave me an anti-imperialist consciousness that serves me well in a university that promotes "structural reform" and other euphemisms for empire. My wife's sharp working-class feminism is a running source of criticism of the pervasive elitism and sexism. Regular work with Cuba shows me vividly that there is an alternative to a competitive, individualistic, exploitative society.

Community organizations, especially in marginalized communities, and the women's health movement raise issues that academia prefers to ignore: the mothers of Woburn noticing that too many of their children from the same small neighborhood had leukemia, the hundreds of environmental justice groups that noted that toxic waste dumps were concentrated in Black and Latino neighborhoods, and the Women's Community Cancer project and others who insist on the environmental causes of cancer and other diseases while the university laboratories are looking for guilty genes. Their initiatives help me maintain an alternative agenda for both theory and action.

Within the university I have a contradictory relationship with the institution and with colleagues, a combination of cooperation and conflict. We may share a concern about health disparities and persistent poverty, but we are in conflict about corporations funding research for patentable molecules and about government agencies such as AID (Agency for International Development) promoting the goals of empire.[2]

I never aspired to what is conventionally considered a "successful career" in academia. I do not find most of my personal validation through the formal reward and recognition system of the scientific community, and I try not to share the common assumptions of my professional community. This gives me wide freedom of choice. Thus when I declined to join the National Academy of Sciences and received many supportive letters praising my courage or calling it a difficult decision, I could honestly say that it was not a hard decision, merely a political choice taken collectively by the Science for the People group in Chicago. We judged that it was more useful to take a public stand against the Academy's collaboration with the Vietnam-American War than to join the Academy and attempt to influence its actions from inside. Dick Lewontin had already tried that unsuccessfully and resigned, along with Bruce Wallace.

Most of my research has objectives at two levels: the particular problem at hand and some major theoretical or polemical issue. The study of temperature adaptation in fruit flies was also an argument for multiple levels of causation. Niche theory was also a foray into the interpenetration of opposites (organism and environment). Biogeography was about multiple levels of ecological and evolutionary dynamics. Ecological pest management was also a claim for whole-system strategies. Work on new and resurgent infectious disease combined biology and sociology. We examined why the public health community was caught by surprise when infectious disease would not go away. It therefore was an exercise in the self-examination of science.

I have always enjoyed mathematics and see one of its tasks as making the obscure obvious. I regularly employ a sort of mid-level math in unconventional ways to promote understanding more than prediction. Much modeling now aims at precise equations giving precise prediction. This makes sense in engineering. In the field of policy, it makes sense to those who are the advisors to the rulers who imagine they have complete enough control of the world to be able to optimize their efforts and investments of resources. But those of us who are in the opposition have no such illusion. The best we can do is decide where to push the system. For this, a qualitative mathematics is more useful. My work with signed digraphs (loop analysis) is one such approach. Rejecting the opposition between qualitative and quantitative analysis and the notion that quantitative is superior to qualitative, I have mostly worked with those mathematical tools that assist conceptualization of complex phenomena.

Political activism, of course, attracts the attention of the agencies of repression. I have been fortunate in that regard, having experienced only relatively light repression. Others did not fare as well, with lost careers, years of imprisonment, violent attacks, intense harassment even of their families, and deportations. Some, mostly from the Puerto Rican, African-American, and Native American liberation movements, as well as the five Cuban anti-terrorists arrested in Florida, are still political prisoners.

Exploitation kills and hurts people. Racism and sexism destroy health and thwart lives. Studying the greed and brutality and smugness of late capitalism is painful and infuriating. Sometimes I have to recite from Jonathan Swift's "Ballad in a Bad Temper":

Like the boatman on the Thames
I row by and call them names.
Like the ever-laughing sage

In a jest I spend my rage
But it must be understood
I would hang them if I could.

For the most part scholarship and activism have given me an enjoyable and rewarding life, doing work I find intellectually exciting, socially useful, and with people I love.

Notes

1 The End of Natural History?
This chapter first appeared in a slightly different form as Richard Lewontin and Richard Levins, "The End of Natural History?" *Capitalism, Nature, Socialism* 7, no. 1 (1996): 95–98.

2 The Return of Old Diseases and the Appearance of New Ones
This chapter first appeared in a slightly different form as Richard Lewontin and Richard Levins, "The Return of Old Diseases and the Appearance of New Ones," *Capitalism, Nature, Socialism* 7, no. 2 (1996): 103–7.

3 False Dichotomies
This chapter first appeared in a slightly different form as Richard Lewontin and Richard Levins, "False Dichotomies," *Capitalism, Nature, Socialism* 7, no. 3 (1996): 27–30.

4 Chance and Necessity
This chapter first appeared in a slightly different form as Richard Lewontin and Richard Levins, "Chance and Necessity," *Capitalism, Nature, Socialism* 8, no. 1 (1997): 65–68.

5 Organism and Environment
This chapter first appeared in a slightly different form as Richard Lewontin and Richard Levins, "Organism and Environment," *Capitalism, Nature, Socialism* 8, no. 2 (1997): 95–98.

6 The Biological and the Social
This chapter first appeared in a slightly different form as Richard Lewontin and Richard Levins, "The Biological and the Social," *Capitalism, Nature, Socialism* 8, no. 3 (1997): 89–92.

7 How Different Are Natural and Social Science?

This chapter first appeared in a slightly different form as Richard Lewontin and Richard Levins, "How Different Are Natural and Social Science?" *Capitalism, Nature, Socialism* 9, no. 1 (1998): 85–9.

1. See chap. 16, "Ten Propositions on Science and Antiscience," in this volume for a brief summary of the "objective" and "subjective" aspects of natural science.

8 Does Anything New Ever Happen?

This chapter first appeared in a slightly different form as Richard Lewontin and Richard Levins, "Does Anything New Ever Happen?" *Capitalism, Nature, Socialism* 9, no. 2 (1998): 53–56.

9 Life on Other Worlds

This chapter first appeared in a slightly different form as Richard Lewontin and Richard Levins, "Life on Other Worlds," *Capitalism, Nature, Socialism* 9, no. 4 (1998): 39–42.

10 Are We Programmed?

This chapter first appeared in a slightly different form as Richard Lewontin and Richard Levins, "Evolutionary Psychology," *Capitalism, Nature, Socialism* 10, no. 3 (1999): 127–130.

11 Evolutionary Psychology

This chapter first appeared in a slightly different form as Richard Lewontin and Richard Levins, "Evolutionary Psychology," *Capitalism, Nature, Socialism* 10, no. 3 (1999): 127–130.

1. Leda Cosmides and John Tooby, "Beyond Intuition and Instinct Blindness: Toward an Evolutionary Rigorous Cognitive Science," in *Cognition on Cognition*, ed. Jacques Mehler and Susana Franck (Cambridge: MIT Press, 1995).

12 Let the Numbers Speak

This chapter first appeared in a slightly different form as Richard Lewontin and Richard Levins, "Let the Numbers Speak," *Capitalism, Nature, Socialism* 11, no. 1 (2000): 63–67.

13 The Politics of Averages

This chapter first appeared in a slightly different form as Richard Lewontin and Richard Levins, "The Politics of Averages," *Capitalism, Nature, Socialism* 11, no. 2 (2000): 111–14.

14 Schmalhausen's Law

This chapter first appeared in a slightly different form as Richard Lewontin and Richard Levins, "Schmalhausen's Law," *Capitalism, Nature, Socialism* 11, no. 4 (2000): 103–8.

15 A Program for Biology

This chapter first appeared in a slightly different form as Richard Levins and Richard Lewontin, "A Program for Biology," *Biological Theory* 1, no. 4 (2006): 1–3.

1. The epidemiological transition is the proposition that as countries develop, infectious disease would decline and be replaced by chronic disease as the major health problem.

16 Ten Propositions on Science and Antiscience

This chapter first appeared in a slightly different form as Richard Levins, "Ten Propositions on Science and Antiscience," *Social Text* 46–47 (1996): 101–12.

1. Nancy Krieger, "Epidemiology and the Web of Causation: Has Anyone Seen the Spider?" *Social Science of Medicine* 30, no. 7 (1994): 887–903.

17 Dialetics and Systems Theory

This chapter first appeared in a slightly different form as Richard Levins, "Dialectics and Systems Theory," *Science & Society* 62, no. 3: 375–99.

1. John Maynard Smith, "Molecules Are Not Enough," review of *The Dialectical Biologist, London Review of Books* 6 (February 1986).

2. Göran Therbom, *What Does the Ruling Class Do When It Rules?* (London: New Left Books, 1978).

3. The term "dialectical materialism" is often associated with the particular rigid exposition of it by Stalin and its dogmatic applications in Soviet apologetics, whereas "dialectical" by itself is a respectable academic term. At a time when the retreat from materialism has reached epidemic proportions it is worthwhile to insist on the unity of materialism and dialectics, and to recapture the full vibrancy of this approach to understanding and acting on the world. Here I use materialist dialectics and dialectical materialism interchangeably.

4. Norbert Wiener, *Cybernetics: Or, Control and Communication in the Animal and the Machine* (Cambridge: MIT Press, 1961).

5. Hubert M. James, Nathaniel B. Nichols, and Ralph S. Phillips, eds., *The Theory of Servomechanisms* (New York: McGraw-Hill, 1947), ix, 2.

6. Donella H. Meadows, Dennis L. Meadows, and Jorgen Randers, *Beyond the Limits* (Post Mills, Vt.: Chelsea Green, 1992).

7. Richard Levins, "The Strategy of Model Building in Population Science," *American Scientist* 54 (1966): 421–31.

8. Ludwig von Bertalanffy, "An Outline of General Systems Theory," *British Journal for the Philosophy of Science* 1, no. 2 (1950): 139–64.

9. William Ross Ashby, *Design for a Brain* (New York: Wiley, 1960).

10. Robert V. O'Neill, D. L. DeAngelis, J. B. Waide, and T. E. H. Allen, *A Hierarchical Concept of Ecosystems* (Princeton, N.J.: Princeton University Press, 1986).

11. Meadows, Meadows, and Randers, *Beyond the Limits*, 2.

12. See Bertell Ollman, *Dialectical Investigations* (New York: Routledge, 1993) for a detailed examination of dialectical abstraction.

13. Viktor G. Afanasev, *The Scientific Management of Society* (Moscow: Progress Publishers, 1971).

18 Aspects of Wholes and Parts in Population Biology

This chapter first appeared in a slightly different form as Richard Lewontin and Richard Levins, "Aspects of Whole and Parts in Population Biology," in *Evolution of Social Behavior and Integrative Levels,* ed. Gary Greenberg and Ethel Tobach (Hillsdale, N.J.: Lawrence Erlbaum Associates Publishers, 1988), 31–52.

1. Caroline Merchant, *The Death of Nature* (San Francisco: Harper & Row, 1983).

2. For a review of this history see H. Waitzkin, "The Social Origins of Illness: A Neglected History," *International Journal of Health Services* 1, no. 2 (1981): 77–104.

3. Ivan Illich, *Medical Nemesis* (New York: Bantam Books, 1976).

4. J. Vandermeer, "The Competitive Structure of Communities: An Experimental Approach Using Protozoa," *Ecology* 50 (1969): 362–71.

5. H. Levene, O. Pavlovsky, and T. Dobzhansky, "Interaction of the Adaptive Values in Polymorphic Experimental Populations of *Drosophila pseudoobscura,*" *Evolution* 8 (1954): 335–49.

6. C. S. Lumsden and E. O. Wilson, *Genes, Mind and Culture* (Cambridge, Mass.: Harvard University Press, 1983).

7. Richard Levins and Richard Lewontin, *The Dialectical Biologist* (Cambridge, Mass.: Harvard University Press, 1985), esp. chap. 4.

8. J. G. Kingsolver and R. S. Moffat, "Thermo Regulation and the Determinants of Heat Transfer in Colias Butterflies," *Oecologia* 53 (1983): 27–33.

9. E. Sober and Richard Lewontin, "Artifact, Cause and Effect," *Philosophy of Science* 49 (1982): 332–38.

10. For the derivation and manipulation of this representation of systems, see C. J. Puccia and Richard Levins, *Qualitative Modeling of Complex Systems* (Cambridge, Mass.: Harvard University Press, 1985).

11. The distinction between inside and outside is not clear-cut. Holists without academic connections and lacking research resources depend on reinterpretation of published reports for their arguments. Scientists in recognized institutions live lives outside as well and have other sources of inspiration than their professional world.

12. We criticize this approach in chap. 4 of Levins and Lewontin, *The Dialectical Biologist.*

13. This line of inquiry was suggested by chap. 1 of Karl Marx's *Capital* (Chicago: Charles Kerr, 1906), in which Marx focuses on the nonequivalence of the terms joined by the equal sign in the equations of value.

19 Strategies of Abstraction

This chapter first appeared in a slightly different form as Richard Levins, "Strategies of Abstraction," *Biology and Philosophy* 21 (2006): 741–755.

1. Bertell Ollman, *Dance of the Dialectic: Steps in Marx's Method* (Chicago: University of Illinois Press, 2003).

2. Bertolt Brecht, "To Posterity," in *Bertolt Brecht: Poetry and Prose*, ed. Reinhold Grimm and Vedia Caroline Molina (New York: Continuum Press, 2003).

3. Richard Lewontin and Richard Levins, "On the Characterization of Density and Resource Availability," *American Naturalist* 134, no. 4 (1989): 513–24.

4. Marxists see contradiction as a process in time, as opposed to the standard view in analytic philosophy that sees contradiction as a static, set-theoretic relation. The relationship between these concepts is discussed in the last chapter of Richard Levins and Richard Lewontin, *The Dialectical Biologist* (Cambridge, Mass.: Harvard University Press, 1985).

5. George Polya, *Mathematics and Plausible Reasoning* (Princeton, N.J.: Princeton University Press, 1990).

20 The Butterfly ex Machina

This chapter first appeared in a slightly different form as Richard Levins, "The Butterfly ex Machina," in *Thinking About Evolution*, ed. Rama Singh, Costas B. Krimbas, Diane B. Paul, and John Beatty (New York: Cambridge University Press, 2001), 529–43.

1. N. Edward Lorenz, "Deterministic Nonperiodic Flows," *Journal of Atmospheric Science* 20 (1963): 130–41.

2. Robert May, "Simple Mathematical Models with Very Complicated Dynamics," *Nature* 261 (1976): 459–67.

3. Peter Carruthers, Interview on *Talk of the Nation*, National Public Radio, January 17, 1994.

4. Deepak Chopra, *Quantum Healing* (New York: Bantam Books, 1989).

5. Michael Hilliard, "The Future of East Germany After the GDR: Interview with Peter Kruger," *Rethinking Marxism* 6, no. 1 (1993): 115–27.

6. Tien-Yien Li and James A. Yorke, "Period Three Implies Chaos," *American Mathematical Monthly* 82, no. 10 (1975): 985–92.

7. Edward A. Grove, G. Ladas, Richard Levins, and C. Puccia, "Oscillation and Stability in Models of a Perennial Grass," *Proceedings of Dynamic Systems and Applications* 1 (1994): 87–93; David Tilman and David Wedin, "Oscillations and Chaos in the Dynamics of a Perennial Grass," *Nature* 353 (1991): 653–55.

8. Richard Levins, "Preparing for Uncertainty," *Ecosystem Health* 1, no. 1 (1995): 48–55.

21 Educating the Intuition to Cope with Complexity

1. The impact of a changed parameter of insulin on glucose is proportional to the feedback of the (E,A) subsystem, which is the strength of the positive feedback loop minus the product of the two self-inhibitions.

22 Preparing for Uncertainty

This chapter first appeared in a slightly different form as Richard Levins, "Preparing for Uncertainty," *Ecosystem Health* 1 (1995): 47–57.

1. M. E. Wilson, Richard Levins, and A. Spielman, "Disease in Evolution: Global Changes and the Emergence of Infectious Diseases," *Annals of the New York Academy of Sciences* (1994).

2. Janet Raloff, "Something's Fishy," *Science News* 146 (1994): 89.

3. Richard Levins and Richard Lewontin, *The Dialectical Biologist* (Cambridge, Mass.: Harvard University Press, 1985).

4. Alfred Crosby, *The Columbian Exchange* (Westport, Conn.: Greenwood Press, 1972).

5. Paul Epstein, "Commentary: Pestilence and Poverty—Historical Transitions and the Great Pandemics," *American Journal of Preventive Medicine* 8, no. 4 (1992): 263–65.

6. Donald R. Strong, Earl D. McCoy, and Jorge R. Rey, "Time and the Number of Herbivore Species: The Pests of Sugar Cane," *Ecology* 58, no. 1 (1977): 167–75.

7. Vandana Shiva, *The Violence of the Green Revolution* (London: Zed Books, 1991); Daniel Faber, *Environment Under Fire* (New York: Monthly Review Press, 1993).

8. M. J. Bouma, H. E. Sandorp, and H. J. van der Kaay, "Climate Change and Periodic Epidemic Malaria," *The Lancet* 343 (1994): 1440.

9. C. Puccia and Richard Levins, *Qualitative Analysis of Complex Systems* (Cambridge, Mass.: Harvard University Press, 1985).

10. Richard Levins and John H. Vandermeer, "The Agroecosystem Embedded in a Complex Community," in *Agroecology*, ed. C. Roland Carroll, John H. Vandermeer, and Peter Rosset (New York: Wiley, 1990).

24 Genes, Environment, and Organisms

This chapter first appeared in a slightly different form as Richard Lewontin, "Genes, Environment, and Organisms," in *Hidden Histories of Science*, ed. Robert B. Silvers (London: Granta Books, 1997), 115–39.

1. The original study in the late 1940s was by C. C. North and P. K. Hatt, "Jobs and Occupations: A Popular Evaluation," in *Sociological Analysis*, ed. Logan Wilson and William Lester Kolb (New York: Harcourt Brace, 1949). Later studies gave essentially identical results.

2. Bernard Barber, *Science and the Social Order* (Glencoe, Ill.: Free Press, 1952), 14.

3. In Diderot's *Le Rêve de d'Alembert*, both the physician Bordeu's waking expositions and d'Alembert's sleeping meanderings center on these issues. "Who knows
   what races of animals preceded us? Who knows what races of animals will succeed ours?" d'Alembert wonders in his sleep. Bordeu asks, "How could we know
   that the man, leaning on his stick, whose eyes are blind, who drags himself along
   with such effort, yet more different on the inside than the outside, is the same man
   who yesterday walked so lightly, shifting with ease the heaviest loads?" These biological questions were such a central concern of philosophy that Diderot has Mlle
   de l'Espinasse remark sarcastically, "There is no difference between a physician
   who is awake and a philosopher who is asleep."

4. For a perceptive and informative view of this debate see Shirley Roe, *Matter, Life
   and Generation: Eighteenth-Century Embryology and the Haller-Wolff Debate*
   (Cambridge: Cambridge University Press, 1981).

5. Jens Clausen, David D. Keck, and William W. Hiesey, "Environmental Responses
   of Climatic Races of *Achillea*," Carnegie Institution of Washington Publication
   581 (1958).

6. The selective theory of the formation of the central nervous system is explicated
   in Gerald M. Edelman, *Neural Darwinism: The Theory of Neuronal Group
   Selection* (New York: Basic Books, 1989).

7. The sophisticated birdwatcher will not recognize any real bird in this composite
   life history.

25 The Dream of the Human Genome

This chapter first appeared in a slightly different form as a chapter in Richard
Lewontin, *It Ain't Necessarily So: The Dream of the Human Genome and Other
Illusions* (New York: New York Review Books, 2000). It was based on a review
essay that first appeared in *The New York Review of Books* on May 28, 1992. The
following were the books under review: Daniel J. Kevles and Leroy Hood, eds.,
The Code of Codes: Scientific and Social Issues in the Human Genome Project
(Cambridge: Harvard University Press, 1992); Joel Davis, *Mapping the Code: The
Human Genome Project and the Choices of Modern Science* (New York: Wiley,
1990); Lois Wingerson, *Mapping Our Genes: The Genome Project and the Future
of Medicine* (New York: Dutton, 1990); David Suzuki and Peter Knudtson,
Genethics: The Ethics of Engineering Life (Cambridge: Harvard University Press,
1990); Committee on Mapping and Sequencing the Human Genome, *Mapping
and Sequencing the Human Genome* (Washington, DC: National Academy Press,
1988); Jerry E. Bishop and Michael Waldholz, *Genome: The Story of the Most
Astonishing Scientific Adventure of Our Time—The Attempt to Map All the Genes in
the Human Body* (New York: Simon and Schuster, 1990); Christopher Wills,
Exons, Introns, and Talking Genes: The Science Behind the Human Genome Project
(New York: Basic Books, 1991); Dorothy Nelkin and Laurence Tancredi, *Dangerous
Diagnostics: The Social Power of Biological Information* (New York: Basic Books,
1989); and Committee on DNA Technology in Forensic Science, *DNA Technology
in Forensic Science* (Washington, DC: National Academy Press, 1992).

1. John Cairn, Gunther S. Stent, and James D. Watson, eds., *Phage and the Origins of Molecular Biology* (Cold Springs Harbor, N.Y.: Cold Spring Harbor Laboratory of Quantitative Biology, 1966).

2. Daniel J. Kevles and Leroy Hood, eds., *The Code of Codes: Scientific and Social Issues in the Human Genome Project* (Cambridge, Mass.: Harvard University Press, 1993).

3. Richard Dawkins, *The Selfish Gene* (New York: Oxford University Press, 1976), 21.

4. Joel Davis, *Mapping the Code: The Human Genome Project and the Choices of Modern Science* (New York: Wiley, 1990).

5. Daniel J. Kevles, *In the Name of Eugenics: Genetics and the Uses of Human Heredity* (Berkeley: University of California Press, 1986).

6. See *New York Times,* April 9, 1992; *Wall Street Journal,* April 17, 1992; and *Nature,* April 9, 1992, 463.

7. Remarks made at the First Human Genome Conference in October 1989. Quoted in Keller, "Nature, Nurture, and the Human Genome Project," in *The Code of Codes.*

8. Pressure was also brought by scientists in the genome sequencing establishment against the editor of the journal in which it was to be published, including one of the contributors to *The Code of Codes.* As a result the editor delayed its publication, demanded changes in galley proof, and asked two defenders of the method to write a counterattack. One report of the scandal is given in Lesley Roberts's "Fight Erupts over DNA Fingerprinting," *Science* (December 20, 1991): 1721–23.

9. Based on *Frye* v. *United States,* 293 F. 2nd DC Circuit 1013, 104 (1923).

10. *DNA Technology in Forensic Science,* report of the Committee on DNA Technology in Forensic Science (Washington, DC: National Academy Press, 1992). The reader should know that I, Richard Lewontin, am not a disinterested party either with respect to the report or to the body that sponsored it. I have twice testified in federal court on the weaknesses of DNA profiles, am the author of a position paper that was a basis for the original very critical version of the NRC report's chapter on population considerations, and am the author, with Daniel Hartl, of a highly critical paper in *Science* that was the object of considerable controversy. I resigned from the National Academy of Sciences in 1971 in protest against the secret military research carried out by its operating arm, the National Research Council.

11. See M. Allison, "The Radioactive Elixir," *Harvard Magazine* (January–February 1992): 73–75.

12. See epilogue to chap. 8 in Lewontin, *It Ain't Necessarily So.*

13. William A. Haseltine, "Life by Design: Gene Mapping, Without Tax Money," *New York Times,* May 21,1998.

14. This is calculated as the present value of the drug when the after-tax income stream is projected for thirty-five years using a long-term estimate of average rates of return. (Genomics II, Lehman Brothers, January 23,1998).

15. National Research Council, *The Evaluation of Forensic DNA Evidence* (Washington, DC: National Academy Press, 1996).

26 Does Culture Evolve?

This chapter first appeared in a slightly different form as Joseph Fracchia and Richard Lewontin, "Does Culture Evolve?" *History and Theory* 38 (1999): 52–78.

1. C. P. Snow, *The Two Cultures and a Second Look* (Cambridge, UK: University Press, 1964), 8–9.

2. Ibid., 70.

3. Herbert Spencer, *The Principles of Biology* [1867] (New York and London: D. Appleton and Company, 1914), 432–33.

4. Leslie White, "Preface," in *Evolution and Culture*, ed. Marshall Sahlins and Elman Service (Ann Arbor: University of Michigan Press, 1960), v.

5. Ibid., vii.

6. E. O. Wilson, *On Human Nature* (Cambridge, Mass.: Harvard University Press, 1978), x.

7. Ibid., 13.

8. Ibid., 34.

9. Alexander Rosenberg, *Sociobiology and the Preemption of Social Science* (Baltimore: Johns Hopkins University Press, 1980), 22–23.

10. Ibid., 4, 158.

11. Leda Cosmides and John Tooby, "The Psychological Foundations of Culture," in *The Adapted Mind*, ed. Jerome H. Barkow, Leda Cosmides, and John Tooby (New York: Oxford University Press, 1992), 22–23.

12. Richard Shelly Hartigan, "A Review of *The Biology of Moral Systems*" (by Richard D. Alexander), *Politics and the Life Sciences* 7, no. 1 (1988), 96.

13. See for example the following essays, all from *Politics and the Life Sciences*: Elliot White, "Self-Selection and Social Life: The Neuropolitics of Alienation—The Trapped and the Overwhelmed" (vol. 7, no. 1, 1989); John H. Beckstrom, "Evolutonary Jurisprudence: Prospects and Limitations on the Use of Modern Darwinism Throughout the Legal Process" (vol. 9, no. 2, 1991); Lee Ellis, "A Biosocial Theory of Social Stratification Derived from the Concepts of Pro/Anti-sociality and r/K Selection" (vol. 10, no. 1, 1991); Hames N. Schubert et al., "Observing Supreme Court Oral Argument: A Biosocial Approach" (vol. 11, no. 1, 1992); Larry Amhart, "Feminism, Primatology, and Ethical Naturalism" (vol. 11, no. 2, 1992).

14. Ibn Khaldun, *The Muqaddimah* (Princeton, N.J.: Princeton University Press, 1958), chap. 2:24.

15. Ibid., chaps. 3:3, 3:46, 3:11.

16. François Furet and Mona Ozouf, eds., *A Critical Dictionary of the French Revolution* (Cambridge, Mass.: Belknap Press, 1989), xvi.

17. Leslie White, *Social and Cultural Evolution*, vol. 3, *Issues in Evolution*, ed. S. Tax and C. Callender (Chicago: University of Chicago Press, 1960), Panel 5.

18. In the introduction to a collection of essays on *History and Evolution*, ed. Matthew Nitecki and Doris Nitecki (Albany: State University of New York Press, 1992), Matthew Nitecki states categorically: "The common element of evolutionary biology and history is the concept of *change* over time." Despite great differences in their definitions of the relation between biology and history, all authors included in the volume (and probably all cultural evolutionists) share Nitecki's definition of history as change over time. This definition allows them to elevate history to scientific status and to subject history to evolutionary explanation, as do, for example, Boyd and Richerson in their categorical claim that "Darwinian theory is both scientific and historical" (in Nitecki, 179–80). Though there are many problems with defining history simply as change over time, we will only make two comments here: this definition almost inevitably results in treating people living in societies in which change is not the norm as people without history; and once change is defined as a transhistorical constant, it is very likely, though not logically necessary, that the next step will be to seek a transhistorical explanatory law—which for cultural evolutionists is that of selection.

19. *Evolution and Culture* (Ann Arbor: University of Michigan Press, 1960) is the work of four authors, each of whom contributed a chapter to the volume: Thomas Harding, David Kaplin. Marshall Sahlins, and Elman Service. The most influential of these has been Sahlins's "Evolution: Specific and General." Sahlins's approach to culture has, of course, evolved considerably since 1960.

20. Marshall Sahlins, "Evolution: Specific and General," in *Evolution and Culture*, 20.

21. E. O. Wilson, "Human Decency Is Animal," *The New York Times Magazine*, October 12, 1975.

22. On the basis of the two questions he asks of each cultural evolutionary theory, William Durham, in *Coevolution: Genes, Culture and Human Diversity* (Stanford, Calif.: Stanford University Press, 1991), is able thoroughly to survey the cultural evolutionist plain, and, in so doing, provides a sense of its paradigmatic unity. His questions are: "Is culture a second inheritance system? What are the best units to use in the study of cultural transmission?" (155). Based on the responses, he establishes in an "approximate[ly] chronological" order (155) a tripartite division of the cultural evolutionist terrain. The earliest theories of cultural evolution tended to be "models without dual inheritance." These conceptualize culture not "as part of the phenotype"; and they explain "phenotypic change in human populations in terms of a single fitness principle, namely, reproductive fitness in one of its guises" (155–56). Examples include both the stronger and weaker versions of sociobiology, David Barash's "genetic determinism," and the "genes on a leash" model of E. O. Wilson's *On Human Nature* (Cambridge, Mass.: Harvard University Press, 1978). The second type, "models with dual inheritance and trait units" also conceptualizes culture "as part of the phenotype" but views it "as

a second, nongenetic inheritance system whose units are defined as culturally heritable aspects of phenotype. These units are recognized as having their own measure of fitness within the cultural system [i.e. 'cultural fitness']. . . . The differential transmission of traits or behaviors within a population constitutes cultural evolution" (156). Examples are: Cavalli-Sforza's and Feldman's cultural transmission model; Richard Alexander's social learning model; Lumsden's and Wilson's gene-culture transmission model; and his own early coevolution model. The third type is "models with dual inheritance and ideational units." These also treat culture "as a separate 'track' of informational inheritance," and focus not on phenotypical traits but the differential transmission of ideas, values, and beliefs in a population" (156). Examples include Alben G. Keller's social selection model, H. Ronald Pulliam's and Christopher Dunford's "programmed learning model," Boyd's and Richerson's "Darwinian culture theory," and Durham's own book, *Coevolution*. The "evolution" of theories of cultural evolution has, as Durham indicated, roughly followed the sequence of his types: from behavioral to (Clifford Geertzian) ideational definitions of culture and from single inheritance, culture-on-a-genetic-leash models to dual inheritance models. This evolution resulted from dissatisfaction with single inheritance models for tying culture to a short genetic leash, and with behavioral definitions of culture for their unreliability, the impossibility of knowing precisely which meme motivates a given behavior.

23. Wilson, *On Human Nature*, 78.

24. Roben Boyd and Peter Richerson, *Culture and the Evolutionary Process* (Chicago: University of Chicago Press, 1985), 292; and "How Microevolutionary Processes Give Rise to History" in *History and Evolution*, ed. Nitecki and Nitecki, 181.

25. Melvin Konner, *The Tangled Wing: Biological Constraints on the Human Spirit* (New York: Holt, Rinehart, and Winston, 1982), 414.

26. Unaware of the implications of their reduction of societies to populations, Boyd and Richerson, much to their surprise, found themselves criticized by David Rindos ("The Evolution of the Capacity for Culture: Sociobiology, Structuralism, and Cultural Selectionism," *Cultural Anthropology* 27 [1986]: 315–16); and William Durham (*Coevolution*, 179ff.) for not having adequately addressed the social. In their direct response to Rindos (included in Rindos, 327) Boyd and Richerson claim, correctly, that they spent an entire chapter of their *Culture and the Evolutionary Process* on "the scale of human social organization" implying, incorrectly, that therewith the matter was resolved. That chapter first develops a taxonomy of biases (direct, indirect, and frequency dependent) and then constructs models to analyze how the frequency of these biases affects the transmission of culture. Though such biases certainly affect social behavior, their origins and persistence are nowhere discussed. Consequently, the authors end up explaining how social biases affect individual choice by transforming clichés into explanatory principles: "When in Rome, do as the Romans do" becomes the law of "frequency-dependent bias" (286) and "keeping up with the Joneses" the law of "indirect bias" (287). The questions of whether all Romans do as some

Romans do or of whether keeping up with the Joneses makes sense in societies not based on commodity production and exchange are crucial questions that disappear in their biases.

Durham makes perhaps the most concerted effort to consider asymmetries of social power and the "imposition" of group values on individual "choice" (*Coevolution*, 198–99). He identifies "reference groups" within a given population, thereby acknowledging "the simple fact that cultural evolution is an intrinsically political process" (211). Because he does not ask the essential questions of why particular "reference groups" exist and what the distinct and discrete social logic is behind particular asymmetries in group power, Durham can only treat any particular set of reference groups and social asymmetries of power as arbitrary and subordinate variable factors affecting individual choice, rather than as constitutive factors of social and cultural forms and their "evolution."

27. The most perspicacious critic of theories that reduce societies to populations was Karl Marx. The hallmark of political economy and the source of its errors, Marx argued, was that it took as its starting point the population without having determined the components of the populations, its "subgroups" or classes *and* the logic of their internal relations. Such an approach would produce not "a rich totality of many determinations and relations," but "ever thinner abstractions" and "a chaotic conception of the whole" (*Marx-Engels Reader,* ed. Robert Tucker [New York: Norton, 1978], 237). Or, as he later summarized it more succinctly: "Society does not consist of individuals, but expresses the sum of interrelations, the relations within which these individuals stand" (247). The analysis of a society reveals much about its population, but the converse is not necessarily true.

28. Wilson, *On Human Nature*, 78; Jerome Barkow, *Darwin, Sex, and Status: Biological Approaches to Mind and Culture* (Toronto: University of Toronto Press, 1989), 142.

29. Bernardo Bernardi, "The Concept of Culture: A New Presentation," in *The Concepts and Dynamics of Culture*, ed. Bernardo Bernardi (The Hague: Mouton: 1977); Martin Stuart-Fox, "The Unit of Replication in Socio-Cultural Evolution," *Journal of Social and Biological Structures* 9 (1986): 67–90.

30. L. L. Cavalli-Sforza and M. W. Feldman, *Cultural Transmission and Evolution:* A *Quantitative Approach* (Princeton, N.J.: Princeton University Press, 1981); Richard Dawkins, *The Selfish Gene* (New York: Oxford University Press, 1976).

31. See Cavalli-Sforza and Feldman, *Cultural Transmission and Evolution,* 10; Boyd and Richerson, *Culture and the Evolutionary Process,* 8ff.; and Boyd and Richerson, "How Microevolutionary Processes Give Rise to History," 182.

32. Peter J. Richerson and Robert Boyd. "A Darwinian Theory for the Evolution of Symbolic Cultural Traits," in *The Relevance of Culture,* ed. Morris Freilich (New York: Bergin & Garvey Publishers, 1989), 121; Boyd and Richerson, "How Microevolutionary Processes Give Rise to History," 181.

33. Martin Daly, "Some Caveats about Cultural Transmission Models," *Human Ecology* 10 (1982): 402–4.

34. Ibid,. 406. See also David Hull, "The Naked Meme," in *Learning, Development and Culture: Essays in Evolutionary Epistemology*, ed. H. C. Plotkin (Chichester, UK: Wiley, 1982); H. Kaufman, "The Natural History of Human Organizations," *Administration and Society* 7 (1975); Timothy Goldsmith. *The Biological Roots of Human Nature: Forging Links between Evolution and Behavior* (New York: Oxford University Press, 1991).

35. Stuart-Fox, "Unit of Replication in Socio-Cultural Evolution," 68.

36. Rosenberg, *Sociobiology and the Preemption of Social Science*, 151.

37. Charles Lumsden and E. O. Wilson, *Genes, Mind and Culture: The Coevolutionary Process* (Cambridge, Mass.: Harvard University Press, 1981), 358, 360, 362. See also E. O. Wilson, *Consilience* (New York: Knopf, 1998).

38. Boyd and Richerson, *Culture and the Evolutionary Process*, 25; "How Microevolutionary Processes Give Rise to History," 203.

39. Laurence J. Peter, *Peter's Quotations: Ideas for Our Time* (New York: Morrow, 1977), 477.

27 Is Capitalism a Disease?: The Crisis in U.S. Public Health

This chapter first appeared in a slightly different form as Richard Levins, "Is Capitalism a Disease?: The Crisis in U.S. Public Health," *Monthly Review* 52, no. 4 (2000): 8–33.

28 Science and Progress: Seven Developmentalist Myths in Agriculture

This chapter first appeared in a slightly different form as Richard Levins, "Science and Progress: Seven Developmentalist Myths in Agriculture," *Monthly Review* 38, no. 3 (1986): 13–20.

29 The Maturing of Capitalist Agriculture: Farmer as Proletarian

This chapter first appeared in a slightly different form as Richard Lewontin, "The Maturing of Capitalist Agriculture: Farmer as Proletarian," *Monthly Review* 50, no. 3 (1998): 72–84.

30 How Cuba Is Going Ecological

This chapter first appeared in a slightly different form as Richard Levins, "How Cuba Is Going Ecological," *Capitalism, Nature, Socialism* 16, no. 3 (2005): 7–25.

1. This paper was prepared for the Latin American Studies Association meetings, October 6–10, 2004.

2. The following references introduce each of the areas of Cuban achievement not discussed in this paper. Jerry M. Spiegel and Annalee Yassi, "Lessons from the Margins of Globalization: Appreciating the Cuban Health Paradox," *Journal of Public Health Policy* 25, no. 1 (2004): 85–110; Anuario Estadistico, Ministerio de Salud Publica de la Republica de Cuba, Candido Lopez Pardo, Miguel Marquez and Francisco Rojas Ochoa, "Desarrollo Humano y Equidad en America Latina y el Caribe," XXV Internacional Congreso of the Latin

American Studies Association, October 2004; Francisco Rojas Ochoa and Candido Lopez Pardo, "Desarrollo Humano y Salud en America Latina y el Caribe," *Rroisra Cubana de Salud Publica* 29, no. 1 (2003): 8–17; United Nations Development Program, "Report on Human Development, 2004," online at http://www.undp.org; Mario Coyula and Jill Hamburg, "Understanding Slums: The Case of Havana, Cuba," *David Rockefeller Center for Latin American Studies* 4 (2004–5); Maria Caridad Cruz and Roberto Sanchez Medina, *Agriculture in the City: A Key to Sustainability in Havana, Cuba* (Kingston, Jamaica: Ian Randle Publishers, 2003), 210; Miren Uriarte, *Cuba Social Policy at the Crossroads: Maintaining Priorities, Transforming Practice* (Boston: Oxfam America, 2002).

3. Minor Sinclair and Martha Thompson, *Cuba Going Against the Grain: Agricultural Crisis and Transformation* (Boston: Oxfam America, 2001), 56.

4. For the purposes of this study, *socialism* refers to a society in which the associated producers (past, present, and future workers) own most of the means of production locally or nationally, in cooperative or state enterprises, and make decisions about the society through combined participatory and representative democracy (both state and non-state). Production is decided on the basis of judgments about human needs, distribution is according to work and need, and labor power is not a marketable commodity. Within this framework many kinds of organization and ways of working have been and are being tested.

5. *Complexity* has become a fashionable buzzword in discussions about science, a recognition that fragmented, reductionist, and ahistorical science has caused major disasters and left us unprepared for the resurgence of infectious disease, for antibiotic resistance in bacteria and pesticide resistance in agriculture, for flooding as a result of flood control engineering, for the emergence of hospitals as foci of infection, food aid leading to hunger, and economic growth giving rise to new forms of poverty. The rediscovery of complexity emphasizes uncertainty, non-linearity, connectedness, the interaction of chance and determinism, and the near orderliness of chaos. In one sense it is a groping toward dialectics without acknowledging the Marxist tradition. See chap. 17, "Dialectics and Systems Theory," and chap. 20, "The Butterfly ex Machina," in this volume.

6. Josefina Toledo Benedit. *La Ciencia y la Técnica en José Martí* (Havana: Editorial Ciencia y Técnica, 1994), 209.

7. The Cuban Speleological Society was organized by teenage boys from Havana who loved exploring the countryside. They began as a Boy Scout troop but broke away when they decided that the Boy Scouts were a conservative, militaristic organization dominated by the United States. The geographer Amonio Nunez Jimenez was a member and introduced Fidel Castro to cave exploration in the early years. The Society continued as a small ecology-minded NGO.

8. Silvia Martinez Puentes, *Cuba Mas Allá de los Sueños* (Havana: Editorial Jose Marti, 2003), 112.

9. Ibid., 128.

10. Fidel Castro, speech at the closing of the Congress of Pedagogy, February 2, 2003.

11. J. D. Bernal, *The Social Function of Science* (London: Routledge, 1939).

12. Jorge Aldoregía Valdes-Briro, Pablo Resik Habib, and Hecror Rodríguez Báster, "Organización y Administración de la Investigación," *Revista Cubana de Administratióm de Salud* 11 (1985): 305–16.

13. Carlos Borroto, discussion in Fidel Castro Diaz-Balart, *Cuba. Amanecer del Tercer Milenio* (Madrid: Publishing Debate, 2002), 257.

14. CITMA (Ministerio de Ciencia, Tecnología y Medio Ambiente), *Taller "Media Ambiente y Desarrollo: Consulta Nacional Rio + 5"* (Havana: CITMA, 1997), 74.

15. Ibid., 22.

16. The Second Biennial International Seminar on the Philosophical, Epistemological and Methodological Implications of Complexity Theory and Parallel Workshop on Complex Biological Systems, Havana. January 7–10, 2004.

17. Agustín Lage, discussion in Fidel Castro Diaz-Balart, 145.

18. Gina Rey, "Cuba, Integral Development, and the Environment," presentation at Dalhousie University, 1989.

19. CITMA, National Environment Strategy, 1997.

20. José Luis Garcia, discussion, in Fidel Castro Diaz-Balart, 306.

21. CITMA, Programa Nacional de Medio Ambiente y Desarrollo, 1995.

22. The people mentioned in this paper are an idiosyncratic sampling of names I happen to remember and people I have worked with. Others whose contributions are just as important have been omitted.

23. Fernando Funes, Luis Garcia, Martin Bourque, Nilda Perez, and Peter Rosset, *Sustainable Agriculture and Resistance: Transforming Food Production in Cuba* (Oakland: Food First Books, 2002), 307.

24. Maria Caridad Cruz and Roberto Sanchez Medina, *Agriculture in the City*.

25. A classic example is the case of Alar, a possible carcinogenic chemical used on apples. Environmentalists were successful in getting Alar removed from the American market in 1989 following a major exposé on the CBS News program *60 Minutes*. But this turned into a major defeat for environmental activism after a food and chemical industry-sponsored public relations campaign successfully spun this legitimate victory into "environmental fear-mongering," a label that has become a potent weapon against environmentalism in general.

31 Living the 11th Thesis

Richard Levins is grateful to Rosario Morales for her assistance in conceptualizing and editing this paper.

1. John Montgomery Brown had been a Lutheran Episcopal bishop of the Missouri Synod, excommunicated when he became a Marxist. In the 1930s he published the quarterly journal *Heresy*.

2. AID carries out programs on health and development in strategically chosen third world countries. Its separate programs are sometimes helpful and their participants motivated by humanitarian concerns. But the agency is also a terrorist organization, supporting counter-revolutionary groups in Venezuela, Haiti, and Cuba. It once sponsored the LEAP (Law Enforcement Assistance Program) that taught torture to Uruguayan and Brazilian police.

Index